Industrial Electronics

Consulting Editor
M. R. Ward
Vice-Principal, South East London Technical College

Related books in the Technical Education Series:

Smithson: Mathematics for Electrical and Telecommunications Technicians, Volumes 1, 2, and 3

Morris: Control Engineering

Stott and Birchall: Electrical Engineering Principles: for Electrical, Telecommunications, and Installation Technicians

Ward: Electrical Engineering Science (01)

Industrial Electronics

N. M. MORRIS, B.Sc., C.Eng., M.I.E.E., M.I.E.R.E.
Principal Lecturer in Electrical and Electronic Engineering
North Staffordshire Polytechnic

McGRAW-HILL · LONDON

New York · Sydney · Toronto · Mexico · Johannesburg
Panama · Düsseldorf · San Francisco · St Louis · Singapore

Published by

McGraw-Hill Publishing Company Limited
MAIDENHEAD · BERKSHIRE · ENGLAND

07 094257 9

PRINTED AND BOUND IN GREAT BRITAIN

Preface

In a world of rapidly developing technology the success of any organization depends on the quality of its technical staff. This, in turn, is dependent upon the dissemination of information throughout industry. It is the purpose of this book to present information which will enable technician and senior technician students to grasp the fundamentals of electronic circuits.

Electronic technology is ever-changing and, with it, courses and syllabi develop to keep abreast of the changes. It was with this in mind that the book was written. The contents can be broadly divided into two sections; basic devices and their characteristics are described in the first five chapters, and the final eight chapters are devoted to circuits and their applications, including switching circuits. The greater part of the book deals with semiconductor devices and circuits, but vacuum and gas-filled devices are not neglected as they fulfil useful engineering functions.

The book is, broadly speaking, aimed at the requirements of the technician engineer and, to this end, a large number of examples have been worked out in the text to supplement basic design principles. In addition, a plentiful supply of problems are given at the end of each of the chapters. While the emphasis of the work is on the practical understanding of the operation of devices and circuits, vital analysis has not been omitted. In many cases both detailed analysis and simplified analysis are given.

A great deal of the information in this book has been derived from industrial sources, and I would like to thank them for the advice and assistance which was freely given. In particular, thanks are due to General Electric, Hewlett-Packard, Texas Instruments, and Mullard. The City and Guilds of London Institute has kindly given me permission to reproduce problems which have appeared in past examination papers; the responsibility for the accuracy of the solutions is mine alone. Members of the electrical and electronic staff of the North Staffordshire

Polytechnic have also made significant contributions to the ideas embodied in the text. In particular I would like to thank Mr V. L. C. Mitchell for his help in the early stages of the book, and also Mr H. Hand and Mr J. W. Machin. Finally, I would like to thank my wife for all the work she has put into the preparation of the manuscript

N. M. MORRIS

Contents

1. Thermionic valves

It was perhaps fate that lead early pioneers of electronics to discover the phenomenon of thermionic emission, and to the logical development of thermionic
devices. It is fortunate that many of the physical principles which govern the
flow of current in valves also govern the flow of current in semiconductor devices,
and in the following an attempt is made to outline these principles.

1.1   The constitution of matter

Elements, compounds, and molecules: There are approximately 100 basic substances known to man and these are called *chemical elements*; elements are
fundamental to nature, and cannot be reduced further by any chemical process.
Chemical elements are made up of a number of smaller particles known as *atoms*.
Two or more different elements can combine to form a *chemical compound,* and
the smallest particle of a compound is called a *molecule.* Each molecule has the
chemical properties of the compound, and comprises one or more of each of the
atoms contained in the compound. The word molecule also describes combined
atoms of the same element, e.g. ozone (O_3) which is *triatomic,* that is it contains
three atoms of oxygen.

Certain elements are very *active* and readily combine with other substances;
an example is sodium (Na), which is very corrosive. One of the reasons for the
combination of elements is that they are either short of electrons, or have too
many for an 'ideal' atomic design. Gases known as *inert gases* have this ideal
design, and do not combine with other elements under normal conditions, or
even with atoms of their own kind. These gases are helium (He), neon (Ne),
argon (A), krypton (Kr), xenon (Xe), and radon (Rn). Molecules of inert gas are
monatomic, and consist of only one atom. Some elements also become
monatomic in their gaseous state, an example being mercury.

1

The atom: From an engineering point of view the atom consists of two types of *charged* particles—*protons* and *electrons*. To complete the basic chemical properties of the atom, a further particle known as the *neutron* is introduced. The proton carries a positive charge of electricity, and the electron a negative charge of electricity. The two charges are equal in magnitude and opposite in polarity, and are given the symbol *e*, where

$$e = \pm 1 \cdot 6 \times 10^{-19} \qquad \text{coulomb}$$

The mass of the proton is $1 \cdot 66 \times 10^{-27}$ kg, which is about 1840 times greater than the mass of the electron. The neutron carries no charge and has a mass equal to that of the proton. The mass of the electron varies with its speed, but at the velocities normally associated with electronic devices it is approximately $9 \cdot 1 \times 10^{-31}$ kg. Atoms do not carry any electrical charge in their normal state, since they contain equal numbers of positive and negative charges. The addition or removal of an electron gives the atom a net negative or positive charge, respectively. An atom carrying an electrical charge is an *ion*, and is said to be *ionized*. Both positive and negative ions exist in nature.

The structure of all atoms is broadly the same. At the centre of the atom is the *nucleus,* which contains the protons and neutrons, and has a positive charge. The nucleus is surrounded by electrons in concentric layers or *shells.* The distance between the nucleus and the orbiting electron is determined by the kinetic energy of the electron. The greater the kinetic energy, the further out it will orbit, subject to the limitation that it must orbit in one of a given number of shells. When an electron moves to a higher orbit, as a result of receiving energy from an external source, the atom is said to be *excited*, and the process is known as *excitation.* For convenience the shells are lettered alphabetically, commencing at the letter K. The numbers of electrons that the shells can possess when filled are: K,2; L,8; M,18; N,32; etc. Some of the shells comprise a number of subshells. For instance, the M shell consists of three sub-shells containing 2, 6, and 10 electrons, respectively, when they have their full complement of electrons.

The atomic structures of different elements differ only in the arrangement and numbers of the three basic particles. For example helium has two protons and two neutrons in the nucleus, and two electrons in the K shell, thereby 'filling' that shell. The nucleus of pure oxygen contains eight protons and eight neutrons, while eight electrons are in orbit, two in the inner or K shell, and six in the L shell. Thus helium is chemically inert since its outer shell is full of electrons, while the L shell of oxygen is incomplete since it can hold eight electrons, and oxygen reacts with other substances to gain extra electrons to fill the shell.

Valency: Atoms with unfilled outer shells combine together to acquire full outer shells. They do so in one of three ways. They either *lose electrons* to other atoms to empty their incomplete outer shell, or *gain electrons* from other atoms, or they *share electrons* with other atoms. In the sharing process, an electron from one atom in the structure orbits around the parent atom and one other atom, binding the two atoms together. This is known as *covalency,* and the atoms are said to be joined by a *covalent bond.* This process is particularly important in semiconductor devices.

Some elements are known as *electro-positive elements* since they readily lose one or more electrons to become positive ions. An *electro-negative element* acquires one or more electrons, becoming a negative ion. Broadly speaking, metals are electro-positive elements, and non-metals are electro-negative elements. The number of electrons which an atom gains, loses, or shares, according to the method of chemical combination, is known as the *valency* of the element. The electrons in the outer shell are called the *valence electrons.* Elements are *mono-, di-, tri-, tetra-, penta-, hexa-,* or *heptavalent,* if the number of valence electrons involved is 1, 2, 3, 4, 5, 6, or 7.

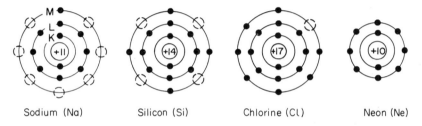

| Sodium (Na) | Silicon (Si) | Chlorine (Cl) | Neon (Ne) |

Fig. 1.1   The atomic structure of some well-known elements.

Figure 1.1 shows the basic atomic structure of some well known elements. Electrons in the shells are shown as black dots, and a shell is said to have a *hole* in it if an electron is missing. Holes are represented by the circles drawn in broken lines in the figure. The element sodium is an electro-positive substance, while chlorine is electro-negative. Silicon is one of those chemical substances which combines with other elements by means of covalent bonds, while neon is an inert gas. A hole may be regarded as the absence of an electron in a position where one would normally be found, and it can be regarded as a *positive charge carrier,* much as an electron is a negative charge carrier.

The broad outline of the structure of matter given here is adequate for the purpose of this book, but a closer investigation into the chemical properties of matter soon shows up the limitations of simple man-made theories.

1.2   Conduction in metals

From a chemical standpoint a metal is an electro-positive element, and readily gives up its valence electrons. A metal may therefore be thought of as a number of fixed positive ions (the atoms themselves), and a number of mobile negative ions (the valence electrons). In an isolated piece of metal the electrons move about in a random fashion, and the net charge on the metal is zero. Since the electrons move in a random fashion, the charge distribution throughout the metal varies with time. If the electrons tend to group at one point in the metal, that region acquires a net negative charge, and other regions a net positive charge. Thus, the region with the negative charge repels further electrons from entering it, and the regions with a positive charge attract electrons to it, resulting in further migration of electrons.

The application of an e.m.f. between the ends of a metal bar results in the free electrons drifting towards the positive pole of the supply source. To maintain the balance of electrons in the bar, electrons enter it from the negative pole of the source of supply. Although the flow of current from one end of the bar to the other appears to be almost instantaneous, as measured by electrons entering at one end and other electrons leaving at the other end, the net drift velocity of an individual electron is very small, being 0·1 mm/s typically.

Electrical resistance of metals: At absolute zero of temperature ($-273°C$), free electrons in metals move about without hindrance, and the electrical resistance is zero (or the conductivity is infinite). When energy in the form of heat is absorbed by the atoms from an external source, they become excited and begin to vibrate. The free electrons begin to lose some of their kinetic energy as a result of the collisions with the atoms, and when an e.m.f. is applied between the ends of the metallic bar a smaller current results than is the case at absolute zero of temperature. As a result, the metal exhibits the property of increasing resistance to flow of current with increasing current. That is to say metals have a *positive resistance-temperature coefficient.*

Work function in metals: Electrons cannot easily escape from the surface of a metallic element since the loss of an electron results in the metal taking on a positive charge. The resulting electrostatic force acting on the electrons which leave the surface of the metal attracts them back to the general body of the element. However, some electrons do acquire sufficient energy to escape from the surface of the metal, and this is the electronic equivalent of water vapour leaving the surface of a pan of heated water. To define the energy required to make an electron leave the surface of the metal a new unit of energy, the *electron-volt,* is used. The electron-volt is defined as the energy gained when an electron falls through a potential difference of one volt.

1 electron-volt (eV) = 1 volt × The charge on an electron in coulombs

$$= 1·6 \times 10^{-19} \text{ joules}$$

The energy required to enable an electron to escape varies from one element to another, and is known as the *work function* of the material. Typical values for the work functions of some common materials are, in electron-volts: caesium (Cs), 1·75; copper (Cu), 4·2; mercury (Hg), 4·5; tungsten (W), 4·55; platinum (Pt), 6·15.

The work function enables the phenomena known as *contact potential* (or *contact e.m.f.*) and *thermo-electric effect* to be explained. When two metals with dissimilar work functions are connected together, electrons from the material with the lower work function escape more easily into the material with the higher work function than is the case in the reverse direction. The material with the lower work function thus acquires a positive potential with respect to the material with the higher work function. The value of the e.m.f. developed in this way may only be a fraction of a volt, and is dependent on the work functions of the materials used.

1.3   Electron emission

At room temperature very few free electrons in metallic bodies attain sufficient energy to escape from the surface. It is necessary to increase the energy of the electrons before escape is possible. The four principal methods of achieving electron emission are

(a) Thermionic emission.
(b) Photoemission.
(c) Secondary emission.
(d) Field emission.

Thermionic emission: When a metal surface is raised to a sufficiently high temperature, a number of free electrons gain sufficient energy to escape from it. This is known as thermionic emission. It is an unfortunate fact that metals with low work functions melt, or even boil, at the temperature at which adequate thermionic emission takes place. Emitting materials in common use today are tungsten, thoriated-tungsten, and oxide coated materials, having work functions in electron-volts of 4·55, 2·7, and 1, respectively.

Tungsten, despite its relatively high work function, is used extensively in high voltage (i.e., over 10 kV) electronic devices and in large transmitting valves. It has a melting point of 3380°C and is worked as an emitter between 2300°C and 2500°C. A *thoriated-tungsten filament* is a tungsten filament which has an atomic layer of the oxide thoria on its surface. The work function of a thoriated-tungsten filament is found to be less than that of the constituent elements, tungsten and thorium. These filaments provide a greater thermionic emission than an equivalent rated tungsten filament and work at a much lower temperature, typically 1500°C. They are commonly used in electronic devices which operate at voltages less than about 10 kV. *Oxide coated emitters* take the form of a coating of oxides of barium and strontium on the surface of a metal sleeve, usually made of nickel. The temperature of the metal sleeve is raised either by passing a current directly through the metal sleeve, or by heating the sleeve indirectly by a heating filament known as a *heater,* the latter method being the most common. The electron emission of oxide coated emitters is much greater than either tungsten filaments or thoriated-tungsten filaments, and the operating temperature is about 700°C. These emitters are restricted to use in electronic devices operating with anode voltage less than about 1 kV. If the temperature

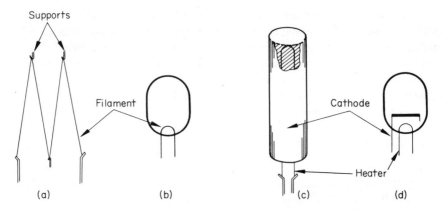

Fig. 1.2   The filament-type emitter shown in (a) is represented by the symbol in (b), and the indirectly heated cathode (c) is represented by symbol (d).

of an oxide coated emitter falls below its rated value, the cathode becomes susceptible to *cathode poisoning*. This term describes the reduction in emission that occurs as a result of surface contamination when small quantities of gas are present.

    An illustration of a typical filament emitter is shown in Fig. 1.2(a), and of an indirectly heated oxide coated cathode in Fig. 1.2(c). The circuit symbols of the two forms of emitter are shown in Figs. 1.2(b) and (d), respectively. The characteristics of materials commonly used as emitters are given in Table 1.1, in which the emission efficiency in mA/W of heating power is given. Owing to the high efficiency and reliability of oxide coated cathodes, the majority of industrial valves employ indirectly heated oxide coated cathodes.

Photoemission: Light is one form of energy, and if the frequency of the light waves impingeing on a metal surface exceeds a threshold value, electrons are emitted from the metal. To obtain the best possible performance, low work function materials are used in the construction of the cathode of *photoemissive cells* or *phototubes*.

Table 1.1

Characteristics of emitting materials

Material	Work function (eV)	Emission efficiency (mA/W)	Operating temperature (°C)
Tungsten	4·55	5-20	2300-2500
Thoriated-tungsten	2·7	50-100	1500-1700
Oxide coated	1·0	100-10,000	700-900

Secondary emission: When a charged particle (an electron or a positive ion) strikes a conducting surface it transfers some, or all, of its kinetic energy to the atomic structure of the metal. If the energy transferred exceeds the work function of the material, an electron is dislodged from its surface.

Field emission: Electrons can be literally torn out of their orbits by the application of a very intense electric field of the order of 1 MV/cm at the surface of the material. This is known as field emission, and is thought to be the principal mechanism of electron emission in mercury-pool devices.

1.4 Thermionic valve circuit notation

Throughout this book a notation generally similar to BS 3363 is adopted. Upper case (capital) letters indicate either steady (d.c.) or r.m.s. quantities, and lower case (small) letters indicate instantaneous quantities. Potentials and currents are identified by a double subscript notation, the second subscript being the reference node, the measurement being made at the point indicated by the first subscript. The anode electrode is designated A or a, and K or k is used to denote the cathode. For example V_{AK} is the average value of the anode voltage with respect to the cathode, and i_{AK} is the total instantaneous value of the anode-to-cathode current. Where the reference terminal is clearly understood, as in the case of the cathode in thermionic devices, the reference subscript is omitted. A supply voltage is defined by repeating the subscript of the electrode to which it is connected. . The notation as applied to the diode is given in Table 1.2. An illustrative example

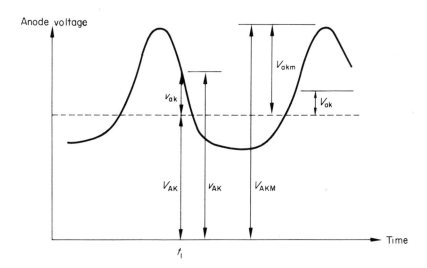

Fig. 1.3   A waveform illustrating the circuit notation used for diodes.

is given in Fig. 1.3, which depicts the anode voltage waveform in which

V_{AK} = Average value of the anode voltage.

v_{ak} = Instantaneous value of the varying component of the anode voltage at time t_1.

v_{AK} = Total instantaneous value of the anode voltage at time t_1.

V_{AKM} = Peak value of the total anode voltage.

V_{akm} = Peak value of the varying component of the anode voltage.

V_{ak} = R.M.S. value of the varying component of the anode voltage.

Table 1.2

Diode valve notation

Symbol	Description
v_{AA}	Total instantaneous value of the anode supply voltage ($= V_{AA} + v_{aa}$)
V_{AA}	Average value of v_{AA}
v_{aa}	Instantaneous value of the varying component of v_{AA}
V_{aa}	R.M.S. value of v_{aa}
v_A or v_{AK}	Total instantaneous value of the anode-to-cathode voltage ($= V_A + v_a$)
V_A or V_{AK}	Average value of v_A
v_a or v_{ak}	Instantaneous value of the varying component of v_A
V_a or V_{ak}	R.M.S. value of v_a
i_A or i_{AK}	Total instantaneous value of the anode-to-cathode current ($= I_A + i_a$)
I_A or I_{AK}	Average value of i_A
i_a or i_{ak}	Instantaneous value of the varying component of i_A
I_a or I_{ak}	R.M.S. value of i_a

Some valves contain control electrodes known as *grids*, and are designated by the symbol G or g. Where more than one grid is employed, the grids are numbered consecutively with G1 nearest to the cathode. In the pentode, which has three grids, G1 is known as the control grid, G2 the screen grid, and G3 the suppressor grid.

Throughout the book arrows shown *on a conductor* represent flow of current in that conductor, the direction of the arrow representing the conventional direction of flow of current (flow of holes in a semiconductor). Arrows drawn *by the side* of a circuit element represent the direction of potential rise in the element, the arrowhead indicating the point which is assumed to have the more positive potential. Examples of current and potential arrows are shown in Fig. 1.6(a).

1.5   The diode

The thermionic diode is a valve containing only two electrodes, a cathode and an anode inside an evacuated container. When the cathode is heated to its working temperature, it releases electrons by thermionic emission and is surrounded by free negative charges. Under steady operating conditions, the negative *space charge* resulting from the presence of the electrons limits further electron emission from the cathode.

In the diode the anode, which is usually made from nickel in the form of a cylinder, surrounds the cathode as shown in Fig. 1.4(b). When the anode is at the same potential as the cathode it is found that some electrons have sufficient energy to cause them to travel to the anode, resulting in a small anode current. This is shown as I_{AO} on the *anode characteristic* or *output characteristic* in Fig. 1.4(c); I_{AO} has a value of the order of a few microamperes, and may be neglected for most practical purposes. I_{AO} is sometimes described as *splash current* and is due to *fast electrons* or electrons which have an energy level which is sufficient to carry them to the anode. In some applications, notably electronic voltmeters (see chapter 11), this current can be troublesome. To reduce the anode current to zero it is necessary to apply a small negative potential to the anode; this repels the electrons which approach the anode. If the anode potential is made progressively more negative, a small reverse leakage current (of the order of a few microamperes) flows from the anode to the cathode. The safe *peak inverse working voltage* that may be applied to the anode is restricted below the value at which a discharge would commence. When the diode operates with a negative anode voltage it is said to be operating in a *reverse blocking* mode, since it blocks the flow of current.

When the anode is positive with respect to the cathode, electrons are attracted to it, resulting in an increase in anode current. At first the anode current increases in an approximately linear fashion with voltage, between I_{AO} and point L on the anode characteristic; this section is described as the *linear* part of the characteristic. Owing to the effect of the cathode space charge, the anode potential can only influence the outer edge of the cloud of electrons, and in the linear region of the characteristic the anode current is said to be *space-charge limited*. As the anode voltage is increased, more electrons are drawn from the outer edge of the electron cloud, depleting the space charge and allowing more electrons to leave the cathode. The electrons in motion between the cathode and anode constitute a current, and are collectively known as the *space current*.

Strictly speaking, the relationship between the anode current and voltage between the points given above is not truly linear, but is given by the relationship

$$i_A = k v_A^{3/2}$$

which is known as the *three-halves power law* or the *Langmuir-Child law*, since the equation is due jointly to the work of Langmuir and Child.

When the anode voltage is increased beyond point L in Fig. 1.4(c), the curve

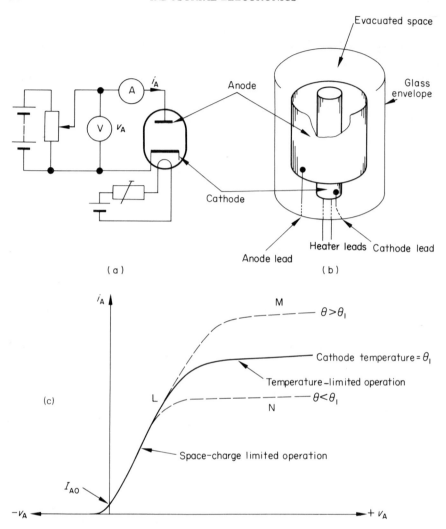

Fig. 1.4   (a) A typical test circuit for use with a thermionic diode. In (b) the
anode structure is cut away to show the electrode structure. The anode
characteristics for various values of cathode temperature are shown in (c).

begins to flatten out because all the electrons emitted by the cathode are collected by the anode. Further increase in anode voltage has negligible effect on the
anode current. In this region of the characteristic, the anode current is said to be
temperature-limited. Only by increasing the cathode temperature is it possible to
increase the anode current, corresponding to characteristic M in Fig. 1.4(c). The
maximum temperature at which the cathode may be operated is limited, of
course, by the temperature at which the oxide coating begins to evaporate. A

reduced cathode temperature reduces the value of the saturation current, shown in characteristic N, in addition to which the cathode becomes susceptible to poisoning (see section 1.3). An additional cause of cathode damage occurs under temperature-limited conditions due to the fact that it is impossible to obtain a perfect vacuum inside valves. The residual gas molecules become ionized at high values of anode voltage, and under temperature-limited conditions the ions fall upon the surface of the cathode, which is no longer protected by its electron shield. At high anode voltages the ions bombard the cathode with sufficient energy to destroy the oxide film.

It is found with increasing values of anode voltage in the temperature-limited region that the anode current increases by a very small amount. This is due to a reduction of the work function of the emissive surface of the cathode when it experiences an electric field. It is known as the *Schottky effect* after the man who first calculated its magnitude.

The curve in Fig. 1.4(c) is also known as the *static characteristic of the diode,* to distinguish it from the *dynamic characteristic of the circuit* containing the diode. The dynamic characteristic of a circuit gives the relationship between the voltage applied to the circuit (i.e., a diode and resistive load connected in series) and the current in the circuit. The static characteristics of both the diode and load must be known in order to predict the dynamic characteristic of the circuit. The construction of this characteristic is dealt with in section 1.7.

1.6 Effective resistance of the diode

The instantaneous resistance of a circuit element is defined as the ratio of the instantaneous voltage applied to the element to the instantaneous current flowing in the element. Owing to the curvature of the diode characteristic, the *instantaneous anode resistance* r_A varies with anode voltage, and is defined as

$$r_A = v_A / i_A$$

In Fig. 1.5 the value of r_A at point M is $22/75 \times 10^{-3} = 293 \ \Omega$, and at point N is $34/132 \cdot 5 \times 10^{-3} = 257 \ \Omega$.

When the anode voltage changes continuously between two points on the characteristic, the average operating point is known as the *quiescent point* or *operating point*. The effective resistance of the diode at the quiescent point is given by the slope of the static characteristic at that point and is known as the *slope resistance* r_a, and is defined by the relationship

$$r_a = \frac{\Delta v_A}{\Delta i_A} \ \Omega$$

where Δv_A and Δi_A are small changes in anode voltage and anode current respectively. The slope resistance of the diode in Fig. 1.5 between points M and N is

$$r_a = \frac{34 - 22}{(132 \cdot 5 - 75) \times 10^{-3}} = 209 \ \Omega$$

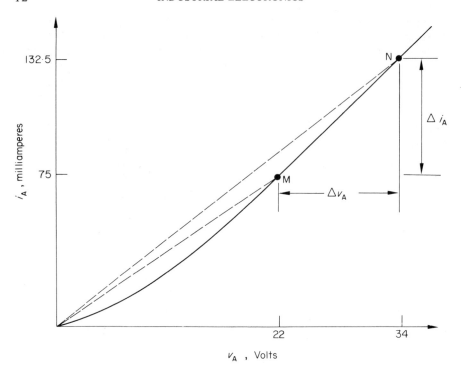

Fig. 1.5   Methods of evaluating r_A and r_a.

When the anode of the diode is positive with respect to the cathode, the diode is said to be *forward biased* and operates in its *forward conducting* mode. The effective resistance of the diode in this mode is generally taken to be equal to r_a. When the anode of the diode is negative with respect to the cathode, the space current of the diode is practically zero and the diode is said to be *reverse biased*. The slope resistance of a reverse biased diode is many megohms and is referred to as the *reverse blocking resistance*.

1.7   Series circuit operation

Many electrical circuits involve the operation of two devices in series. When one, or both, of the devices has a non-linear characteristic, a mathematical solution of the circuit equations is difficult. The problem, in the case of a diode with a resistive load, is illustrated in Fig. 1.6(a). The total voltage V_{AA1} applied to the network is divided between the non-linear and linear devices according to their characteristics. In the diagram, quantities with suffix A refer to the diode, and those with suffix L refer to the load.

Steady operating conditions are reached when the supply voltage is equal to the sum of the p.d.'s across the two devices, and when the diode current is equal

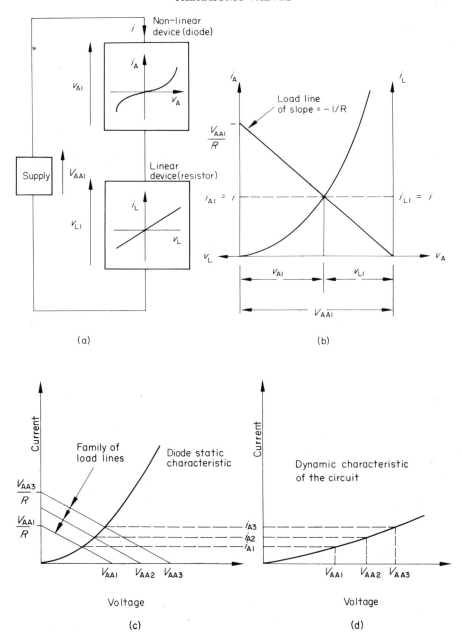

Fig. 1.6   The voltage distribution between the two circuit elements in (a) and the circuit current can be evaluated by the graphical construction in (b). One method of determining the dynamic characteristic of the circuit is illustrated in (c) and (d).

to the current flowing in the resistor. Expressed in mathematical form

$$V_{AA1} = v_{A1} + v_{L1}$$

and

$$i_{A1} = i_{L1}$$

These equations are best solved graphically using the construction in Fig. 1.6(b). The linear resistive characteristic is plotted in the reverse direction on the diode characteristic, from a false zero corresponding to the supply voltage V_{AA1}. The intersection of the two characteristics gives both the voltage distribution and the circuit current. When a linear characteristic is plotted in this way it is known as a *load line*. The equation of the load line is deduced from the mesh equation of the circuit as follows

$$V_{AA1} = v_A + i_A R$$

Where R is the resistance of the load. Rewriting in terms of i_A yields the load-line equation

$$i_A = (V_{AA1} - v_A)/R$$

The load line intersects the vertical axis of the diode characteristic when $v_A = 0$, i.e., when

$$i_A = (V_{AA1} - 0)/R = V_{AA1}/R$$

Since there is one load line (with a given value of R) for each value of supply voltage, there are a family of load lines as shown in Fig. 1.6(c), resulting in a family of anode currents i_{A1}, i_{A2}, i_{A3}, etc., corresponding to supply voltages V_{AA1}, V_{AA2}, V_{AA3}, etc. The *dynamic characteristic of the circuit* for a given load resistance is obtained by plotting the values of circuit current so obtained to a base of supply voltage, shown in Fig. 1.6(d).

The construction used here can be extended to deal with two non-linear devices in series*. It can also be used to determine the voltage distribution under reverse bias conditions by plotting the load line on the reverse blocking characteristic of the diode. Generally speaking, the dynamic characteristic of the circuit under reverse biased conditions is a straight line on the zero current axis.

Example 1.1: Plot the dynamic characteristic for the circuit in Fig. 1.7(a) if the diode has the static characteristic in Fig. 1.7(b). Hence estimate the current in the circuit for supply voltages of (a) 10 V, (b) 25 V, and (c) 40 V.

Solution: To illustrate the basic simplicity of the method, the circuit dynamic characteristic is constructed on the static characteristic of the diode, Fig. 1.7(b). For a supply voltage of 10 V, the load line corresponding to a 200 Ω resistor terminates at $i_A = 10/200 = 0.05$ A or 50 mA. The current in the circuit is given by the intersection of the load line and the diode static characteristic, at a

---

* See *Control Engineering* by N. M. Morris, published by McGraw-Hill (1969).

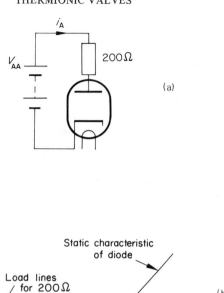

Fig. 1.7   Figure for example 1.1.

current of 17 mA. A similar construction is used for supply voltages of 25 V and 40 V, and the required solutions are: (a) 17 mA, (b) 48 mA, and (c) 83 mA.

It should be noted that the dynamic characteristic is much more linear than the static characteristic of the diode.

1.8   Equivalent circuit of the diode

The equivalent circuit of an electronic device is a circuit which has the same static characteristics as the device over a specified range of operating conditions.

In the case of the diode the characteristic is non-linear, and a true equivalent circuit is very complex. The problem is simplified if only the linear parts of the characteristic are considered.

A simplified *linear equivalent circuit* of the diode is shown in Fig. 1.8(a). The diode is regarded as a single-pole double-throw switch which is open (OFF) when the diode is reverse biased, and is closed (ON) when the diode is forward biased. The effective resistance in the forward biased condition is taken to be the slope resistance r_a. A more accurate representation is shown in Fig. 1.8(b), where R is the leakage resistance of the diode in the reverse biased condition. Since the anode and cathode of the diode are concentric metallic conductors which are separated by a dielectric, a capacitance exists between them. This is shown as C in the equivalent circuit in Fig. 1.8(c).

An example of the use of the equivalent circuit technique is illustrated in

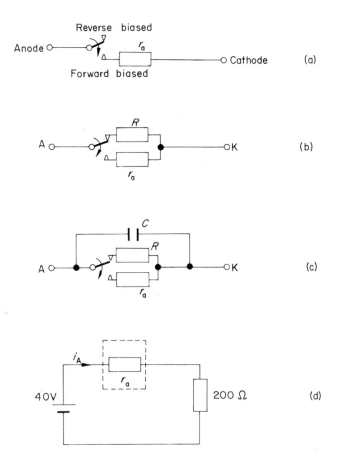

Fig. 1.8   The linear equivalent circuit of the diode. In (d) it is used to determine the current i_A.

Fig. 1.8(d). The circuit current with a load resistance of 200 Ω and V_{AA} = 40 V is

$$i_A = 40/(r_a + 200) \text{ A}$$

Using the diode in example 1.1, which has a slope resistance of 209 Ω at a current of 100 mA, the circuit current is

$$i_A = 40/409 = 0.0978 \text{ A} \quad \text{or} \quad 97.8 \text{ mA}$$

The true circuit current, from Fig. 1.7(b), is 83 mA. This example illustrates the limitation on the accuracy of calculation which is imposed by using a linearized version of the non-linear characteristic.

1.9 Voltage and current sources

The principal requirement of an ideal constant voltage source is that its terminal voltage must be unvarying over a wide range of load conditions, i.e., from no-load to a value in excess of its rated power output. This is illustrated in Fig. 1.9(a).

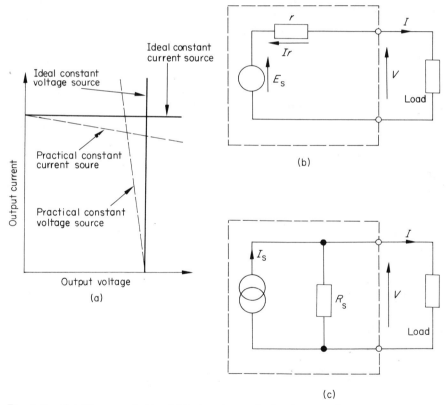

Fig. 1.9   (a) Characteristics of ideal and practical voltage and current sources. The equivalent circuit of a practical voltage source is shown in (b), and that of a practical current source is shown in (c).

No-load operation is defined as the condition in which zero power is delivered into the load. In the case of a constant voltage source it occurs when the load current is zero, i.e., the load is disconnected.

An ideal constant current source must be capable of supplying an unvarying current to the load, from no-load to a value in excess of its rated power output. The no-load condition of a constant current source corresponds to the case where it delivers current into a load of zero resistance, i.e., when its terminals are short-circuited.

To satisfy the above conditions, the internal impedance or *output impedance* of the ideal constant voltage generator is zero, so that there is no reduction in terminal voltage with increase of load current. The output impedance of an ideal constant current generator is infinity, so that whatever load impedance is connected it does not alter the current delivered to the load. Ideal conditions are rarely achieved in practice, and a compromise must be made to obtain equivalent circuits which represent practical versions of the generators. The practical limitations are accounted for by including a resistor r in series with the ideal constant voltage source E_S, as shown in Fig. 1.9(b) (often known as a *Thevenin equivalent voltage source*), and by shunting the ideal current source I_S in Fig. 1.9(c) by resistor R_S (often known as a *Norton equivalent current source*). The characteristics of the two practical circuits are then as shown by the broken lines in Fig. 1.9(a).

In circuit design it is often convenient to use the constant voltage generator concept in some applications, and the constant current concept in others. The relationship between the two is deduced as follows. For the constant voltage source

$$V = E_S - Ir$$

or

$$I = \frac{E_S}{r} - \frac{V}{r}$$

therefore

$$\frac{E_S}{r} = I + \frac{V}{r} \tag{1.1}$$

For the constant current generator

$$I_S = I + \frac{V}{R_S} \tag{1.2}$$

For equivalence of eqs. (1.1) and (1.2)

$$I_S = E_S/r \quad \text{and} \quad R_S = r$$

Thus, a constant voltage source with E_S = 50 V and r = 100 Ω is equivalent to a constant current source having I_S = 50/100 = 0·5 A and R_S = 100 Ω. Both develop an open-circuit voltage of 50 V, and pass 0·5 A into a short-circuit.

1.10   The triode

Since space current in high vacuum thermionic diodes is in the form of electron flow, it can be modified by introducing a third electrode between the anode and the cathode, which is arranged to alter the electrostatic field distribution in the valve. The third electrode is known as the *control grid* or G1, which is in the form of a spiral wire mesh close to, and around the cathode, and is supported on metal rods. The geometry of a typical triode is shown in Fig. 1.10(a) together with its circuit symbol. The anode is cut away to display the electrode structure.

If the control grid is negative with respect to the cathode, some of the electrons in transit to the anode are repelled by the grid and return to the cathode. This has the effect of reducing the space current. Since the grid is physically close to the cathode, a small grid potential exerts a considerable influence on the space current of the triode. Increasing the negative potential on the grid reduces the space current further, and high values of negative grid voltage result in the anode current being reduced to zero. In this event, the anode current is said to be *cut-off* or simply OFF.

A typical triode test circuit is shown in Fig. 1.10(b), and the *static anode characteristics* or *output characteristics* corresponding to various values of v_{GK} are shown in Fig. 1.10(c). With negative grid potentials, electrons are not attracted to the grid, and the grid current for all practical purposes is zero. The

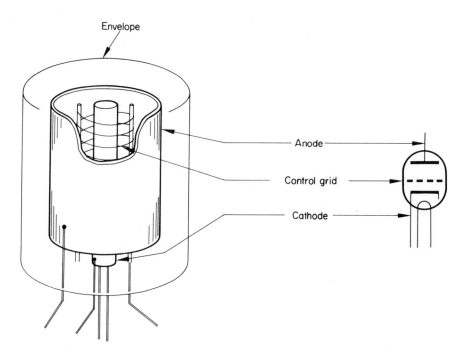

Fig. 1.10(a)

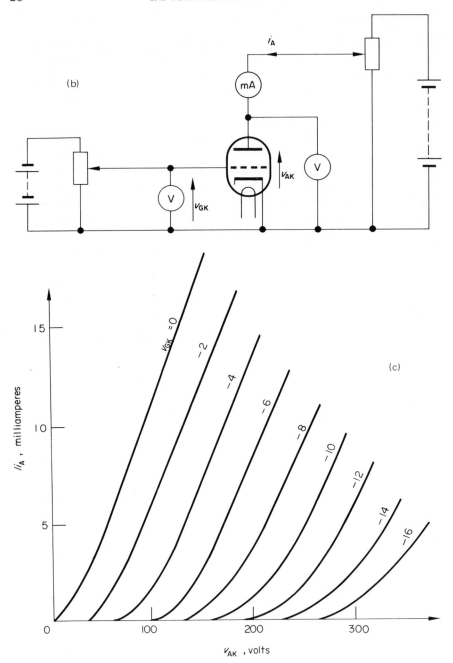

Fig. 1.10 The construction of a triode valve and its circuit symbol are shown in (a), the anode being cut away for clarity. The *static anode characteristics* in (c) are obtained from the test circuit in (b).

input impedance measured between the grid and cathode (expressed by v_{GK}/grid current) is, therefore, very high indeed.

If the grid becomes positive with respect to the cathode it acts as a second anode, and the net space current increases. Since the grid is in the form of an open wire mesh, most of the increased space current passes through it and is collected by the anode, so increasing the anode current. However, not all the additional electrons escape, and those that reach the grid return to the cathode via the external grid circuit. Since the grid now draws current, the input impedance of the valve is reduced to a value of the order of 1 kΩ. If the grid potential is to be maintained positive, the grid structure must be designed to dissipate heat, otherwise it may be damaged. In most applications the grid is maintained at a negative potential with respect to the cathode.

The triode output characteristics may be regarded as a family of diode characteristics, each characteristic corresponding to a particular control grid voltage. If the anode voltage is raised to a sufficiently high level, temperature-limited operation occurs as in the diode. In all normal applications the triode is operated in the space-charge limited region of its characteristics.

1.11   Triode parameters

The valve constants or *parameters* are defined in terms of the characteristics of the valve, Fig. 1.11. The total change δi_A in anode current is dependent upon the change δv_A in the anode voltage, and upon the change δv_G in the grid voltage. Each of these changes may occur independently or simultaneously, and the parameters enable the effects of these changes to be predicted.

In Fig. 1.11(a) the grid voltage is maintained constant at v_{G1} while the anode voltage is increased by an amount δv_A, resulting in an increase δi_A in the anode current. The relationship between the two changes is given by the *slope resistance* r_a, where

$$r_a = (\delta v_A / \delta i_A)_{\delta v_G = 0} \; \Omega$$

The suffix $\delta v_G = 0$ implies that there is no change in grid voltage, and this statement is usually omitted from the equation. From an inspection of Fig. 1.11(a), it appears that the equivalent circuit of the valve for a change in i_A at constant v_G is that of a resistor connected between the anode and cathode, as shown in Fig. 1.11(b).

If the grid voltage is made more positive, by being increased from v_{G1} to v_{G2} at constant anode voltage, as in Fig. 1.11(c), the current flowing into the anode increases by δi_A. The relationship between the changes (at constant v_A) is given by the *mutual conductance* or *transconductance* g_m, where

$$g_m = (\delta i_A / \delta v_G)_{\delta v_A = 0} \; mA/V \text{ or } mS$$

The change in anode current is, therefore, $\delta i_A = g_m \delta v_G$. The equivalent circuit of the valve, based on Fig. 1.11(c), is as shown in Fig. 1.11(d), the change in anode current being generated by a current generator.

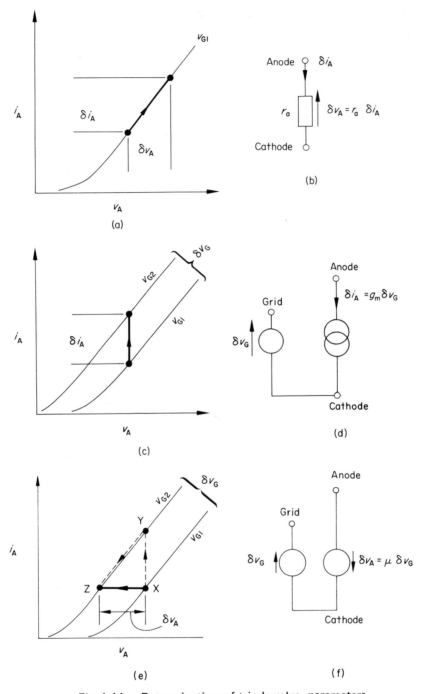

Fig. 1.11   Determination of triode valve parameters.

Changes in v_G and v_A can occur simultaneously to give zero net change in anode current if v_G is increased by δv_G, causing a change along the line XY in Fig. 1.11(e), and v_A is decreased by δv_A causing a change along the line YZ. The relationship between the two changes necessary to maintain a constant value of i_A is given by the *amplification factor* μ, where

$$\mu = (\delta v_A/\delta v_G)_{\delta i_A = 0}$$

which is a dimensionless quantity (having dimensions of V/V). The magnitude of the change in v_A, at constant i_A, is $\delta v_A = \mu \delta v_G$. It is seen from Fig. 1.11(e) that it is necessary to reduce v_A by δv_A when v_G is increased by δv_G in order to maintain a constant anode current. The equivalent circuit representing Fig. 1.11(e) is therefore as shown in Fig. 1.11(f), in which the equivalent generators in the grid and anode circuits are of opposite polarity to one another.

The changes δi_A, δv_A, and δv_G considered here are small changes in current and voltage, and the equivalent circuits so derived are known as the *small-signal*

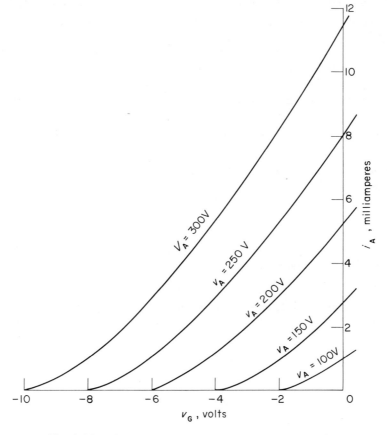

Fig. 1.12   Static mutual characteristics of the triode.

equivalent circuits. The complete small-signal equivalent circuit of the valve is developed in section 1.12.

The relationship between the small-signal parameters is obtained as follows. For a given change in anode voltage, the following equations hold good

$$\delta v_A = r_a \delta i_A \quad \text{and} \quad \delta v_A = \mu \delta v_G$$

hence

$$\mu v_G = r_a \delta i_A$$

or

$$\mu = r_a(\delta i_A / \delta v_G) \tag{1.3}$$

$$= r_a g_m$$

Owing to the curvature of the static characteristics, the value of the small-signal parameters depends upon the point at which they are measured. It is normal practice for the parameters to be specified at particular values of anode and grid potentials. Valves are often operated at a point on the characteristics which is well away from the point at which the parameters are measured in order to obtain some particular advantage, e.g., a greater value of μ or g_m. Also, owing to physical imperfections in manufacturing processes, parameters are subject to variation in value, even in valves of the same type. This is known as *parameter spread*, which is more marked in semiconductor devices than in valves.

Another useful set of curves, known as the *mutual characteristics* or *transfer characteristics* of the valve, are shown in Fig. 1.12. These show the variation of i_A to a base of v_G, at various values of v_A. Much the same information can be obtained from these curves as is obtained from the anode characteristics.

Example 1.2: To determine the small-signal parameters of a triode the following tests were made. With an initial grid voltage of -6 V and an anode voltage of 175 V, the anode current was found to be 4·6 mA. The anode voltage was then increased to 200 V, at constant grid voltage, when the anode current rose to 7 mA. The anode voltage was then maintained at 200 V and the grid voltage was altered to -4 V, when the anode current was found to be 11·4 mA. Compute the values of the valve parameters.

Solution: The first change corresponds to that in Fig. 1.11(a), where $\delta v_A = 200 - 175 = 25$ V and $\delta i_A = 7 - 4·6 = 2·4$ mA.
Hence

$$r_a = 25/2·4 \times 10^{-3} = 10,420 \ \Omega \text{ or } 10·42 \text{ k}\Omega$$

The second change corresponds to that in Fig. 1.11(c), where $\delta i_A = 11·4 - 7 = 4·4$ mA and $\delta v_G = -4 - (-6) = +2$ V.
Hence

$$g_m = 4·4 \text{ mA}/2 \text{ V} = 2·2 \text{ mA/V or mS}$$

From eq. (1.3),

$$\mu = r_a g_m = 10{\cdot}42 \text{ k}\Omega \times 2{\cdot}2 \text{ mA/V} = 22{\cdot}9$$

1.12 The small-signal equivalent circuit of the triode

A change in the characteristics from condition X to condition Y, in Fig. 1.13(a) and (c), can occur in two basic ways, giving two types of equivalent circuit. In

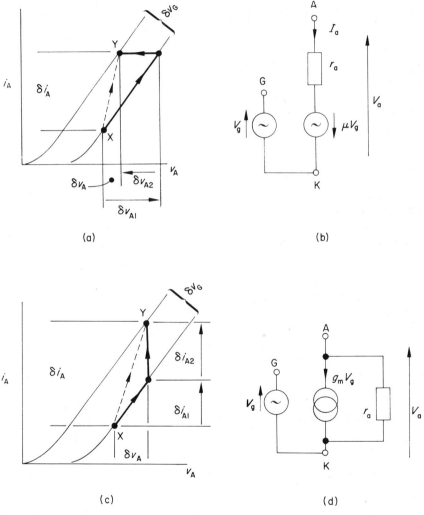

(a)   (b)

(c)   (d)

Fig. 1.13   The *a.c. small-signal constant-voltage equivalent circuit* of the triode (b) is derived from characteristic (a); equivalent circuit (d) is deduced from characteristic (c).

Fig. 1.13(a), the net change in anode voltage is

$$\delta v_{\mathbf{A}} = \delta v_{\mathbf{A}1} - \delta v_{\mathbf{A}2}$$

Change $\delta v_{\mathbf{A}1}$ occurs at constant grid voltage, while change $\delta v_{\mathbf{A}2}$ occurs at constant anode current. The equation can thus be modified to

$$\delta v_{\mathbf{A}} = r_{\mathbf{a}} \delta i_{\mathbf{A}} - \mu \delta v_{\mathbf{g}}$$

From this equation, the equivalent circuit of the valve is that of a resistor $r_{\mathbf{a}}$ carrying a current $\delta i_{\mathbf{A}}$, in series with a voltage generator with an output voltage $-\mu \delta v_{\mathbf{G}}$, i.e., a generator which makes the anode less positive when the grid becomes more positive. If the grid signal is an alternating voltage of r.m.s. value $V_{\mathbf{g}}$, then the anode current has an r.m.s. value $I_{\mathbf{a}}$, and the voltage generator develops an output $-\mu V_{\mathbf{g}}$ with respect to the cathode. The small-signal constant-voltage generator equivalent circuit of the valve for alternating input signals is therefore as shown in Fig. 1.13(b).

If the total change $\delta i_{\mathbf{A}}$ in anode current occurs as shown in Fig. 1.13(c), then

$$\delta i_{\mathbf{A}} = \delta i_{\mathbf{A}1} + \delta i_{\mathbf{A}2} = \frac{\delta v_{\mathbf{A}}}{r_{\mathbf{a}}} + g_{\mathbf{m}} \delta v_{\mathbf{G}}$$

For an alternating input signal $V_{\mathbf{g}}$, this equation becomes

$$I_{\mathbf{a}} = \frac{V_{\mathbf{a}}}{r_{\mathbf{a}}} + g_{\mathbf{m}} V_{\mathbf{g}}$$

giving a small-signal constant-current generator version of the *a.c. equivalent circuit* shown in Fig. 1.13(d). The two equivalent circuits developed here are seen to be directly related to the basic equivalent circuits in section 1.9.

The concept of the small-signal equivalent circuit is a particularly powerful tool when the design principles of electronic amplifiers are considered. In the above, the parameters have been determined from the static characteristics of the device. In practice, it is more convenient to evaluate the parameters by means of a *dynamic test*, in which the valve is excited by an alternating signal, and the small signal changes are measured directly.

1.12.1   Input capacitance of the triode

Owing to the very nature of the construction of the triode, a capacitance $C_{\mathbf{gk}}$ exists between the grid and cathode of the valve, and a capacitance $C_{\mathbf{ga}}$ exists between the grid and the anode of the valve. Capacitor $C_{\mathbf{ga}}$ causes current to be *fed back* from the anode circuit to the grid circuit when the valve is used in an amplifier. If the *gain of the amplifier* is m (as distinct from the triode amplification factor μ), then an increase in grid voltage of $v_{\mathbf{g}}$ causes the anode voltage to change by $m v_{\mathbf{g}}$, and the net result is a change in p.d. across $C_{\mathbf{ga}}$ of $(v_{\mathbf{g}} - m v_{\mathbf{g}}) = v_{\mathbf{g}}(1 - m)$. This implies that the input circuit 'sees' a capacitor between the input terminals with an apparent value of $C_{ga}(1-m)$, in addition to C_{gk}. The effective input capacitance of the valve is therefore

$$C_{in} = C_{gk} + C_{ga}(1-m)$$

The increase in C_{ga} due to voltage amplification is known as the *Miller effect*, and is used to advantage in some oscillator circuits. Unfortunately, the increased input capacitance reduces the input impedance at high frequencies, so limiting the usefulness of triodes at high frequencies. This is illustrated in the following. If C_{ga} = 2 pF, C_{gk} = 10 pF, and $m = -19$, then C_{in} = 10 + (2 x 20) = 50 pF. At a frequency of 31·8 Hz the input reactance is 100 MΩ, and at 31·8 MHz it is 100 Ω. To overcome this defect of the triode, other devices such as the tetrode and the pentode have been developed.

1.13   The tetrode

In the tetrode, a fourth electrode known as the *screen grid* or G2 is interposed between the control grid and the anode. The circuit symbol is shown in Fig. 1.14(a), and typical output characteristics for one value of control grid voltage are in Fig. 1.14(b).

The screen grid potential is usually fixed at about two-thirds of the anode supply voltage, so that when the anode potential is zero, i_A is also zero and all the space current (i_K) flows to the screen grid. As the anode potential is increased, the anode begins to collect some of the electrons which pass through the coarse mesh of the screen, and i_A begins to increase rapidly at first, as shown in Fig. 1.14(b).

The screen grid provides an effective electrostatic screen between the anode and the cathode, and for any given control grid potential the total space current

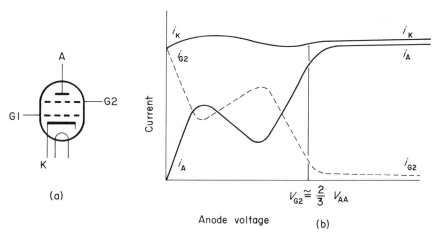

Fig. 1.14   (a) The circuit symbol of the tetrode, and (b) a typical set of characteristics for a constant value of control grid voltage.

is set by the potential of the screen grid. The anode potential has little influence on the total space current so effective is the 'screening' of G2. As a result, an increase in i_A causes a reduction in i_{G2}, and the screen grid current characteristic is almost a mirror image of the anode current characteristic (for a given value of v_{G1}).

Following the region in which i_A rises rapidly, there is a region in which it reduces with increasing anode voltage. This is equivalent to the tetrode having a negative slope resistance. This occurs because of secondary emission from the anode due to the impact of primary electrons. Since the screen grid potential is still higher than that of the anode, the secondary electrons which leave the anode are attracted to the screen grid, reducing i_A and increasing i_{G2}.

When the anode potential is increased further, some of the secondary electrons are attracted back to the anode, causing the characteristic to begin to flatten again. When $v_A > v_{G2}$, practically all of the space current flows to the anode. In this region of the characteristics the anode current, at a set value of control grid voltage, remains substantially constant over a wide range of anode voltages, the value of the current being limited by the screen grid voltage.

The tetrode is normally operated with an anode voltage greater than the screen voltage to avoid the kinks in the characteristic. Under these conditions, the anode current is almost independent of the actual anode voltage used, resulting in a slope resistance of the order of 10^5 to 10^6 Ω. Typical values for g_m and μ are 0·5 mA/V to 5 mA/V and 100 to 1500, respectively.

Owing to the inherent screening introduced between the control grid and the anode, the grid-anode capacitance of the tetrode is much less than that of the triode, being typically 0·001 pF to 0·01 pF.

1.14   The beam tetrode

The kink in the tetrode characteristics results in a limitation on the range of anode voltage available for useful purposes. One method of overcoming this problem resulted in the evolution of the beam tetrode. Historically, the beam tetrode was not developed until some years after the production of the pentode, which also overcame the same problem. But, from a theoretical point of view, it can be regarded as the logical successor to the tetrode. The essential features of the beam tetrode are shown in Fig. 1.15(a), and are listed below.

(a) A flattened cathode is used to permit high emission from the flat faces.
(b) *Beam forming plates,* which are connected to the cathode, are contained within the structure.
(c) The shape of the anode is designed to minimize secondary emission.
(d) Each turn of wire on the screen grid is aligned with each turn of wire on the control grid.

By aligning the two sets of grid wires the screen grid current is kept to a low value, and most of the space current is collected by the anode at all voltage levels.

This alone, however, does not explain the reason for the removal of the kinks in the tetrode characteristic, since secondary emission still occurs when primary electrons with sufficient energy strike the anode.

   The effect of the beam forming plates, which are at cathode potential, is to form the space current into beams of high electron density. At low anode potentials, the resulting charge cloud between the anode and the screen gives rise to a situation which is analogous to that existing in a space-charge limited diode. The secondary electrons emanating from the anode are repelled by the charge cloud,

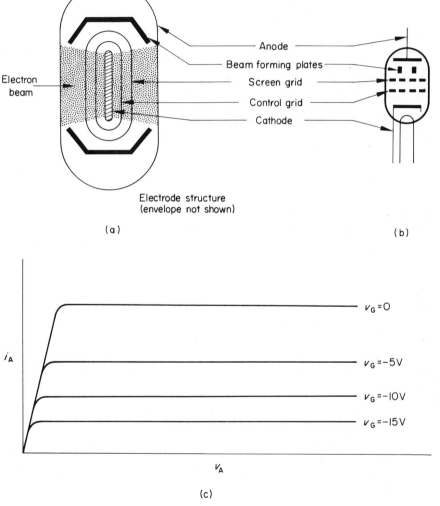

Fig. 1.15   (a) A plan view of the section of a beam tetrode valve, and (b) its circuit symbol. A family of anode characteristics is shown in (c).

and return to the anode. This results in the characteristics in Fig. 1.15(c) with a sharp 'knee' on each curve.

The beam tetrode depends for its operation on a high density electron beam, and therefore a high cathode current. For this reason, beam tetrodes are generally used in power amplifiers as output devices. In addition the beam tetrode, in common with other high current devices, is inherently 'noisy' in operation (see section 1.17). This arises from the fact that each time an electron arrives at the anode it gives up its charge to the anode circuit, resulting in an electronic 'ping' in that circuit.

1.15   The pentode

The pentode, or five electrode valve, has three grids between the cathode and the anode. The third grid, the *suppressor grid* or G3 is a coarse mesh grid between the screen grid and the anode. The circuit symbol of the pentode is shown in Fig. 1.16(a), the suppressor grid usually being connected to the cathode either internally or externally, although in some circuits it is used as an additional control grid. In its normal operating mode, the principal function of the suppressor grid is to form a region of zero potential between the anode and the screen grid. When secondary electrons are emitted by the anode, they are repelled by the zero potential region and are returned to the anode.

The anode characteristics are generally similar to those of the beam tetrode, the pentode characteristics having a 'softer' or more curved knee than those of the beam tetrode. The pentode is worked on the flat parts of the characteristics which are characterized by a high slope resistance. Consequently, the current-source equivalent circuit in Fig. 1.15(b) is preferred to the voltage-source equivalent circuit; with normal types of construction, g_m has a maximum value of about 8 mA/V.

Since the grid/cathode geometry of a pentode is very similar to that of a triode, the mutual conductances of the two valves are much the same. However, the

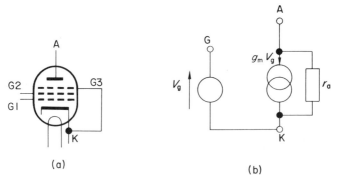

Fig. 1.16   (a) The circuit symbol of the pentode, and (b) the most commonly used a.c. small-signal equivalent circuit.

slope resistance of the pentode is much greater, with the result that the amplification factor of the pentode may have a value of the order of several thousand, compared with a value of about 20 in the triode.

1.16   The variable-mu pentode

If the spacing between the turns of wire on the control grid is not uniform, or the spacing between the control grid and the cathode varies along its length, or the diameter of the control grid wires is not uniform, then the amplification

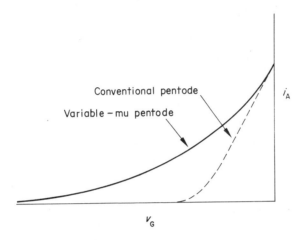

Fig. 1.17   The mutual characteristic of a variable-mu pentode.

factor of the valve varies with the control grid voltage. The general effect of a variable-pitch grid on the mutual characteristics is shown in Fig. 1.17, the curve shown in the full line corresponding to a *variable-mu pentode* or *remote cut-off pentode,* while that shown in the broken line corresponds to a conventional pentode (*sharp cut-off pentode* or *short grid-base pentode*).

These valves were developed for communications systems where they had to deal with a tiny input signal at one moment, and a much larger signal at some other time. In both cases, the output signal was required to be of much the same order of magnitude. It is arranged that the grid bias voltage applied to the valve is closely related to the incoming alternating signal, a smaller signal giving a smaller bias voltage. As the grid bias voltage increases the amplification factor is reduced, hence the name variable-mu.

When compared with conventional pentodes, variable-mu pentodes are at a disadvantage since they have to be worked at a higher quiescent anode current to achieve a comparable amplification factor.

1.17   Noise in valves

Noise is defined as all spurious signals, random or otherwise, that are no part of the input information. Noise is generally regarded as being either *narrowband* or *broadband*. Narrowband noise is of low frequency origin, typically in the range 10 Hz to 1 kHz, and is often harmonically related to the power supply frequency, e.g., mains 'hum'. Broadband noise comprises signals above a frequency of about 1 kHz, and in valves is generated by the movement of electrons.

A great deal of the noise generated in valves is due to the random flow of electrons between the anode and the cathode, and is known as *shot* noise. The rate at which electrons arrive at the anode varies very rapidly, giving the apparent effect of the presence of a broadband signal at the control grid. An ammeter connected in the anode of the valve does not show these variations, since it reads the average anode current. As a result of shot noise, a valve carrying a large current is 'noisier' than one carrying a small current. This is one reason why beam tetrodes are not used in stages which are to amplify small signals. This problem also affects the variable-mu pentode, since a high amplification factor is only achieved with a high anode current; for this reason variable-mu pentodes are not used in the first amplifying stage in communications receivers, since the noise introduced would have a detrimental effect on the performance of the amplifier.

The division of the space current between positively charged electrodes introduces another component of noise known as *partition noise*. This is a random phenomenon, which is not related to the random variations in electron emission from the cathode. Multi-electrode valves are inherently noisy, and in applications where low noise value is important, triodes are often used in preference to pentodes.

Another source of noise in valves is *microphony*, which results from the vibration of electrodes when the valve is subject to a mechanical shock. Microphony can be reduced by mounting the valve in a resilient holder, and acoustically shielding the valve.

1.18   Valve ratings

Published data often give *absolute maximum rating* values of operating conditions, which should not be exceeded under the worst possible operating conditions. Care should be taken in interpreting these ratings, since the maximum anode voltage and the maximum anode current cannot be achieved simultaneously. The interpretation of some maximum values is illustrated in Fig. 1.18. Two other maximum values are of interest. Firstly the *peak inverse voltage* (p.i.v.), which usually applies to diodes, and is the peak repetitive inverse voltage that may be applied without damage to the valve. Secondly, the heater-to-cathode voltage rating of indirectly heated valves is limited by the breakdown voltage of the insulation between the heater and the cathode.

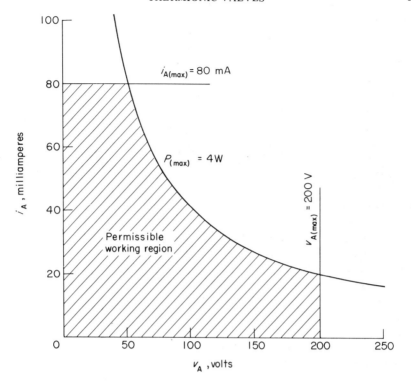

Fig. 1.18   Illustrating the maximum ratings of electronic devices.

There is, in addition, a *design centre* system of ratings which gives the average figures for valves of that type. Most published data give design centre values, which allow for variations in supply voltage, component tolerance, etc.

1.19   Valve construction

The majority of industrial valve electrodes are made of nickel or iron, which are easily workable and are chemically stable materials. Electrodes are insulated from one another by spacers, usually made from mica, which are treated to remove surface impurities before being sealed. The electrode assembly is then connected to the valve base and inserted into the glass envelope, together with a small magnesium pellet, known as the *getter.*

The whole assembly is then heated to a temperature just below the softening temperature of the envelope, which is about 400°C for glass and 600°C for Pyrex, and the interior of the valve is evacuated by a vacuum pump. At the same time, the electrodes are heated by induction. The two processes have the effect of removing practically all traces of gas and water vapour from the inner surfaces of the valve.

Finally the getter, which is a chemically active material, is fired by induction heating. The getter combines with any further small quantities of gas that may be liberated during the lifetime of the valve. The getter deposits itself on the glass envelope of the valve, forming the familiar silvery film on the inside of the valve.

A valve which has a high vacuum inside it is known as a *hard valve*, while one containing gas is described as a *soft valve*. Soft valves are considered in detail in chapter 2. Since many gases are electro-negative, i.e., they acquire electrons, the presence of such gases (oxygen is an example) seriously affects the performance of the valve at high frequencies. Gas particles also increase the noise level introduced by the valve.

1.20   Signal distortion

Owing to the curvature of the characteristic curves of valves (and transistors for that matter!), the output signal from amplifiers may be a distorted image of the input signal under certain operating conditions.

The principal factors influencing the degree of distortion are (1) the point on the characteristics at which the valve is operated, and (2) the magnitude of the input signal. If the operating point (set by the bias circuit) is close to the curved parts of the characteristic, then even a small input signal causes the anode current to be a distorted version of the input signal. Large input signals invariably cause distortion.

In triodes, distortion is largely *second harmonic distortion* due to the introduction of signals of twice the frequency of the input signal by the valve. In pentodes, distortion can be either second harmonic or *third harmonic* (i.e., due to the introduction of frequencies three times greater than the signal frequency). The reason for the distortion in the triode is that its mutual characteristic is curved only at its lower end (assuming that temperature-limited operating conditions are not reached), whereas in the pentode the mutual characteristic is curved at both the lower and upper ends. Generally speaking, for least distortion, the valve should be biased to the centre of the linear part of the mutual characteristic, and the input signal should have small value.

Problems

1.1 Explain the following terms: valence electron, covalent bond, work function.

1.2 Explain the following terms in relation to the high vacuum valve: space charge, Schottky effect.

1.3 Distinguish between temperature-limited operation and space-charge limited operation in a vacuum diode.

1.4 Describe a test made to obtain the static characteristic of a vacuum diode. Draw the circuit arrangement used; list the equipment necessary and indicate the ranges of instruments used. Sketch the characteristics, and show how the forward conduction resistance and reverse blocking resistance can be calculated.

1.5 Describe the construction of an indirectly heated cathode suitable for use in a vacuum diode. Discuss the relative merits of filamentary and indirectly heated cathodes.

1.6 When the anode current and the anode voltage of a certain vacuum diode are 5 mA and 70 V, respectively, the relationship between the anode current and anode voltage is given by $i_A = kv_A^{1.5}$. Determine the value of k, and calculate (a) the anode voltage for an anode current of 4 mA, and (b) the anode current for an anode voltage of 75 V.

1.7 A diode has the following static characteristic:

v_A	(V)	5	10	15	20	25	30	35
i_A	(mA)	8	22·5	42	65	88	113	137·5

Plot the characteristic, and determine (a) the instantaneous slope resistance r_A, and (b) the slope resistance r_a at anode voltages of (i) 10 V, and (ii) 30 V.

1.8 For the diode in question 1.7, plot the dynamic characteristics of the circuit for loads of (a) 120 Ω, (b) 200 Ω, (c) 700 Ω, and (d) 1·4 kΩ. The maximum supply voltage is 50 V.

1.9 Determine the voltage across the diode in question 1.7 when the load resistance is 1 kΩ and the supply voltage is 40 V.

1.10 Discuss the relative merits of the constant voltage (Thevenin) generator circuit and the constant current (Norton) generator circuit in electronic circuit analysis.

1.11 A certain generator provides an output voltage of 9 V when delivering 10 mA, and 8 V when delivering 20 mA. Derive both the Thevenin and Norton equivalent circuits of the generator.

1.12 (a) Describe a test made to obtain the static characteristic of a vacuum triode. Sketch the circuit arrangement used.

(b) List all the equipment required giving typical ranges.

(c) Sketch the form of curves obtained by plotting typical test results and explain how (i) the anode slope resistance, (ii) the mutual conductance, and (iii) the amplification factor may be derived from these curves.   (C & G)

1.13 In a triode valve, $\mu = 20$ and $r_a = 5$ kΩ. Draw the equivalent constant current (Norton) generator circuit.

1.14 In a certain triode, the anode voltage and current were found to be 140 V and 12 mA, respectively, when the grid bias voltage was −2 V. When the anode voltage was increased by 10 V, at constant grid voltage, the anode current was found to be increased by 3 mA. The anode conditions were then returned to their initial values, and the grid bias voltage was increased to −3 V at a constant anode current of 12 mA. To maintain the anode current constant in the latter test, the anode voltage had to be increased by 10 V. Evaluate the parameters of the valve.

1.15 The following characteristics refer to a triode:

v_A(V)	i_A (mA) at grid voltages of				
	−2	−3	−4	−6	−8
130	10·5	8	5·6	2·7	0·9
150	13·5	10·6	8	4·3	2
170	16·8	13·5	10·6	6·2	3·3

Plot the characteristics and estimate the parameters of the valve at a grid bias of −3 V and an anode voltage of 150 V.

1.16 The following characteristics were obtained from a thermionic device. State if it is a triode, a tetrode, or a pentode.

v_A (V)	50	70	90	110	130	150	170	190	210
i_A (mA) (v_G = −3 V)	0·6	1·8	2·4	5·6	7·8	10·6	13·4	16·6	20·0
i_A (mA) (v_G = −4 V)	0·2	0·8	2·0	3·8	5·5	7·9	10·5	13·5	16·8

Draw the characteristics and estimate the values of the parameters μ, r_a, and g_m at each point given in the table. Plot the results to a base of i_A.

1.17 Discuss the reasons for the development of tetrode and pentode valves.

1.18 The following characteristics refer to a pentode valve.

v_A(V)	i_A (mA) at grid voltages of			
	−0·5	−1·0	−1·5	−2·0
150	13·0	10·5	8·25	6·25
200	13·2	10·7	8·3	6·3
250	13·3	10·8	8·4	6·4

Plot the characteristics and estimate the slope resistance on each curve for v_A = 200 V. Estimate the value of μ when v_A = 200 V, v_G = −1·5 V.

1.19 In a pentode valve, g_m = 6 mA/V and r_a = 200 kΩ. Draw the voltage generator (Thevenin) equivalent circuit.

1.20 Draw diagrams to show the construction of a beam tetrode valve.

Explain what is meant by secondary emission and how the effects of it are overcome in a beam tetrode.

Compare the anode characteristics of the beam tetrode with those of a similarly rated pentode. Comment on the relative fields of applications of these valves and explain why one is preferred to the other for power amplification. (C & G)

1.21 Define the *amplification factor* of a valve and describe one method by which it may be measured. State the factors which govern the amplification factor of a valve: thence explain (i) how a *variable-mu* (i.e., variable amplification factor) characteristic is obtained, (ii) why the amplification factor of a pentode is much greater than that of a triode valve. (C & G)

1.22 Describe the sources of noise in:
　(a) thermionic valves,
　(b) transistors,*
and mention, where appropriate, how these vary with frequency. (C & G)
　　* See chapter 4.

1.23 (a) Sketch typical anode current/anode voltage static characteristics for a tetrode and a pentode valve.

Using these, explain (i) the function of the earthed grid in the pentode, and (ii) why the tetrode can behave as a negative resistance to a.c.

(b) Show that the anode circuit of a pentode can be regarded as a constant current generator. (C & G)

1.24 What is the Miller effect and how can it be reduced? Show that the effective input capacitance at the grid of a resistance loaded amplifying stage is given by

$$C_{gk} + C_{ga}(1 - m)$$

where

C_{gk} is the grid-cathode interelectrode capacitance
C_{ga} is the grid-anode interelectrode capacitance
m is the magnitude of the stage gain.

When the stage gain is −8, the effective capacitance is 15 pF and rises to 27 pF when the gain is doubled. Calculate the two valve capacitances.   (C & G)

1.25 The characteristic of a certain valve is given by $i_A = a + bv_G + cv_G^2$, where i_A is the anode current and v_G is the grid voltage. Draw a graph showing how mutual conductance varies with grid bias.

Given that $a = 18$, $b = 6$, $c = 0 \cdot 5$, calculate the mutual conductance at a bias of 3V.

(Based on a C & G question)

2. Gas-filled and vapour-filled valves

2.1 Ionization in gases

Under normal conditions, gas molecules are constantly being ionized by atmospheric radiation. For example, cosmic rays are a source of ionization. Thus, if a gas is enclosed in an evacuated tube which has two electrodes in it, the application of a potential between the electrodes causes the electrons and positive ions in the gas to be attracted to opposite electrodes. At low pressure* there are few gas molecules in the tube, and the space current is very small (of the order of 10^{-14} A, depending on the magnitude of the applied voltage).

Free electrons are accelerated in the tube by the electric field, and acquire *kinetic energy*. When the applied potential is low the kinetic energy of the electrons is small, and if an electron collides with a gas molecule it has little chance of ionizing it. As the applied potential is increased, the mean velocity and kinetic energy increase. A critical condition is achieved when the energy reaches what is known as the *ionization potential* of the atom or molecule of gas. When this happens, the electron has a finite chance of detaching an electron from any gas molecule it collides with. If an electron is detached, the molecule is said to be *ionized by collision*. Ionization potentials of commonly used gases are listed in Table 2.1

Table 2.1

Atom or molecule	Helium	Neon	Argon	Krypton	Mercury
Ionization potential (eV)	24·5	21·5	15·7	14·0	10·4

* Pressure, in valve technology, is given in *millimetres of mercury* (mm Hg) and atmospheric pressure is 760 mm Hg. *High pressure* refers to a pressure which is approximately atmospheric pressure, while *low pressure* is a fraction of atmospheric pressure.

A voltage is finally reached at which the current in the tube continues to flow, even if the external source of radiation is removed. The voltage at which this occurs is known as the *breakdown voltage,* and the discharge is said to be *self-maintained.* When breakdown occurs, the potential distribution along the tube is as shown in Fig. 2.1(a). The majority of the tube p.d. occurs close to the cathode, and is known as the *cathode fall.* The magnitude of the cathode fall varies from about 60 V to 400 V, depending upon the gas and the work function

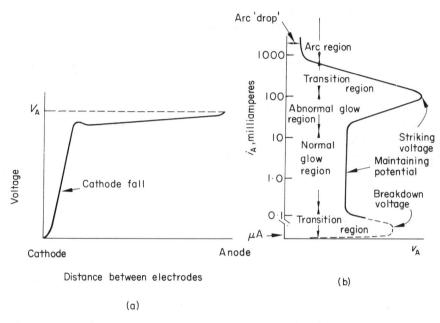

(a)

(b)

Fig. 2.1   (a) Potential distribution in a glow discharge tube, and (b) the *V/I* characteristic of a disruptive discharge.

of the cathode coating. After breakdown, the positive ions move toward the cathode and enter the region of the cathode fall. The high field in this region causes them to be accelerated towards the cathode, and they strike it with sufficient energy to release electrons from the cathode by secondary emission. Since the gas ions themselves release electrons for the purpose of ionizing further molecules, the process becomes self-maintained.

Under this condition, the gas emanates visible radiation in the region between the anode and the cathode. This is known as *normal glow,* and is associated with a constant voltage drop (known as the *maintaining potential*) across the tube over a wide range of values of current, as shown in Fig. 2.1(b). The colour of the glow is characteristic of the gas used, and the p.d. across the tube is dependent upon the gas used and the work function of the cathode. A feature of normal glow operation is that the cathode current density is substantially constant. An increase in current results in a proportional increase in the area of the cathode which is covered by the glow.

The limit of the normal glow characteristic is reached when the cathode glow cannot increase further in area. Further increase in current results in the current density increasing. This results in an increase in the p.d. across the tube, leading to ions striking the cathode with greater energy, so raising its temperature. This is shown as the *abnormal glow* region of the characteristic in Fig. 2.1(b), and has no significant engineering application.

Increasing the current density further results in a transition from the secondary emission characteristic to one which is dependent on some other mechanism. In the *thermionic arc,* electrons are supplied by a heated cathode. The heat may be generated either by positive ion bombardment or by means of a heating circuit. In *mercury-pool devices,* the electron emission mechanism is thought to be due to field emission.

The *arc region* of the characteristic is characterized by a high current and a low voltage drop, in contrast with the normal glow region which is characterized by a low current and a high voltage drop. An arc may be distinguished from a glow discharge in terms of the cathode fall. If the fall is less than the breakdown potential of the gas, then the discharge is described as an arc.

2.2   Cold-cathode diodes

In the normal glow region the cathode remains cool, and diodes which are designed to operate in this part of the characteristic are known as *cold-cathode diodes* or *glow tubes.* Since the p.d. across the tube is reasonably constant over

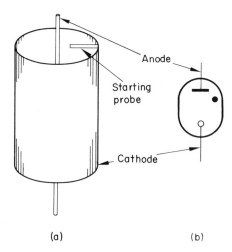

(a)   (b)

Fig. 2.2   (a) Electrode structure of one form of cold-cathode diode, and (b) its circuit symbol.

the working current range (1 mA to 50 mA), cold-cathode tubes are frequently used in simple shunt voltage regulators (see chapter 13).

Resulting from the fact that the cathode current density is constant, the surface area of the cathode must be large in order to handle the higher values of current. One method of construction is shown in Fig. 2.2(a), in which the cathode is the inner surface of a cylinder which surrounds an anode wire. The cathode surface may be coated with a low work-function material to reduce the cathode fall. Gases in common use in these tubes include argon, helium, and neon, the usual range of breakdown potentials being 75 V to 150 V. Cold-cathode lamps are also used as indicator lamps and test lamps. Some tubes have a *starting probe* (shown in Fig. 2.2(a)) to reduce the breakdown potential of the tube. To distinguish between hard valves and gas-filled tubes, a dot is placed inside the circuit symbol of gas-filled tubes; in some diagrams a gas-filled tube is indicated by cross-hatching the circuit symbol.

Since the cathode of the tube has a much larger area than the anode and the work function of the cathode surface is less than that of the anode, normal glow discharge occurs only when the anode is positive with respect to the cathode. This property permits the cold-cathode tube to be used as a diode. In its usual applications, a resistor known as a *ballast resistor* is coupled in series with the cold-cathode diode to limit the current through the tube to a safe value once the discharge has commenced.

2.3   Hot-cathode gas-filled diodes

Hot-cathode gas-filled devices have a thermionic cathode, and contain either an inert gas at low pressure or mercury vapour or both. For the purpose of this book, mercury vapour fulfils a similar function to that of a gas. The current ratings of these devices range from 50 mA to over 20 A, and are chiefly used as rectifiers. The cathode may be in the form of an oxide-coated ribbon filament or an indirectly heated oxide-coated emitter.

After the gas has been ionized, positive ions surround the cathode and partially neutralize the electronic space-charge. The overall effect is to minimize the space-charge limitation which is imposed on the cathode current in other-wise equivalent high vacuum valves. As a direct consequence of this, a small increase in anode voltage results in a large increase in the current through the valve. The anode current remains largely due to the electrons released by thermionic emission from the cathode, the positive ions contributing by the reduction in the cathode space-charge.

When operated under normal conditions, the anode-to-cathode p.d. of the gas-filled valve is 10 V to 17 V, typically. When the valve is overrun, the cathode becomes completely stripped of its protective coat of electrons and the cathode surface is bombarded by positive ions. This can only result in the destruction of the emissive surface of the cathode in a very short time. Hot-cathode gas-filled devices should, therefore, never be operated outside the manufacturer's ratings.

The cathode construction used in a hot-cathode gas-filled tube differs from that used in hard valves, due mainly to the principle of operation of the valve. In high vacuum valves, the cathode cannot be allowed to have pockets or recesses, otherwise high concentrations of electrons accumulate in them, cutting off emission from the lower parts of the pocket. In gas-filled valves, much of the space-charge is eliminated by positive ions, allowing the use of specially shaped electrodes. One form of *heat-shielded cathode* used in these tubes is illustrated in Fig. 2.3. The cathode, which is oxide coated, comprises an inner cylinder

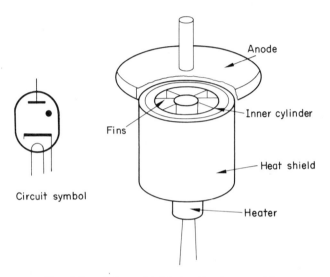

Circuit symbol

Anode

Inner cylinder

Fins

Heat shield

Heater

Fig. 2.3   A heat-shielded cathode assembly.

with fins heated by a central concentric heater. The whole is surrounded by several concentric polished metallic heat shields, which reflect heat back to the cathode. By this means, the thermal efficiency of the cathode is increased above that of the type employed in hard valves. Owing to the relatively massive cathode structure, it may take a considerable length of time for the cathode to reach its working temperature; large cathodes may take as long as 30 min. It is inadvisable to apply the anode voltage in these valves before the cathode has reached its working temperature, otherwise ionic bombardment may damage the cathode. The top of the cathode structure in Fig. 2.3 is open to allow electron emission, and the anode in the form of a cap is placed over the open end.

2.4   The thyratron

The name thyratron is given to gas-filled triodes and tetrodes. It was originally a trade name of a specific device, but the name became so popular that it was adopted to describe all gas-filled triodes and tetrodes. The grid of the triode version is a massive structure, shown in Fig. 2.4, and serves only to control the initiation of the arc between the anode and cathode. Once this has occurred, the grid circuit loses control over the anode current. This is explained in the following paragraphs. Assume for the moment that the grid is held at a negative potential, and that the anode is at a small positive potential. Some of the electrons emitted by the cathode are captured by the anode, in much the same way as in a thermionic triode. An increased anode voltage results in an increased anode current, but it is still of the order of a few microamperes. As the anode voltage is increased further there will be a point at which electrons

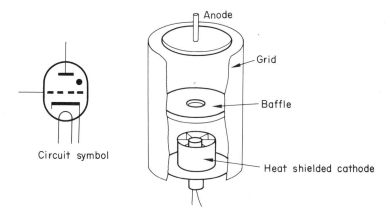

Fig. 2.4   One form of electrode structure of a thyratron valve.

acquire sufficient energy to ionize gas molecules. For a given anode voltage, the grid voltage at which this occurs is known as the *critical grid voltage.*

The *turn-on time* of the thyratron depends on the rate at which ionization proceeds, and is usually 0.1 μs to 10 μs, depending on the gas pressure, the temperature, and the type of gas. To turn the arc off it is necessary either to reduce the anode voltage to zero (or to make it negative) or to divert the anode current through another path for a short time. The grid must also be taken to a large negative potential. During this period of time the positive ions and free electrons *recombine,* to provide an insulating medium between the anode and the cathode. The *turn-off time* or *deionization time* lies between 10 μs and 1 ms for mercury-filled valves and 5 μs and 10 μs for hydrogen-filled valves. As a result of the relatively long turn-off time, thyratrons are limited to applications in which the maximum supply frequency is a few hundred hertz.

Once the arc has become established, the negative charge on the grid repels electrons and attracts positive ions to itself. This results in the formation of a positive *ion sheath* on the surface of the grid structure. This has two principal effects. Firstly, the positive ions neutralize the negative charge on the grid and prevent it from exercising control over the anode current. Secondly, the potential of the grid may no longer be negative. If the grid is raised to a potential which is a few volts above that of the cathode, it will act as a second anode, and an arc will be established between it and the cathode. To prevent the possibility of this happening, a *grid stopper* resistor R_s (Fig. 2.5(a)) of value 10 kΩ to 100 kΩ is connected in the grid circuit.

Thyratrons fall into one of two distinct groups; *negative control valves* and *positive control valves.* Both types have baffles in the grid structure, and in negative control thyratrons the baffles have relatively large holes in them, and a negative grid potential is required to *prevent* the arc from striking. In positive control valves, the baffle has a number of very small holes in it, and the electrostatic shielding between the anode and the cathode is almost complete. In these

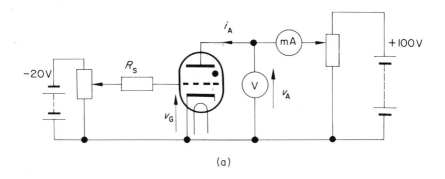

(a)

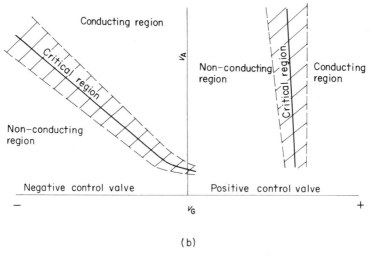

(b)

Fig. 2.5　(a) A test circuit to obtain the control characteristics (b) of the thyratron valve.

valves, a positive grid potential is required to *initiate* the arc. Generally speaking, negative control valves are used in industrial applications, and the magnitude of the linear part of the critical grid characteristic (Fig. 2.5(b)) is known as the *control ratio*.

Valves filled with mercury have a family of characteristics as shown in Fig. 2.5(b), dependent upon the condensed-mercury temperature, while those filled with an inert gas have a single characteristic. Circuits must be designed to accommodate the changes in the characteristics of mercury-filled devices.

Because of the increasing use of semiconductor devices, the thyratron is becoming obsolescent, particularly in low voltage applications. In the higher voltage range they are still holding their own, and a typical specification of one such thyratron is

Peak inverse voltage (kV)	15
Peak forward voltage (kV)	15
Peak anode current (A)	75
Mean anode current (A)	15

Example 2.1: Two points on the critical grid characteristic of a negative control thyratron are found to be $v_A = 170$ V when $v_G = -8$ V, and $v_A = 90$ V when $v_G = -4$ V. Determine (a) the control ratio, (b) the critical grid voltage to cause conduction when $v_A = 150$ V, and (c) the minimum value of v_A to cause conduction when $v_G = -12$ V.

Solution: The characteristic is shown in Fig. 2.6.
(a) The control ratio is given by the modulus of the slope of the characteristic.

$$\text{Control ratio} = (170 \sim 90)/(8 \sim 4) = 80/4 = 20$$

(b) The simplest method of solution is to obtain the equation of the linear part of the characteristic. The vertical intercept of the linear part of the characteristic is given by C in Fig. 2.6, where

$$C = 90 - (4 \times 20) = 10 \text{ V}$$

and the equation of the linear part of the characteristic becomes

$$v_A = -20v_G + 10 \tag{2.1}$$

Since $v_A = 150$ V, then

$$v_{G2} = -(v_A - 10)/20 = -(150 - 10)/20 = -7 \text{ V}$$

(c) From eq. (2.1)

$$v_{A2} = -20(-12) + 10 = 250 \text{ V}$$

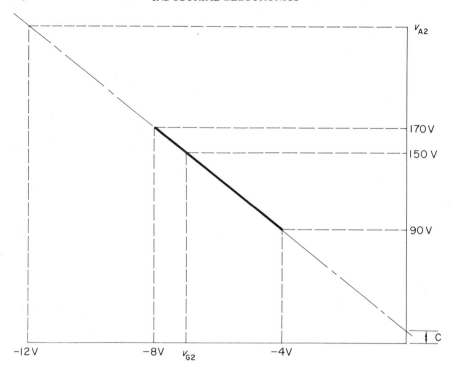

Fig. 2.6   Figure for example 2.1.

2.5   Shield-grid thyratrons

Prior to breakdown, the grid current of a thyratron is of the order of a few tenths of a microampere. Although this is a minute current, it can cause trouble in circuits which have a very high source impedance, e.g., photoelectric cell circuits. To limit the pre-breakdown grid current to a very small value, a fourth electrode known as the *shield grid* is introduced. The general form of construction of the shield-grid thyratron is shown in Fig. 2.7. The shield grid is a large structure which surrounds the anode and the cathode; the control grid is in the form of a ring or loop, which is opposite the hole in the baffle in the shield grid. The value of the pre-breakdown control grid current in the shield-grid thyratron is reduced to the order of 1 nA (10^{-9} A) to 10 nA.

The shield-grid potential clearly has an effect on the characteristic of the thyratron. For either a positive or slightly negative shield-grid voltage, the thyratron has a characteristic which is generally similar to that of a negative control valve. When the shield grid is at a large negative potential, the characteristic is not unlike that of a positive control thyratron.

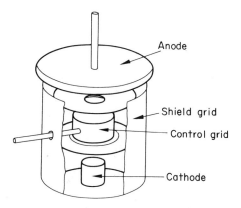

Fig. 2.7   Electrode assembly of a shield grid thyratron.

2.6   The ignitron

The ignitron is a controlled diode with a mercury-pool cathode. Control is provided by an electrode known as an *ignitor*. The ignitor, shown in Fig. 2.8, has the shape of a tapered cone and is made of silicon carbide or boron carbide which is not wetted by mercury. The tip of the ignitor dips into the mercury, and when an impulsive current is passed through it a spark is produced between

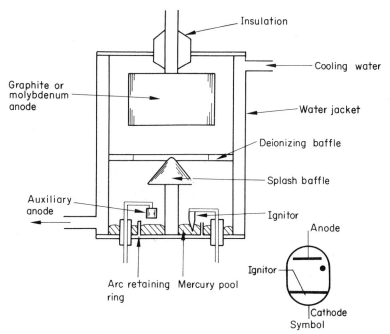

Fig. 2.8   Cross-section of an ignitron.

the ignitor and the mercury. If the anode is at a positive potential with respect to the cathode, it takes up the arc a fraction of a microsecond after the first spark at the ignitor. The signal pulse applied to the ignitor rod has an amplitude of 10 A to 30 A at 100 V to 400 V for a period of 100 μs to 1 ms. The pulse is generated by a control circuit in each half-cycle in which conduction is required.

The *auxiliary anode* or *excitation anode* is supplied from an alternating voltage source, and acts as an intermediate stage in the transfer of the arc to the main anode. When the main anode becomes negative, the arc is automatically extinguished.

The ignitron is generally used to control the flow of current in a.c. circuits; the peak inverse voltage applied to the anode must be less than the breakdown voltage of the gas, otherwise the anode may act as a cold cathode to give a glow discharge. This may develop into an arc to the cathode, known as a *backfire,* possibly resulting in damage to the ignitron and connected equipment. As a precaution against backfire, the anode is made of a high work-function material which is not wetted by mercury. Additionally, there is a deionizing grid between the anode and the cathode, which is normally connected to the case, and a splash baffle is mounted below the anode.

Principal applications of ignitrons include resistance welders, and capacitor dischargers where the peak current may be many hundreds of kiloamperes.

Example 2.2: An ignitron is to be used to control the current in a resistive circuit, the supply being 250 V r.m.s., single-phase. If the peak current in the circuit is 20 A, estimate the value of the circuit resistance. What is the average current in the circuit if the conduction angle is 90 degrees? Neglect the effects of arc drop.

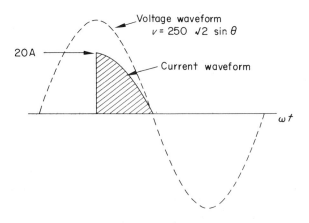

Fig. 2.9   Figure for example 2.2.

Solution: The circuit waveform is shown in Fig. 2.9. Since the peak voltage is $250\sqrt{2}$ V, and the peak current is 20 A, then the circuit resistance is

$$R = 250\sqrt{2}/20 = 17 \cdot 7 \ \Omega$$

If the ignitron conducts for the complete positive half-cycle, the average current is I_m/π. Since conduction is initiated after 90 degrees in this case, the average value is $I_m/2\pi$, hence

$$\text{Average current} = 20/2\pi = 3 \cdot 18 \text{ A}$$

2.7   The mercury-arc convertor

Where continuous control of very large values of direct current (e.g., several thousands of amperes) is required, the multi-anode mercury-arc convertor is used. Figure 2.10 shows a three-anode mercury-arc convertor, in which anodes a1, a2, and a3 are energized from a three-phase alternating supply. The arc is initiated by means of an *ignition mechanism*, and the arc is immediately taken

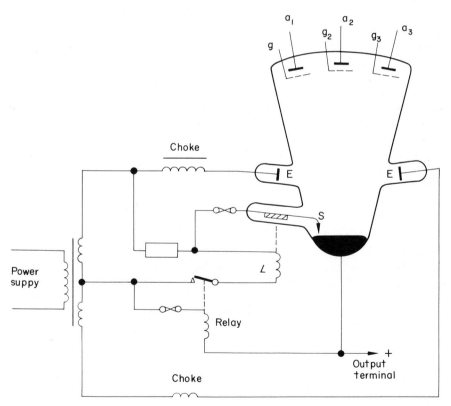

Fig. 2.10   The mercury-arc convertor, showing the starting and auxiliary anode circuits.

up and maintained by *auxiliary anodes* E. In the basic circuit shown, the main arc transfers or *commutates* to the anode with the highest positive potential. When the potential of that anode falls below that of another anode, the arc automatically transfers to the other anode. Thus, each anode carries the full current for one-third of the total period of the supply waveform.

When the supply is switched on, solenoid L is energized via the normally-closed relay contact, and the *starting electrode* or *ignition electrode* S is momentarily brought into contact with the mercury pool. This instantaneously short-circuits the solenoid and de-energizes it; the release of the starting electrode causes a spark at the surface of the mercury. An arc is then initiated between the mercury cathode and the auxiliary anodes, which supply inductive loads. The chokes serve to prevent the current in the auxiliary circuit from falling to zero, so maintaining the 'spot' on the mercury pool to enable conduction to the main anodes to take place at any time. Once current commences flowing in the auxiliary circuit, the relay is energized and the supply to the starting circuit is broken.

Control over the load current is exercised by means of the control grids, not shown in Fig. 2.10, each grid controlling the point at which its anode commences conduction. This is described in section 12.9. By means of grid control, it is possible to prevent the arc transferring to a more positive anode, since a negative grid potential greater than the critical grid potential prevents conduction.

The number of anodes used is an integral multiple of the number of phases of the power supply system. The most common are three-, six-, and twelve-anode convertors. In some instances, groups of anodes can be operated independently of the others in the same convertor, so that more than one anode is conducting at any instant. Such convertors are described as *multiplex convertors*. This mode of operation has the effect of increasing the transformer utilization and improving the overall efficiency, at the expense of an increase in circuit complexity and capital cost.

Factors limiting the rating of a mercury-arc convertor are the *maximum average anode current,* the *peak inverse voltage,* and the *peak current.* If the maximum average anode current is exceeded, the anode overheats and it may function as a thermionic emitter, resulting in the anode acting as a second cathode. Excessive inverse voltage applied to the anode may result in an anode acting as a cold cathode, giving a glow discharge. Either of the above effects may result in either a backfire or a *crossfire* (an arc between two or more anodes). If the maximum peak current is exceeded, the cathode may be damaged.

2.8   Inverted operation

A gas-filled device acts as a *rectifier* pure and simple if it does not have a control grid (i.e., it converts an alternating signal into a unidirectional signal). If it has a control grid it can be used either as a rectifier, or as an *invertor* (which converts a unidirectional signal into an alternating signal). This statement generally applies to any device which has an ON-OFF characteristic. The general principles of inversion are illustrated by reference to the thyratron circuits in Fig. 2.11.

When the instantaneous supply voltage V_s is greater than the voltage across the load (which is a battery in Fig. 2.11(a)), the thyratron can be triggered into conduction by the application of a control signal between the grid and the cathode. A power generating state is seen to occur when the voltage and current

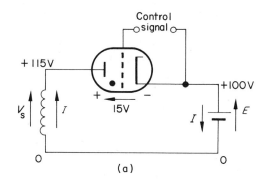

(a)

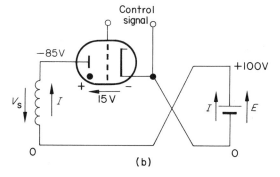

(b)

Fig. 2.11   Illustrating the process of (a) rectification, and (b) inversion.

arrows on the figure act in the same direction, and a power consuming state occurs when the arrows are in opposite directions. Figure 2.11(a) thus corresponds to the process of rectification. If the connections to the battery are reversed, and the gate signal is applied when the transformer supply to the anode is negative, as shown in Fig. 2.11(b), the anode of the thyratron is still positive with respect to its cathode, and current flows from the battery to the secondary winding of the transformer. The potential and current arrows now show the battery to be in a 'generating' state, and the transformer to be in a load consuming state. This corresponds to the condition inversion, when power is supplied from the d.c. source to the a.c. system.

The simple system depicted in Fig. 2.11 is subject to practical limitations, and a poly-phase arrangement would normally be adopted.

2.9   Visual read-out devices

In electronic counting networks, it is often desirable to be able to read the state
of a counter at a particular stage. For this purpose, special indicating devices
have been developed. Early devices were used to give a numerical read-out, and
are generally described as *Nixie tubes,* although this is a trade name.

Devices considered here are gas-filled, and contain an anode and a number of
cathodes of various shapes, either in an alphabetical or numerical configuration.
When any one of the cathodes is taken to a sufficiently negative potential, it
glows and gives the desired numerical or alphabetical indication. *Alpha-numeric
indicator tubes* are manufactured in which there are a large number of linear
cold-cathode segments. A numerical or alphabetical character is formed whenever a negative potential is applied to an appropriate group of segments. Other
units have a plate which is perforated with characters, each perforation being
located over an individual cold cathode. When the cathode glows, the appropriate
character on the plate is illuminated.

One of the earliest read-out devices was the *Dekatron tube,* which combined
both counting and read-out functions. The physical arrangement of one type of
Dekatron tube is shown in Fig. 2.12(a). The tube operates on the principle that
the breakdown potential of a gas discharge is reduced if positive ions already
exist in the vicinity of the cathode. A series of cathodes are distributed con-
centrically around a central anode. These are numbered zero to nine in the
figure. There are ten other *guide cathodes* marked *a,* and ten more marked *b.*
All the *a* guide cathodes are connected in parallel, as are the *b* guide cathodes.

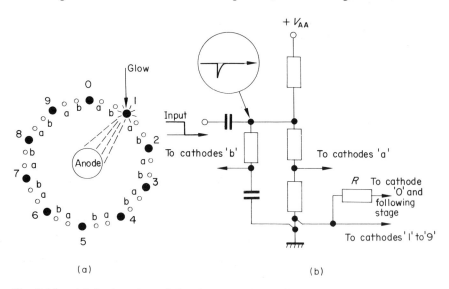

Fig. 2.12   (a) A plan view of the electrodes in one form of Dekatron tube.
(b) A control circuit suitable for use with a Dekatron tube.

A typical control circuit for the Dekatron is shown in Fig. 2.12(b).

With zero input signal, the *a* and *b* guide cathodes are at a positive potential, and the discharge terminates on one of the main cathodes, cathode 1 in the figure. The application of a large negative-going pulse to the input of the control circuit causes guide cathodes *a* to fall instantaneously to a negative potential, causing the discharge to commutate to the nearest *a* guide cathode. The fall in the voltage of the *b* guide cathodes is delayed by the charging time of capacitor *C*. The values of the circuit components are such that the peak potential applied to the *b* guide cathodes is even more negative than that applied to the *a* guide cathodes, resulting in the transfer of the discharge to the *b* cathodes. When the input pulse dies away, the glow automatically transfers to the closest main cathode, i.e., cathode 2. Successive input pulses cause the discharge to transfer in the sequence 1, 2, . . . , 8, 9. The tenth pulse causes it to transfer to the 'O' cathode, and the voltage developed across resistor *R* is used to trigger the next Dekatron in the sequence. The pulse must be applied for about 0·15 ms for the transfer to be effective, limiting the counting rate to about 5000 pulses per second. However, these tubes have a high reliability and their capital cost is low, and they will remain in service for many years to come.

Higher counting rates can be achieved by high vacuum decade switching tubes, and counting rates up to 10^6 per second have been achieved. Semiconductor read-out devices are now coming into more common use, and these are discussed in chapter 5.

Problems

2.1 Write a short essay on the electrical breakdown mechanism in gas discharge tubes.

2.2 How is it possible to differentiate between a glow discharge and an arc discharge? State a typical application of each type of discharge.

2.3 Account for the shape of the characteristic curve of a cold-cathode discharge tube and hence state why a limiting resistor must be used with this type of device.
Sketch the electrode assembly and explain the operation of a dekatron tube.

(C & G)

2.4 Draw a circuit diagram of a laboratory test circuit to obtain the characteristic of a cold-cathode discharge tube. Describe the test procedure and indicate the ranges of the instruments required. What precautions must be taken with conventional moving-coil instruments to ensure accurate results?

2.5 In a laboratory test on a cold-cathode discharge tube, the tube is shunted by a voltmeter of resistance 3 MΩ. Just prior to ionization the applied voltage is 150 V, and the current drawn from the supply is 51·5 μA. What is the forward blocking resistance of the tube?
After conduction has commenced, the tube voltage is found to be 155 V at a current of 1 mA, and is 156 V at 11 mA. Calculate the slope resistance of the tube.

2.6 Draw a connection diagram for a laboratory test circuit to obtain the d.c. control characteristic of a hot cathode gas-filled triode. Outline, briefly, the test procedure.

Sketch, to convenient scales, a sine wave representing an alternating voltage of 300 V peak applied between anode and cathode of such a triode and the critical bias curve. The control ratio is 20.

What bias voltage will just prevent conduction?

(C & G)

2.7   Explain the principles underlying the action of a gas-filled triode. How and why does its behaviour differ from that of the vacuum diode?

Give an example of a practical application of the gas-filled triode.

(C & G)

2.8   What is meant by the process of ionization as applied to a thyratron?
Hence explain why:
(a) the grid has no control over the anode current once ionization has started,
(b) a limiting resistor must be included in the anode circuit and
(c) the anode voltage must not be applied before the cathode has reached its working
    temperature.
State the advantage of the shield grid (tetrode) thyratron over the triode type.

(C & G)

2.9   (a) Draw typical control and anode characteristics for a small-current thyratron and explain its operation.

(b) A thyratron conducts when $v_A = 200$ V with $v_G = -8$ V and when $v_A = 120$ V with $v_G = -4$ V. Determine, (i) the control ratio, (ii) the minimum value of v_A to give conduction when $v_G = -11$ V, and (iii) the critical grid voltage when $v_A = 100$ V.

2.10 Explain, with the aid of characteristic curves and appropriate circuit diagrams, how the d.c. grid control of a vacuum triode is different from that of a hot cathode gas-filled triode.

Outline an industrial application of a hot cathode gas-filled triode, giving details of the grid control system and its effect.

(C & G)

2.11 (a) Describe with the aid of a sketch, the construction of a cold-cathode counting tube (Dekatron).

(b) Draw a diagram of the basic associated circuitry and explain how the discharge moves on one count when an input pulse is applied.

(c) Show how to couple the output from one counting tube to the input of the next.

(C & G)

2.12 Sketch a diagram showing the construction of the ignitron. Clearly label each part on the diagram. Explain the principle of operation of the ignitron.

An ignitron is to be used to control the current in a resistive circuit, and the average current is to be 10 A. The supply is from a 220 V r.m.s. single-phase alternating supply. Calculate the value of the circuit resistance and the peak value of the current if the conduction angle is 90 degrees and the arc voltage drop can be neglected.

3. Semiconductor diodes

3.1 Semiconductors

There are many ways of classifying materials, a common method being by their electrical conductivity. Broadly speaking, materials may be divided into conductors, semiconductors, and insulators. *Conductors* are defined as materials having a low resistivity, in the range zero to 10^{-4} Ωm. *Insulators* have conductivities greater than 10^3 Ωm, while *semiconductors* have conductivity values at room temperature in the range 10^{-4} Ωm to 10^3 Ωm.

There are many semiconductor materials available, but very few of them have a practical application in electronics. The two most useful materials are silicon (Si) and germanium (Ge), both of which are tetravalent. The atoms in a single crystal of either pure silicon or pure germanium are joined together by covalent bonds, and the valence electrons are shared between the atoms. At absolute zero temperature (0 K), all the electrons are at their lowest energy levels, and the valence electrons are taken up in the covalent bonds of the structure. As a result there are no free electrons at this temperature and, in theory, both silicon and germanium are perfect insulators. This does not occur in practice, since it is not possible to manufacture a perfect crystal, and impurities exist within the crystal (a crystal is said to be pure if the impurity content is less than one part in 10^{10}).

At room temperature, some valence electrons acquire sufficient energy to break away from the parent atom to become free. Simultaneously, a *hole* is said to be generated in the crystal structure (it has already been stated that a hole is equivalent to a positive charge carrier). This process is known as *thermal generation of an electron-hole pair*. Electron-hole pairs are generated at many points within the crystal, with the result that an electron from one part of the crystal lattice soon *recombines* with a hole from another part of the structure. The average time that either charge carrier exists is known as the *lifetime* of the carrier, which is typically 10^{-2} to 10^{-4} s. When a hole and an electron recombine,

55

they cease to exist as charge carriers. Thermally generated electron-hole pairs give rise to *intrinsic conduction* in the semiconductor when an electrical potential difference is applied between the ends of the sample.

Electron hole pairs can also be generated by other sources of energy, such as light. Light can be considered to consist of a series of small packets of energy, known as *photons,* whose energy is proportional to frequency. Thus, blue light is more 'energetic' than red light. Atomic radiation can also generate electron-hole pairs in semiconductor materials.

3.2   Extrinsic conduction

The deliberate introduction of a small quantity of impurity (of the order of 1 part in 10^8) modifies the electrical properties of semiconductors. It results in a new type of conduction, known as *extrinsic conduction.*

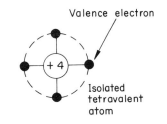

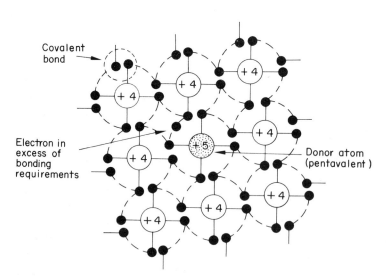

Fig. 3.1   The lattice structure of an *n-type* semiconductor material which has tetravalent 'host' atoms.

If the impurity atom has a valency of five, i.e., it is *pentavalent,* it has one more electron in its outer shell than have the host atoms (assuming them to be either silicon or germanium), as shown in Fig. 3.1. The atomic structure in the figure is simplified by representing it in planar form, or flat form, without loss of accuracy. Since there are only four electrons in the valence shell of the parent tetravalent atoms, the net charge on these atoms if the valence electrons were stripped off would be + 4 electronic units. The additional electron orbiting the pentavalent atom is surplus to the covalent bonding requirements, and can be removed more readily from the crystal lattice than can the valence electrons of the parent atoms. This type of impurity is known as *n-type* impurity (*n* for *n*egative

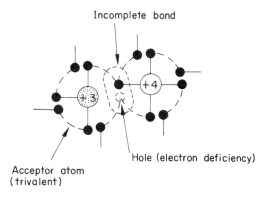

Fig. 3.2   The structure of a *p-type* semiconductor.

charge carrier), and the impurity atom is a do*n*or atom since it donates an electron for the purposes of conduction.

In n-type semiconductors, free electrons are the major means of conducting charges through the lattice, and electrons are described as *majority charge carriers.* Holes generated by thermal or other means also act as charge carriers but, since these are relatively few in number when compared to the majority charge carriers, they are known as *minority charge carriers.* Typical donor elements include phosphorous (P), arsenic (As), and antimony (Sb).

If a germanium or silicon atom in the crystal is replaced by a trivalent atom (valency 3), as shown in Fig. 3.2, a *hole* is generated in the lattice due to incomplete bonding. This hole can be 'filled' by any free electron in the crystal lattice, causing a hole to appear at some other point. This type of impurity is known as *p-type* impurity (*p* for *p*ositive charge carrier), and the impurity atom is known as an acce*p*tor atom, since it results in the acceptance of an electron into the lattice. Typical acceptor elements include boron (B), aluminium (Al), gallium (Ga), and indium (In). The mobile hole has all the characteristics of a positive charge carrier, and holes are the majority charge carriers in p-type semiconductors. Electrons generated by thermal and other effects are minority charge carriers.

Gallium arsenide (GaAs) is another material which is finding increasing use in electronic devices. Gallium has a valency of three, and arsenic has a valency of five. The same general rules for doping apply to gallium arsenide as to silicon and germanium. If the valency of the impurity atom is less than the lowest valency of the host atoms, then the crystal becomes a p-type semiconductor as, for example, if zinc is used, which has a valency of two. If the valency of the impurity atom is greater than the highest valency of the host material, then the crystal becomes an n-type semiconductor. Selenium, which has a valency of six, is a suitable donor impurity element. Tetravalent impurity atoms, e.g., germanium, can be used if the process is correctly controlled. If a germanium atom replaces a gallium atom, an n-type semiconductor is formed, whereas a p-type semiconductor results if it replaces an arsenic atom.

The simple theory outlined here is adequate to explain the operation of most of the devices we come across in electronics, but some devices require a more sophisticated theory. One such theory, the *energy-band theory* is given in simplified form below.

In an atom, electrons can only orbit in certain *levels* or *bands,* in much the same way that a car in a multi-storey car park can only park on one of the floors, and not between the floors. In semiconductors, the highest energy band containing electrons is the valence band, the next higher being an empty band known as the *conduction energy band.* Since electrons are forbidden to appear at energy levels in the range between the two levels, the gap between them is known as the *forbidden energy gap.* In pure semiconductor materials (those exhibiting only intrinsic conduction), the breaking of a covalent bond is equivalent to an electron gaining sufficient energy to leap the forbidden energy gap. This leaves an electron in the conduction band and a hole in the valence band.

The reverse leakage current of a semiconductor device is related to the size of the forbidden energy gap, since a large gap results in a smaller intrinsic leakage current at a given temperature. Since the forbidden energy gap in silicon is greater than in germanium, the leakage current in silicon devices is much less than in otherwise similar germanium devices.

At this point, we reach the first significant difference between the simple covalent bond theory and the energy-band theory. Since an electron in the conduction band is at a higher energy level in the atom than is the hole in the valence band, we find that the electron is more *mobile* than the hole. The difference in the *mobilities* of electrons and holes is of the order of 2:1 in the more usual semiconductors. This significant fact is not apparent from the simple theory.

In the energy-band theory, the effect of adding impurities is to introduce additional available energy levels in the forbidden energy gap. In the case of n-type materials, a 'full' level (i.e., full of electrons) is introduced close to the conduction band. This means that electrons can be transferred into the conduction band by only acquiring a small amount of energy, much less than is necessary to jump across the forbidden energy gap. In p-type semiconductors, an 'empty' level (i.e., one containing no electrons) is introduced into the gap close to the valence band. As a result, electrons can leave the valence band and enter the new energy level with much less energy than is required to cross the forbidden gap, leaving a hole in the valence band.

3.3   Diffusion and drift

Charge carriers move through semiconductor materials by two distinct and separate mechanisms: diffusion and drift. *Diffusion* occurs when there is a concentration of free charge carriers in one part of the crystal. The mutual repulsion between the charge carriers (assuming that they are of the same kind) results in a net movement of charges from an area of high concentration to one of low concentration.

Drift is an effect resulting from the application of an electric field (i.e., the application of a potential difference across the material) to the semiconductor. In p-type semiconductors, the drift current consists of holes which move towards the end of the semiconductor at a more negative potential. In n-type semiconductors, the drift current consists of electrons drifting toward the more positive end of the material. In both cases, the charge carriers are replenished by the supply source which sets up the electric field.

3.4   The p-n junction diode

A p-n junction diode is formed from a *single crystal* containing an n-region and a p-region, as shown in Fig. 3.3. When the p-region is connected to the negative pole of a battery and the n-region to the positive pole, as in Fig. 3.3(a), the mobile holes in the p-region are attracted away from the junction by the negative pole, as are the electrons in the n-region by the positive pole. As a result, the number of charge carriers in the region of the junction is depleted, and is known as the *depletion layer*. The greater the applied voltage, the greater the width of the depletion layer. Theoretically, the flow of current between the two regions should be zero since no mobile charge carriers are available at the junction. In fact, a small *leakage current* flows due to thermal generation of minority carriers in both halves. These are indicated by arrows in Fig. 3.3(a). The leakage current is of the order of a few nanoamperes in a silicon signal diode, and slightly greater in a germanium diode. In this operating state, the diode is said to be reverse biased, and is in its *reverse blocking mode*. The effect of an increase in temperature is to increase the number of minority carriers generated, so increasing the leakage current. The effect on the diode characteristic of an increase in temperature is shown in Fig. 3.4. The leakage current under reverse bias conditions, known as the *reverse saturation current,* is substantially constant at any given temperature. As the temperature increases, the number of

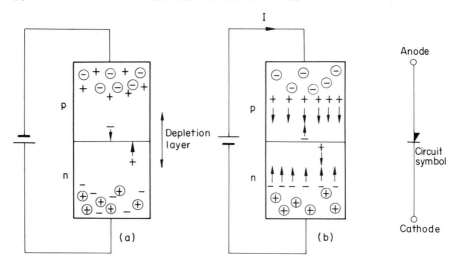

Key

⊖   Fixed negative charges

⊕   Fixed positive charges

−   Mobile negative charges

+   Mobile positive charges

Fig. 3.3   The operation of the p-n junction diode under (a) reverse bias and (b) forward bias.

thermally generated minority carriers increases, and with it the reverse leakage current.

The application of a reverse voltage in excess of the *peak inverse voltage* rating of the diode can lead to an electrical breakdown of the diode. It then operates in the *reverse breakdown* or *reverse conducting* mode. In this phase of its operation, the diode current increases rapidly for a very small increase in applied voltage. Unless the current is limited by some means, breakdown can result in the destruction of the p-n junction. Diodes known as *Zener diodes* are designed to operate in the reverse breakdown mode, and are described in section 3.9.

By reversing the polarity of the applied voltage, see Fig. 3.3(b), the majority carriers are attracted towards the junction under the influence of the electric field. In the region of the junction, the numbers of charge carriers diminish due to recombination of charge carriers. Thermally generated minority carriers also contribute to the current flow. The net result is that the potential drop across the diode in the forward conduction mode is small, being typically 0·8 V to 1 V in a silicon diode, and one-half this value in a germanium diode. An increase in temperature gives rise to the generation of more minority carriers, giving an

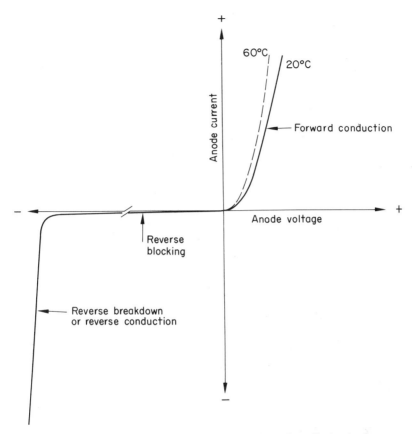

Fig. 3.4   The static anode characteristic of a p-n junction diode. An increase in temperature causes the characteristic to change in the manner shown.

increased forward current for a given forward p.d. The junction is said to be *forward biased* in this mode of operation.

Due to its characteristics, the p-n junction diode finds greatest application as a rectifying device; the rating of one of the larger present-day junction rectifiers is 500 A, 1800 V (p.i.v.), with a maximum reverse leakage current of 15 mA.

3.5   Capacitance of the p-n junction diode

When a reverse voltage is applied to a p-n junction diode, a depletion layer is formed whose width is dependent on the applied voltage. An increase in the reverse voltage causes the majority charge carriers to move further away from the junction, and a reduction in reverse voltage causes them to move closer to the junction. Movement of charges within the diode manifests itself in the external circuit as a current.

Thus, a reverse biased p-n junction diode appears to the external circuit as a capacitor whose capacitance is a non-linear function of the applied voltage. The capacitance usually has a value in the range 5 pF to 20 pF, and is known as the *transition capacitance* of the diode. Diodes which are deliberately operated under reverse biased conditions to give a variable reactance effect are known as *varactor diodes.* Another version, the *step recovery diode,* is a variant of the varactor diode with an abrupt junction.

In addition, another effect known as *storage capacitance* occurs under forward biased conditions. It was stated earlier that holes and electrons diffuse throughout the crystal in order to unify the charge density. This effect is independent of the drift current. Hence, holes from the p-region diffuse into the n-region, and electrons diffuse in the reverse direction. This charge is effectively stored, and returns when a reverse bias is applied, as shown in Fig. 3.5. The storage current is principally hole flow, since the p-type anode region is doped more heavily with impurity atoms than is the n-type cathode. As a result, this phenomenon has become known as *hole storage.* The storage

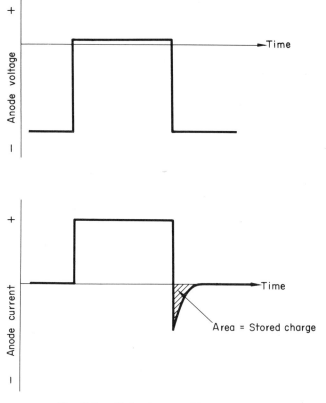

Fig. 3.5   Hole storage effect.

capacitance may have a value of several hundred picofarads. It is a transient effect and should not be confused with the transition capacitance, which is always present under reverse biased conditions.

The combined effects of transition capacitance and storage capacitance limit the useful high frequency range of p-n junction diodes below the gigahertz (10^9 Hz) frequency range. Special diodes are manufactured which overcome the high frequency limitations of the junction diode; two of these are described in sections 3.6 and 3.7, respectively.

3.6 The point-contact diode

The point-contact diode was one of the first types of semiconductor diode to be used, and is the modern version of the old 'cat's whisker' crystal detector used in early radio receivers. The diode consists of an n-type semiconductor against which a fine tungsten spring wire presses. During manufacture, a pulse of current is passed down the wire, and is believed to form a p-region at the point of contact with the semiconductor. The operation then follows normal p-n junction theory. Owing to the small size of the junction, the diode capacitance is very small—of the order of a few picofarads—allowing the diode to be used in the gigahertz frequency band. The small area of contact unfortunately limits the anode current to a very low value.

A variant of the point-contact diode is the *gold-bonded diode*. In this diode, the wire spring is made of gold, resulting in an improved current rating when compared with the point-contact diode.

3.7 The Schottky barrier diode

A Schottky barrier diode or *hot carrier diode* is a rectifying metal-to-semiconductor junction diode. Several metals may be used, including gold, molybdenum, titanium, chromium, nickel, nichrome, and aluminium in conjunction with either n-type or p-type silicon. Since the mobility of electrons is greater than that of holes, n-type silicon is nearly always used as the semiconductor since it ensures better high frequency performance.

Current flow in the Schottky barrier diode differs from current flow in conventional p-n junction diodes in that minority carriers (holes in n-type semiconductors) do not take any part in the process. This has the effect of eliminating charge storage effects, enabling switching speeds less than 0·1 ns to be achieved. As a consequence of this, Schottky barrier diodes can be used at frequencies up to about 40 GHz.

The electron flow results from the property of solids under the influence of a strong electric field to emit electrons. The electrons have energies which are normally associated with electrons which leave a 'hot' body, hence the name hot carrier diode.

In the early days of the development of Schottky barrier diodes, one limitation was that of the reverse breakdown voltage, which was typically −5 V to −10 V. The introduction of a diffused p-type guard ring (*see also* section 4.4) around the edge of a metal-to-n-type Schottky barrier junction has overcome this problem, and reverse breakdown voltages of the order of several hundred volts can be achieved.

3.8   The tunnel diode

The tunnel diode, introduced in 1958, is a junction diode in which both regions are very heavily doped with impurity atoms. The ratio of the doping in tunnel diodes to that in conventional diodes is approximately $10^5:1$. The net effect is to reduce the 'reverse' breakdown voltage to a point where it occurs at a positive voltage, as shown in Fig. 3.6. Typical values are shown in the figure.

The name *tunnel diode* is derived from the *quantum theory* which is used to explain its operation. According to the quantum theory, there is a probability of an electron 'tunnelling' through the thin potential barrier existing between the regions of a reverse biased diode. If, according to the classical theory, the electron does not have sufficient energy to overcome the potential barrier, it can sometimes appear on the other side of the barrier by tunnelling through it.

Tunnel diodes have been constructed from a number of semiconductor materials including germanium, silicon, gallium arsenide, and gallium antimonide.

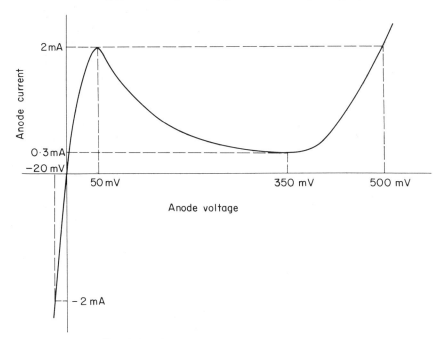

Fig. 3.6   Static characteristic of a tunnel diode.

Applications include its use in amplifiers, switches, and logic circuits.

A feature of heavily doped diodes which is worthy of note is the low reverse voltage drop. Versions of heavily doped diodes, though not sufficient to produce the negative resistance region of the tunnel diode, are operated in the 'reverse' direction to give a smaller voltage drop than is achieved with conventional diodes. The diodes are described as *backward diodes* or *back diodes,* the name implying that the diode conducts more heavily in the reverse direction than in the forward direction. As a consequence of using the diode in a 'backward' or reverse connection, the 'forward' voltage drop when carrying the rated current is about 0·1 V. 'Reverse' breakdown occurs at about 0·7 V. To realize the full rectification efficiency, the backward diode is used with signals with a peak voltage less than about 0·7 V.

3.9   Zener diodes

One of the first useful theories on the breakdown of dielectrics was put forward in 1934 by C. Zener. He suggested that, under the influence of very intense electric fields (several million volts per centimetre), electrons can tunnel through the forbidden energy gap. This theory is certainly true in semiconductor p-n junction diodes with a relatively thin depletion layer, when breakdown occurs with anode voltages in the region of −2·6 V to −5 V. This type of breakdown is known as *Zener breakdown.*

With wider depletion layers, obtained by modifying the impurity doping, breakdown occurs at a much higher voltage than is expected by Zener's theory. In such cases, the breakdown mechanism is thought to be due to *avalanche breakdown,* which is the solid-state equivalent of ionization in gas-filled valves. The breakdown mechanism is explained as follows. Leakage current in p-n junction diodes is largely due to minority carriers drifting across the junction under the influence of the reverse bias. If the electric field in the region of the junction is sufficiently great, the electrons are accelerated and acquire enough energy to ionize atoms by collision, i.e., they strip off valence electrons to generate electron-hole pairs in the depletion layer. The effect is multiplicative, causing a very rapid transition from a reverse blocking state to a reverse conducting state. Both avalanche and Zener breakdown diodes are traditionally known as Zener diodes.

Typical static anode characteristics of various types of Zener diodes are shown in Fig. 3.7(a), together with one form of circuit symbol (b).

An approximate equivalent electrical circuit of the Zener diode is shown in Fig. 3.7(c). Diode D1 allows conduction in the conventional manner when the anode is positive with respect to the cathode. Conduction takes place through D2 when the cathode potential exceeds the breakdown potential V_Z. Resistance r_Z represents the reverse breakdown slope resistance of the diode, which has a value between 0·5 Ω and about 150 Ω, depending on the operating point and the diode rating.

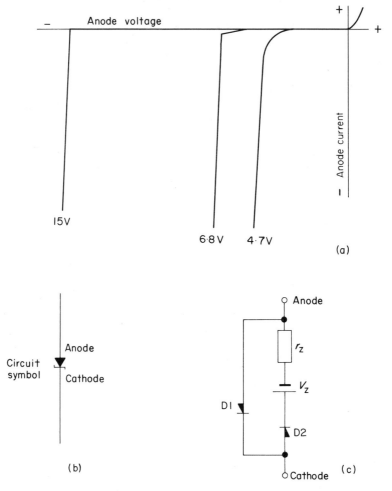

Fig. 3.7   (a) Reverse breakdown characteristics of various types of Zener diode. (b) Shows one form of circuit symbol, and (c) an equivalent electrical circuit.

The anode current of the 4·7 V diode in Fig. 3.7(a) is largely composed of carriers which have tunnelled through the potential barrier, and the characteristic is typified by a smooth 'knee'. Avalanche breakdown diodes have a characteristic with a sharp 'knee', which is due to the sudden transition from reverse blocking to reverse breakdown, and the slope resistance is generally higher than that of a Zener breakdown diode. This is illustrated in the 15 V characteristic. When breakdown occurs between 5 V and 7 V, there is a range of diodes whose breakdown mechanism is partly due to Zener effect and partly due to avalanche breakdown. The slope resistance of this range of diodes is the lowest of all. The forward characteristics of both Zener and avalanche break-down devices are generally similar to a conventional p-n junction diode.

It is found that Zener and avalanche breakdown mechanisms react in different ways to a change in temperature. In the Zener breakdown, the tunnelling effect increases with temperature, resulting in a greater flow of current for a given anode voltage, as illustrated in Fig. 3.8. In avalanche breakdown devices, the increased temperature results in increased activity in the atomic structure, and the free charge carriers find it more difficult to acquire the critical energy to ionize other atoms. As a result, the breakdown voltage of avalanche diodes increases with temperature. Between the two, in the region of −5 V, the two variations balance each other out, giving diodes with characteristics which are relatively insensitive to temperature. The latter are used as *voltage reference diodes,* having voltage-temperature coefficients of the order of zero (at one point on the characteristic only), combined with a very low slope resistance.

Present-day ratings of Zener diodes range from breakdown voltages of 2·6 V to over 200 V, at power ratings up to 75 W and over. They are most commonly

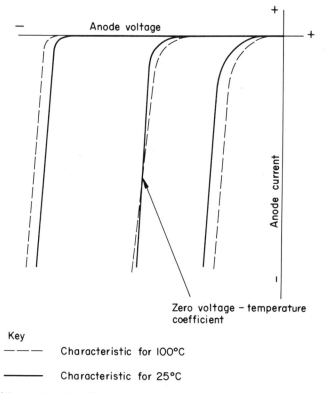

Fig. 3.8   Illustrating the effect of temperature change on the reverse breakdown characteristics of Zener diodes.

used as voltage reference sources, and other applications include meter protection circuits, instrument scale expanders, and low speed computing circuits.

3.10   Unijunction transistors

The unijunction transistor (UJT) or *double-based diode* has two ohmic connections to a bar of n-type material, shown in Fig. 3.9(a). These connections are known as *base-one* (B1) and *base-two* (B2), respectively. A p-n junction is formed between a p-type *emitter* and the bar. In the absence of a signal at the emitter, the interbase resistance is of the order of 5 kΩ to 10 kΩ, so that the leakage current is small. The bar acts as a potential divider, and a potential of about $0.4V_{BB}$ to $0.8V_{BB}$ appears between the emitter and base-one, where V_{BB} is the interbase voltage. The coefficient of V_{BB} above is known as the *intrinsic stand-off ratio* η.

If V_E, the emitter voltage, is less than ηV_{BB}, the emitter-to-bar junction is reverse biased, and the emitter current is very small. When the emitter voltage is increased to the point where the junction is forward biased, the UJT turns on, and the resistance between the emitter and base-one falls to a low value.

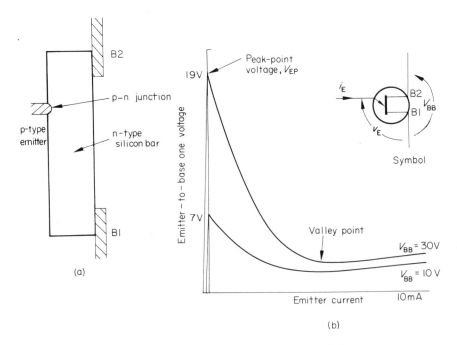

Fig. 3.9   (a) Construction of a unijunction transistor, and (b) its static characteristic curves.

This occurs at the *peak-point voltage* V_{EP}, which is given by

$$V_{EP} = V_j + \eta V_{BB}$$

where V_j is the forward voltage drop across the p-n junction, and is of the order of 0·7 V. It is frequently possible to neglect V_j when compared with ηV_{BB}.

The UJT characteristics shown in Fig. 3.9(b) are typical only. For a general-purpose UJT, the peak-point current is of the order of $1\mu A$ at V_{BB} = 20 V, with a valley-point voltage of about 1·5 V at 6 mA. Complementary UJT's with p-type bars are also manufactured. UJT's are commonly used in relaxation oscillator circuits, in which they are used to periodically discharge a capacitor. In these circuits, the peak emitter current may be several amperes, which falls to a very low value after a few microseconds when the capacitor is discharged.

3.11   Copper oxide and selenium rectifiers

A copper oxide rectifier consists of a copper disc on one side of which is a film of p-type cuprous oxide (Cu_2O), the impurity atoms being oxygen which are introduced during manufacture. Between the p-type oxide and the copper is an insulating *barrier layer* of pure copper oxide about 1 μm (0·04 thou.) thick. Since copper has a lower work function than the p-type semiconductor, electrons more readily leave the copper and pass to the oxide than is true in the reverse direction. Hence, the metal acts as the cathode of the rectifier, and the oxide as the anode.

The theories put forward to explain the operation of the selenium rectifier are more complex than those given hitherto, and what follows is a simplified explanation. A layer of selenium compound is formed on a plate of steel or aluminium, resulting in the formation of a barrier layer between the compound and the plate. The selenium compound then acts as the anode and the metal plate as the cathode.

Owing to the barrier layer in both types, a small forward voltage must be applied before conduction commences, which is about 0·25 V per element in a copper oxide rectifier and 0·7 V in a selenium rectifier. The inverse voltage that a single element of either type can support without excessive flow of reverse current is much less than in p-n junction diodes, being about 6 V per element in copper oxide rectifiers and 10 V to 60 V in selenium rectifiers; in order to operate at high voltages, elements are connected in series.

Copper oxide rectifiers are still used as instrument rectifiers, although they are now being superseded by semiconductor devices. Before the advent of semiconductors,.selenium rectifiers were used in almost every industrial rectifier application, ranging from applications requiring thousands of amperes at a few volts to a few amperes at thousands of volts. Each year semiconductor devices make inroads in areas which traditionally belong to metal rectifiers.

Problems

3.1   Explain the following terms: (a) donor atom, (b) acceptor atom, (c) valence electron, (d) n-type material, (e) p-type material, (f) covalent bond, (g) intrinsic conductivity, (h) extrinsic conductivity.

3.2   What are minority charge carriers and majority charge carriers in semiconductor materials, and what is their function?

3.3   Show how tetravalent semiconductors can be converted into p-type and n-type semiconductor materials. Illustrate your answer with a lattice diagram of the crystal structure. Give the names of commonly used impurity atoms.

3.4   Explain how it is possible for substances like gallium arsenide to be converted into extrinsic conductors.

3.5   Give a simple explanation of the energy-band theory of semiconductors.

3.6   What is meant by the term *mobility* in connection with charge carriers?

3.7   What is meant in semiconductor theory by (a) drift current, and (b) diffusion current. Using these concepts, describe the principle of operation of the p-n junction diode.

3.8   Explain, with the aid of a characteristic curve, the principle of operation of a Zener diode. What is meant by Zener breakdown and avalanche breakdown? What are the essential differences between them?

3.9   If you were selecting a Zener diode for a voltage reference source from the following range, which one would you select and why would you select it? 2·5 V, 4·7 V, 15 V. Explain the effects of temperature variation on the reverse breakdown voltage of the three types.

3.10   Sketch the characteristic curve of a tunnel diode, and explain the principle of operation of the diode. Suggest a practical application for the device.

3.11   Draw a diagram of a unijunction transistor, and explain how the device operates.

3.12   Two copper oxide rectifier elements have the following forward conduction characteristics:

Voltage	0·4	0·6	0·8	1·0	1·2
Current (rectifier 1)	0·4	1·8	5·5	12·5	23
Current (rectifier 2)	1·5	4·5	11	21	—

If the two rectifiers are connected in series to a 1·2 V d.c. supply, estimate the value of the voltage developed across each diode, and the circuit current.

3.13   Explain briefly the principle of the Zener diode and sketch a typical voltage-current characteristic.

Draw a diagram of a circuit which makes use of the properties of a Zener diode, and briefly explain its operation.

(C & G)

4. Transistors and other semiconductor devices

4.1   The junction transistor

The junction transistor* was one of the first types of transistor to come into commercial use, and many thousands are now used in electronic circuits. The junction transistor is a *bipolar transistor*, that is, it employs both p-type and n-type semiconductor regions. Early transistors were made from germanium, and were mainly of the p-n-p type, shown in Fig. 4.1. That is, they comprise a single crystal which contains two p-regions and an n-region. One of the p-regions, known as the *emitter*, is of low resistivity material with a heavy impurity doping. The emitter is the source of charge carriers in the transistor. The central n-region is known as the *base region* of the transistor, and is a relatively pure semiconductor of high resistivity. The base region is the control electrode or control

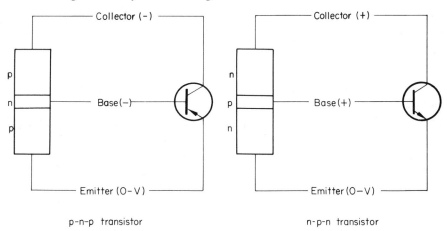

p-n-p transistor   n-p-n transistor

Fig. 4.1   Transistor symbols.

* The name transistor is derived from the expression *trans*fer re*sistor*.

71

region of the transistor. The second p-region, known as the *collector region,* has a lower conductivity than the emitter region, and is the region in which the mobile charge carriers are finally collected.

With the advent of silicon devices, n-p-n transistors have come into more common usage. In an n-p-n transistor, the two extreme n-regions are the emitter and collector, respectively, while the central p-region is the base region. The rules for doping the regions of the p-n-p type apply to the n-p-n type.

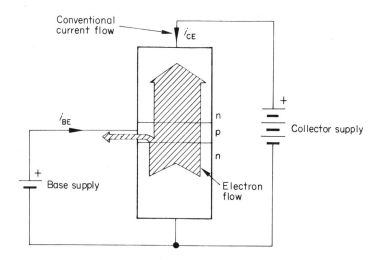

Fig. 4.2   The common-emitter circuit.

In both transistor types, the junction between the emitter and the base is known as the *emitter junction*, and that between the collector and the base as the *collector junction.*

A simplified circuit diagram of an n-p-n transistor in the *common-emitter mode* or *common-emitter connection* is shown in Fig. 4.2. The circuit is known by this name since the emitter is common to the input (base) and output (collector) supplies.

Under normal operating conditions, the emitter junction is forward biased, and charge carriers are emitted by the low resistivity emitter region into the high resistivity base region, the current taking the form of charge carrier drift. The collector supply voltage is normally much greater than the base circuit voltage, and the collector junction is reverse biased. However, the high concentration of charge carriers injected into the base region from the emitter diffuse throughout the crystal, most of them into the lower resistivity collector region. In practice, it is found that more than 95 per cent of the charge carriers leaving the emitter arrive at the collector. The majority charge carriers which leave the emitter become minority charge carriers when they arrive in the base region, and for this reason the bipolar transistor is regarded as a minority charge carrier controlled device.

The minority charge carriers in the base region which do not diffuse into the collector combine with majority charge carriers in the base region and disappear. The majority charge carriers in the base region are then replenished by the current i_{BE} from the external base circuit.

It is clear from the above explanation that the operation of bipolar transistors consists of *drift* of charge carriers from emitter to base, followed by *diffusion* of charge carriers from the base to the collector.

The ratio of the small-signal change δi_{CE} in collector current to the small-signal change δi_{BE} in base current in the common-emitter mode is given by the *hybrid parameter h_{fe}*, where

$$h_{fe} = \frac{\text{Change in } i_{CE}}{\text{Change in } i_{BE}}$$

The value of h_{fe} in present-day transistors usually lies between 20 and 600. Transistor hybrid parameters are discussed in detail in section 4.8.

So far, the transistor has been regarded as a current controlled device. In applications where they are used as switches, it is sometimes convenient to regard them as charge controlled devices, in which the charge built up in the base region by minority carriers is the primary control mechanism. The constants of the transistor, when considered as a charge controlled device, are known as *charge parameters*. Charge parameters are very successful in describing the operation of alloy-junction transistors (see section 4.2), since the charge is held in the base region. It is, however, less successful in describing the operation of planar devices, in which the charge may be stored in parts other than the base region. This introduction to charge parameters is adequate for the purpose of this book, more detailed information being available elsewhere*.

4.2 Types of transistor construction

The early types of transistors were *grown-junction transistors*, illustrated in Fig. 4.2, the p-regions and n-regions being grown within the single crystal. These were followed by *alloy-junction transistors*, in which suitable n-type impurity pellets are alloyed, under controlled conditions, on opposite sides of a p-type wafer which forms the base region. This type of transistor provided improved high frequency performance, when compared with the grown-junction transistor. Even so, the high frequency limit was below 20 MHz, leaving high frequency circuits firmly in the hands of thermionic devices. *Micro-alloy transistors*, having extremely thin base regions, were developed to improve the performance of transistors, but, unfortunately, these had limited collector-to-emitter voltage ratings.

* See Morris, N. M., *Logic Circuits*, McGraw-Hill (1969).

In both the grown-junction and the alloy-junction types, the diffusion mechanism is employed to transfer charge carriers to the collector. This is a relatively slow process, and in *alloy-diffused transistors* the base width is reduced to a few thousandths of a millimetre, and a *drift field* is introduced into the base region. This is achieved by grading the impurity doping between the emitter and the collector, forming a graded base region. The drift field causes the charge carriers to be accelerated towards the collector, causing a more rapid transition of charge carriers to the collector. This increases the frequency response of this type of transistor to the order of 100 MHz.

More recently *planar transistors* have been introduced, and these are described in the following section.

4.3   Planar transistors

The transistors so far described suffer from a limited frequency response, and relatively high leakage current. Leakage current is affected by the ambient temperature, particularly in germanium transistors where the forbidden energy gap is less than in silicon devices, the leakage current approximately doubling for each $8°C$ rise in temperature. In addition, the physical size of the three regions in the older type of transistors is large, when compared with modern transistors, resulting in charge storage problems. Also, in order to achieve high reverse breakdown voltages between the base and collector regions, it is necessary to use relatively high resistivity semiconductor material in the collector. This results in transistors having a high *saturation resistance*, that is a high resistance between emitter and collector when the transistor is fully *turned-on* or *saturated.*

The development of the *planar epitaxial transistor* has overcome many of these problems. In the n-p-n version, shown in Fig. 4.3, a heavily doped (low resistivity) *substrate*, marked n^+, is used as the foundation of the collector. Because of its low resistivity, the drawback of high saturation resistance is overcome. A thin *epitaxial*, lightly-doped n-type layer is then grown on the substrate. The epitaxial layer is of higher resistivity than the substrate, giving the high reverse breakdown voltage between the collector and the base regions.

The resulting piece of silicon, known as a *chip*, has its surface oxidized to give an insulating or *passivating layer* on its surface. It is arranged that the insulating layer has 'windows' left in it at suitable points, and a gas carrying a p-type impurity is allowed to come into contact with the chip. The impurity diffuses into the epitaxial layer, forming the base region of the transistor. A second diffusion of n-type material generates the emitter. Contacts are then made to the exposed areas of the transistor by evaporating metallic conducting material, usually aluminium, over the surface.

Complete circuits, known as *monolithic integrated circuits*, are constructed by these methods. In monolithic integrated circuits, resistors, diodes, and transistors are laid down in a common epitaxial layer, and are connected to one another

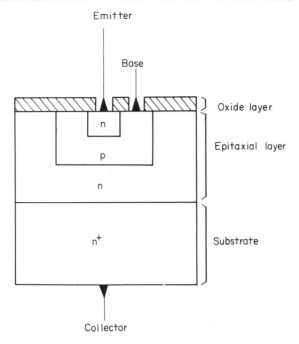

Fig. 4.3   A section of a planar epitaxial transistor.

by metallic interconnections, which are evaporated on to the oxide layer. Special isolating techniques are required in monolithic integrated circuits to prevent random leakage currents between elements on the same substrate from affecting the performance of the circuit. The size of a typical monolithic logic element for a computer may be as small as 1 mm square, and the only connections that need be made to it are those to the power lines and input and output signals. Unfortunately, the external connections to the integrated circuit are often the weakest link in an otherwise ideal circuit element.

4.4   The annular structure

The oxide passivating layer provides protection for the transistor against atmospheric impurities, it reduces the leakage current to a very small value, and it improves the long-term stability of the device. In the early days of the development of passivated transistors, the collector-to-emitter voltage was limited to less than about 30 V, due to a sudden and unpredictable change in the transistor characteristics when the collector region resistivity was increased to try to achieve high voltage working.

The problem can be understood by reference to the p-n-p structure in Fig. 4.4. The passivating layer is generated by converting a thin layer of the p-type epitaxial layer into an oxide. The p-type collector contains a large number of

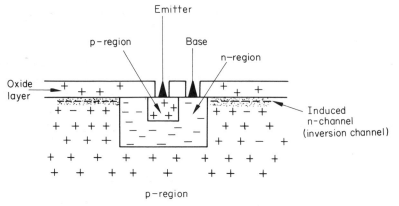

Fig. 4.4   The formation of an induced inversion channel in an epitaxial transistor.

positive charge carriers, and a small number of free electrons, which are generated by thermal effects. The positive charges on, or in, the oxide layer cause the minority carriers in the collector region to concentrate near to the surface, reducing the charge density at the interface between the semiconductor and the oxide. If the collector material has a low impurity concentration (high resistivity), the minority carriers attracted to the surface can convert the surface of the p-type semiconductor into an n-type semiconductor. This results in a shallow *induced inversion channel* at the collector surface. In the case of a p-type collector, this becomes an n-type channel. The net result is a spreading of the collector junction, with consequent modification of the characteristics in an unpredictable fashion.

All passivated devices are subject to channelling, which is used to good effect in some types of *field-effect transistors* (see section 4.14), and in *pinch-effect resistors*. The only way to eliminate channelling is to remove its cause, i.e., the oxide layer, but this solution would lead to an increase in leakage current and a reduction in reliability. Alternatively, methods of limiting the channelling effect

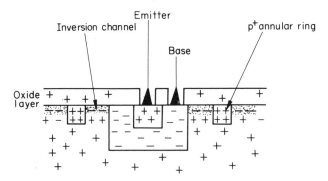

Fig. 4.5   One method of overcoming the effects of the inversion channel.

can be adopted. One such method is illustrated in Fig. 4.5, in which a shallow ring of low resistivity p^+ semiconductor material is diffused around the base region. Owing to the large number of positive charge carriers in the annular ring, relative to the number in the epitaxial layer, it is not converted into an n-type material, and the channelling is effectively stopped by the ring. An inversion layer exists beyond the ring, but this does not affect the operation of the transistor. As a result of this and similar techniques, high resistivity epitaxial collector regions can be used, permitting the development of high voltage transistors.

4.5   Bipolar transistor common-emitter characteristics

The common-emitter circuit is the most popular transistor connection, since it provides both high power gain and high current gain. A typical test circuit to obtain the static characteristics of an n-p-n transistor is shown in Fig. 4.6. Typical characteristics for a low power transistor are shown in Fig. 4.7. In the test circuit, the 500 Ω potentiometers are used to vary the base and collector supply voltages.

The *output characteristics*, Fig. 4.7(a), are particularly useful from a circuit design point of view, since they show how the collector current i_{CE} varies with the collector-to-emitter voltage v_{CE} over the normal working range of voltages. Initially, with the base circuit disconnected, the collector current is equal to the intrinsic leakage current in the common-emitter mode I_{CEO}. The subscript CE indicates that current is flowing from collector to emitter, and the subscript O gives the value of the external base current. In a low power silicon planar transistor, I_{CEO} has a value of a few tens of nanoamperes. Under this condition, the base voltage is of the order of 100 mV, due to the flow of current through the transistor. Other values of leakage current of interest in the common-emitter mode are I_{CES} and I_{CEX}. I_{CES} is the collector leakage current with the base short-circuited to the emitter, and I_{CEX} is the collector leakage current with specified base circuit conditions.

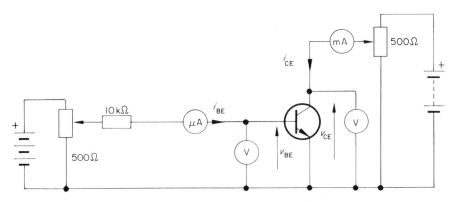

Fig. 4.6   Transistor test circuit.

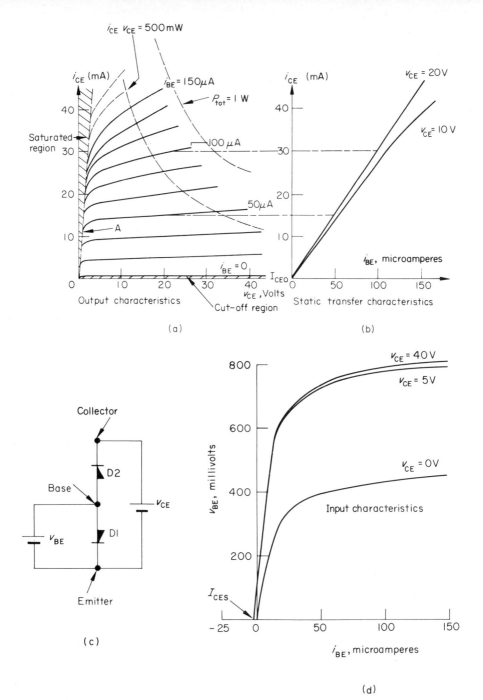

Fig. 4.7 (a) Static output characteristics, and (b) static transfer characteristics of a transistor in the common-emitter mode. A simplified equivalent circuit in the saturated operating mode is shown in (c), and the common-emitter input characteristics are shown in (d).

When the collector voltage v_{CE} is held at a very low value (ideally zero), and voltage v_{BE} is applied to the base, the transistor is said to be operated in the *saturated region*. A very simple explanation of this is afforded by Fig. 4.7(c). To a first approximation, the transistor may be represented by the circuit in the figure, where diode D1 is the base-emitter junction diode, and D2 is the collector-base junction diode. When v_{BE} is greater than v_{CE}, both diodes are forward biased, and current flows from the base region to the emitter and collector regions. When this occurs, both diodes are 'saturated' with current carriers, hence the name saturated operation. This picture of the operation of the transistor is an imperfect one, and modifications are made to it as the chapter proceeds. When the transistor is used as a switching device, e.g., in electronic logic circuits, the transistor is frequently worked in the saturated region of the characteristic, and a parameter of some importance is the static value of the *saturation resistance* r_{CEsat}. This is the effective resistance between the emitter and collector at some specified point on the output characteristics. In Fig. 4.7, this may be specified at some point A, when the *saturated collector voltage* V_{CEsat} may be 0·8 V at a collector current of 10 mA. In this case,

$$r_{CEsat} = \frac{V_{CEsat}}{I_C} = \frac{0·8}{0·01} = 80 \ \Omega$$

To achieve this condition, the *base-emitter saturation voltage* V_{BEsat} is equal to, or greater than, the collector saturation voltage. In the above case, V_{BEsat} would be of the order of 1 V. When the transistor is working in its saturated region it is said to be turned ON, in much the same way that a switch is said to be turned ON when it carries a large current and has a small p.d. across it.

When the collector voltage is greater than the base voltage, the collector junction becomes reverse biased, and the transistor is said to be operating in its *unsaturated region*. In this region, collector current continues to flow by reason of the diffusion process, as explained in section 4.1, and the greater the base current, the greater the collector current. When used in a *linear amplifier*, the transistor is biased to operate in the unsaturated region, when it has a pentode-like characteristic, with a characteristically high slope resistance.

Under zero base-emitter current conditions, the transistor is said to be OFF, in as much as a switch is said to be OFF when it passes only leakage current and supports the whole of the supply voltage across itself. This region of the output characteristics is known as the *cut-off region*. If the polarity of v_{BE} is reversed, i.e., it it is made negative in Fig. 4.6, the emitter junction becomes reverse biased, and the collector current falls even further, to a value known as I_{CBO}. In this event, the emitter is effectively open-circuited, and the collector leakage current flows to the base circuit. The subscript CB designates collector-to-base current, and subscript O indicates the value of the external emitter current. In a low power silicon transistor, I_{CBO} is usually of the order of a few nanoamperes at normal values of collector current; for example, in the 2N929 it is less than

10nA. In transistors which handle several amperes, I_{CBO} may have a value of the order of $100\ \mu A$.

At values of v_{CE} well above the rated collector voltage, there is a sudden transition in the characteristics when the collector current rises very rapidly. This is due to *avalanche breakdown* of the reverse biased collector junction. In addition, another phenomenon known as *punch-through* or *reach-through* causes a sudden increase in collector current. Punch-through is peculiar to alloy devices in which the base region is very thin. An increase in the collector voltage causes an increase in the width of the collector junction depletion layer. At some value of voltage, designated V_{pt}, the collector and emitter depletion layers meet, giving a base region of zero width. As a result, there is an abrupt increase in collector current. When this happens, the collector voltage is said to have 'punched-through' or penetrated the base region.

After avalanche breakdown in transistors, there is thought to be a 'second' breakdown which leads to catastrophic failure of the device. One reason for this is as follows. Any single transistor can be regarded as a number of smaller transistors in parallel, and when avalanche breakdown occurs in one of the smaller units the heat generated in it results in a reduction of resistance between the emitter and collector at that point. As a result, regions of the transistor exhibit a 'current hogging' effect, and local overheating occurs which leads to catastrophic failure of the device.

It is difficult to specify the collector-to-emitter breakdown voltage, unless the conditions pertaining to the base-emitter circuit are known. The condition leading to the lowest value of breakdown voltage occurs when the base is open-circuited. This breakdown voltage is known as $V_{(BR)CEO}$ or V_{CEO}, and this is the value usually quoted by manufacturers. If the base is connected to the emitter by a resistor, the breakdown voltage is specified as $V_{(BR)CER}$, the value of the resistance being given in the specification sheet. The breakdown voltage is higher when the base and emitter are short-circuited together, and is specified as $V_{(BR)CES}$.

The *static transfer characteristic*, at a given value of v_{CE}, is predicted from the output characteristics by noting corresponding values of i_{CE} and i_{BE}. Thus, when $v_{CE} = 20$ V, $i_{CE} = 30$ mA when $i_{BE} = 100\ \mu A$, and $i_{CE} = 15$ mA when $i_{BE} = 50\ \mu A$, etc. These points are then plotted in Fig. 4.7(b) to give the static transfer characteristic for $v_{CE} = 20$ V. The static transfer characteristic for $v_{CE} = 10$ V is also given in the figure; it is curved due to its nearness to the curvature of the collector characteristics in that region.

The *static value of the forward current transfer ratio* h_{FE} (which is also known as the *large-signal common-emitter current gain*) of the transistor is given by the slope of the static transfer characteristic, and is quoted at a particular value of v_{CE}. In the case of Fig. 4.7(b), its value at $v_{CE} = 20$ V is $h_{FE} = 30$ mA/$100\ \mu A = 300$. The numerical value of the current gain usually lies between about 20 and 600. The *parameter spread* of h_{FE} is about 4:1; that is, transistors of one type may have quoted minimum and maximum current gains of 150 and

600, respectively, while a typical value may be 350. Circuits must be designed to take account of the parameter spread. A moment's consideration of Figs. 4.7(a) and (b), indicates that h_{FE} falls in value as v_{CE} falls. The lowest value of current gain occurs when the transistor is working in its saturated mode, when the current gain is described by h_{FEsat}. In the case of the 2N929 transistor, one saturated condition is $V_{CEsat} = 1$ V when $i_{BE} = 0.5$ mA and $i_{CE} = 10$ mA. This corresponds to a current gain of 20, compared with a value of about 400 when working in the unsaturated region.

From the output characteristics, it appears that when a current flows in the base circuit, a much larger current flows in the collector circuit. The relationship between the two currents is given, neglecting leakage current, by the relationship

$$i_{CE} = ki_{BE}$$

where k is another parameter of the transistor. The simple equivalent circuit of Fig. 4.7(c) is modified to take account of this fact, by shunting the collector diode by a constant current generator, as shown in Fig. 4.8. For most low frequency applications, the equivalent circuit of Fig. 4.8 is adequate.

Since the base-emitter circuit of a transistor corresponds roughly to a diode, the input characteristic, Fig. 4.7(d), has a diode-type shape. When $i_{BE} = 0$, i.e., the base circuit is disconnected, there exists a small potential at the base terminal due to the leakage current through the transistor. Only by short-circuiting the base and emitter regions together externally can v_{BE} be reduced to zero. In this

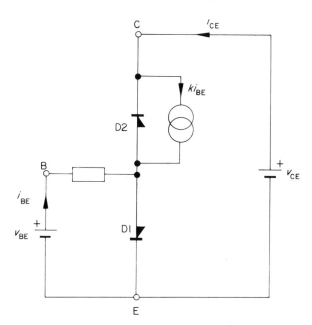

Fig. 4.8   One form of equivalent circuit.

event, leakage current I_{CES} flows out of the base region. In a low power silicon transistor, I_{CES} has the value of a few nanoamperes.

The input resistance of the transistor in the common-emitter mode is approximately equal to the sum of the forward conducting resistance of the diode and the ohmic resistance of the base region. The latter is represented by the resistance in series with the base lead in Fig. 4.8.

4.5.1   Transistor rating limitations

As with valves, the regions on transistor characteristics which may be used are subject to certain limitations. In addition to the maximum collector voltage, which has already been discussed, there is a limit to the *maximum emitter reverse voltage*. Under certain conditions, e.g., in switching circuits, the emitter junction is reverse biased; due to the high impurity doping in the emitter region, the reverse breakdown voltage of the emitter junction is generally lower than the collector junction breakdown voltage.

The *maximum collector current* rating is rarely limited by the current carrying capacity of the device itself. It is often determined either by the power dissipation of the transistor, or by the reduction in current gain at high values of current.

The *maximum power* rating P_{tot} is related to the *maximum allowable junction temperature* T_{jmax}, which is about 85°C in germanium devices and 200°C in silicon devices. The *thermal resistance* of a transistor is an indication of the ability with which the transistor can dissipate heat, and is the junction temperature rise per unit power dissipated (C deg. per watt (or milliwatt)). For instance, a silicon transistor with a rating of 300 mW at 25°C and a maximum permissible junction temperature of 175°C, may have a thermal resistance of 0·5 C deg./mW. This results in a derating curve with a slope of 2 mW/C deg. The rating of this transistor at an ambient temperature of 50°C is $[300 - (50 - 25) \times 2]$ mW = 250 mW. Power derating curves for specific types of transistors are generally supplied by manufacturers.

When a transistor is energized by a sinusoidal signal of constant r.m.s. value but of varying frequency, it is found that the current gain has a constant value at low frequencies, but at higher frequencies its magnitude reduces. There are two principal reasons for this effect. Firstly, the minority charge carriers in the base region take a finite time, known as the *transit time*, to reach the collector. This time reduces with reducing base width, and planar transistors with a base width of about 0·5 μm (0·02 thou.) exhibit a uniform gain up to frequencies of several hundred megahertz. The second effect is due to the capacitance of the collector junction. The *cut-off frequency* of a transistor in any given mode of operation is defined as the frequency at which the current transfer ratio is reduced to 0·707 of its initial value. The common-base cut-off frequency (the *alpha cut-off frequency* f_α or f_{hfb}) is much higher than the cut-off frequency of the same transistor operating in the common-emitter mode. For this reason common-base circuits are often used in applications which are beyond the frequency range of common-emitter amplifiers.

It is now customary for manufacturers to quote the *transition frequency* f_T, which is the frequency at which the common-emitter current gain has fallen to unity. It is usual to determine f_T by measuring the common-emitter current gain at a frequency above the common-emitter cut-off frequency (f_β or f_{hfe}), and extrapolating the result at a slope of -6 decibels per octave* or -20 decibels per decade. For example, if a transistor has an f_{hfe} of 1 MHz and $h_{fe} = 10$ (equivalent to 20 decibels) at 10 MHz, then $f_T = 100$ MHz.

4.6   Common-base characteristics

During the development of the transistor, a great deal of work was done on the common-base configuration, Fig. 4.9(a). The output characteristics and transfer characteristics are shown in Figs. 4.9(b) and (c), respectively. In the common-base mode, the input circuit carries slightly more current than is carried in the collector circuit. Since the emitter current (the input current in this mode) is large, the input impedance of the common-emitter stage is low. Since the collector current is practically constant over the working range of v_{CB}, the output impedance of this configuration is very high, making it difficult to cascade common-emitter stages, since the output impedance of the driver stage is not matched to the input impedance of the driven stage.

The ratio i_{CB}/i_{EB} is known as the *static value of the forward current transfer ratio* h_{FB}. For convenience, in transistor circuit analysis, all currents are assumed to flow into the transistor and, as will be seen from the following, this leads to h_{FB} having a negative value. In Fig. 4.9(a) it is seen that both i_{CB} and i_{EB} appear to flow into the transistor, but a moment's consideration will show that, in fact, one of the two currents must flow out of the transistor. Thus, if a collector current of 98 mA flows *into* the transistor, and the base current is 2 mA, then an emitter current of 100 mA flows *out* of the transistor.

Hence,

$$h_{FB} = \frac{+98}{-100} = -0.98$$

The value of h_{FB} usually lies between -0.98 and about -0.998. The *small-signal common-base current gain* α is the magnitude of the ratio i_c/i_e, without regard to sign, and

$$\alpha = -h_{fb}$$

The leakage current of the transistor in the common-base circuit is the collector-to-base current with the emitter (input circuit) open-circuited. This leakage

---

* See chapter 8

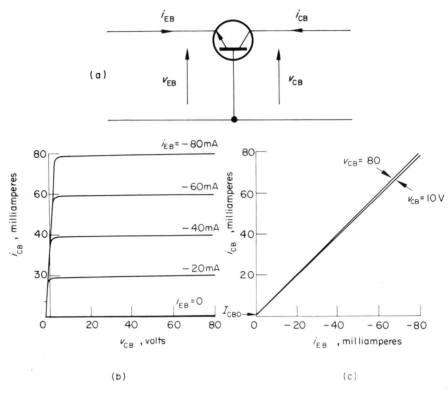

Fig. 4.9   The common-base circuit (a), with (b) the output characteristics, and (c) the static transfer characteristics.

current is designated I_{CBO} for the same transistor. To a first approximation, the relationship between the two values of leakage current is given by

$$I_{CBO} = I_{CEO}(1 + h_{FB})$$

4.7   The common-collector configuration

Transistors are frequently used in this configuration as buffer stages between circuits with widely differing impedances, but common-collector parameters are rarely quoted in manufacturers' literature. The common-collector amplifier will be dealt with at length in the chapter on feedback amplifiers, and no further comment will be made at this point.

4.8   Transistor h-parameters

Many different types of parameters have been used to define the operation of the transistor, but few are as popular as the *hybrid parameters* or *h-parameters*.

They are so named because the dimensions of the parameters are mixed. One of the h-parameters has the dimensions of resistance, one has the dimensions of conductance, while two are dimensionless. The principal advantage of the hybrid parameters is the ease with which they can be measured, and, additionally, they can be simplified to provide ease of calculation.

The small-signal h-parameters are defined by the equations

$$V_1 = h_i I_1 + h_r V_2 \qquad (4.1)$$

$$I_2 = h_f I_1 + h_o V_2 \qquad (4.2)$$

where V_1 is the r.m.s. value of the small-signal voltage applied to the input of the transistor, I_1 is the r.m.s. input current, V_2 is the r.m.s. value of the small-signal output voltage, and I_2 is the r.m.s. output current. The subscripts represent the relationships

i = Input parameter, having the dimensions of resistance.

r = Reverse parameter, which is dimensionless.

f = Forward parameter, which is dimensionless.

o = Output parameter, having the dimensions of conductance.

Depending on the circuit configuration, i.e., common-emitter, common-collector, or common-base, other subscripts are given to the parameters, viz.

e = Common-emitter configuration.

b = Common-base configuration.

c = Common-collector configuration.

Thus, for the common-emitter configuration, eqs. (4.1) and (4.2) become

$$V_{be} = h_{ie} I_{be} + h_{re} V_{ce} \qquad (4.3)$$

$$I_{ce} = h_{fe} I_{be} + h_{oe} V_{ce} \qquad (4.4)$$

where h_{ie} is the input parameter in the common-emitter mode, etc., V_{be} is the r.m.s. input voltage, I_{be} is the r.m.s. input current, and V_{ce} and I_{ce} are the corresponding output quantities, respectively.

To be strictly accurate, h_{ie} and h_{fe} should both be measured with $V_{ce} = 0$, i.e., the output should be short-circuited. This condition can be approximated to under small-signal conditions in a test circuit. Similarly, h_{re} and h_{oe} should be measured with $I_{be} = 0$, i.e., the input should be open-circuited to signal frequencies. In the following, a procedure is outlined for determining the parameters from the static characteristics, but in practice they are obtained by dynamic tests.

In determining the static characteristics of transistors, difficulty is experienced in taking reliable measurements close to the maximum power limit, and a practical solution is to use a transistor curve tracer in which the base current is applied impulsively so that the transistor does not operate for long periods of time at or near to its maximum power rating.

When the transistor is used as a switch, we are primarily interested in the large-signal parameters, which are defined in the common-emitter case by the equations

$$v_{BE} = h_{IE} i_{BE} + h_{RE} v_{CE} \tag{4.5}$$

$$i_{CE} = h_{FE} i_{BE} + h_{OE} v_{CE} \tag{4.6}$$

A qualitative explanation of the hybrid equations for the common-emitter mode of operation follows from an inspection of Fig. 4.8. When the transistor is conducting, the emitter diode D1 carries the collector current, resulting in a p.d. across the diode. The proportion of the collector signal voltage which appears

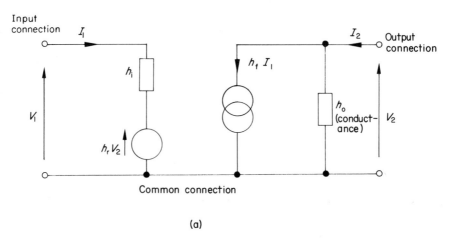

(a)

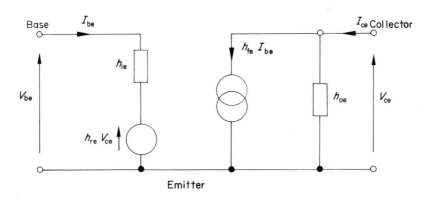

(b)

Fig. 4.10   (a) The general hybrid equivalent circuit, and (b) the common-emitter hybrid equivalent circuit.

between the base and emitter due to the collector current flowing in D1 is $h_{re}V_{ce}$. The net signal voltage acting in the input circuit is, therefore, $V_{be} - h_{re}V_{ce}$. Since the base and emitter regions are coupled together, there is an ohmic connection between them. If this resistance has a value $h_{ie}\Omega$, then the voltage drop in the base-emitter circuit when it carries current I_{be} is $h_{ie}I_{be}$. That is

$$V_{be} - h_{re}V_{ce} = h_{ie}I_{be}$$

or

$$V_{be} = h_{ie}I_{be} + h_{re}V_{ce}$$

which corresponds to eq. (4.3)

The total collector current I_{ce} is due in part to the collector voltage V_{ce} appearing across the ohmic connections within the transistor. If h_{oe} is the effective conductance of the conducting path between the collector and the emitter, then the proportion of the collector current which flows in h_{oe} is $h_{oe}V_{ce}$. The remaining part of the collector current, which is usually the greater part, is due to the amplification of the base current. This is $h_{fe}I_{be}$. The total collector current is, therefore, the sum of the two components

$$I_{ce} = h_{fe}I_{be} + h_{oe}V_{ce}$$

which corresponds to eq. (4.4).

The equivalent circuit satisfying the general h-parameters, eqs. (4.1) and (4.2), is shown in Fig. 4.10(a), and that for the common-emitter equations, eqs. (4.3) and (4.4), is shown in Fig. 4.10(b). It is seen that both equivalent circuits contain two active elements, a voltage generator in the input circuit and a current generator in the output circuit. The parameters for any given configuration are obtained either from the static characteristics or by means of a dynamic test *in that particular configuration*. Typical h-parameter values for a silicon planar transistor in both the common-emitter and common-base modes are given in Table 4.1.

Table 4.1

Typical hybrid parameters of a silicon planar transistor

	Common emitter	Common base
h_i	5 kΩ	30 Ω
h_r	2·5 x 10^{-4}	6 x 10^{-4}
h_f	200	−0·994
h_o	15 μS	1 μS

A method of estimating the common-emitter parameters from the static characteristics is shown in Fig. 4.11. Using the input characteristics, Fig. 4.11(a)

$$h_{ie} = \delta v_{BE1}/\delta i_{BE} \qquad \text{at constant } v_{CE}$$

$$h_{re} = \delta v_{BE2}/\delta v_{CE} \qquad \text{at constant } i_{BE}$$

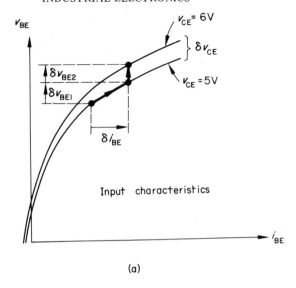

Input characteristics

(a)

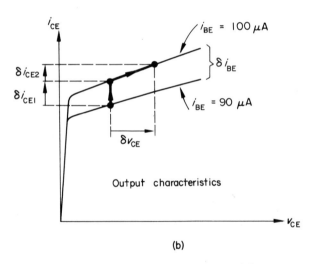

Output characteristics

(b)

Fig. 4.11 (a) One method of evaluating h_{ie} and h_{re}, and (b) determination of h_{fe} and h_{oe}.

and from the output characteristics, Fig. 4.11(b),

$$h_{fe} = \delta i_{CE1}/\delta i_{BE} \qquad \text{at constant } v_{CE}$$

$$h_{oe} = \delta i_{CE2}/\delta v_{CE} \qquad \text{at constant } i_{BE}$$

The common-base parameters are determined in much the same way from the common-base input and output characteristics, respectively.

The h_f parameters are most frequently quoted in technical literature, and the relationship between h_{fe} and h_{fb} may be deduced as follows. Since current is always assumed to flow into the transistor from the external circuits then

$$I_b + I_c + I_e = 0$$

or

$$I_e = -(I_b + I_c)$$

hence,

$$\frac{1}{h_{fb}} = \frac{I_e}{I_c} = \frac{-(I_b + I_c)}{I_c} = -\left(\frac{I_b}{I_c} + 1\right) = -\left(\frac{1}{h_{fe}} + 1\right)$$

therefore,

$$h_{fb} = \frac{-h_{fe}}{1 + h_{fe}} = -\alpha \qquad (4.7)$$

$$h_{fe} = \frac{-h_{fb}}{1 + h_{fb}} = \frac{\alpha}{1 - \alpha} \qquad (4.8)$$

For example, if h_{fe} = 99, then h_{fb} = −99/100 = −0·99, or α = 0·99, and if the transistor has a value of −0·98 for h_{fb}, then h_{fe} = −(−0·98)/[1 + (−0·98)] = 49. This calculation illustrates the fact that a small change in h_{fb} gives a much greater change in h_{fe}; a change in h_{fb} from −0·99 to −0·98 gives a change in h_{fe} of 99 to 49.

4.9   Simplified hybrid equivalent circuit

In some instances, it is possible to make simplifying assumptions to reduce the complexity of the equivalent circuit. Generally speaking, h_r is very small in both the common-emitter and common-base modes (see Table 4.1), so that the voltage generator $h_r V_2$ can be omitted from the input circuit without significant loss of accuracy. This gives the equivalent circuit in Fig. 4.12(a). In other cases, conductance h_o is small when compared with the conductance of any load connected to the output terminals. In this case, h_o may be neglected, giving the ultimate in simplicity in Fig. 4.12(b). It must be emphasized that the simplified equivalent circuits yield accurate results only when the assumptions are justified.

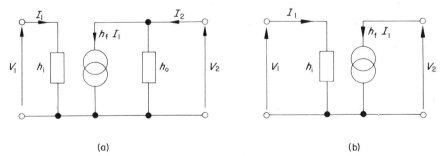

(a)   (b)

Fig. 4.12   Simplified hybrid circuits.

4.10 The 'tee' equivalent circuit

A difficulty associated with the complete h-parameter equivalent circuit of Fig. 4.10 is that one has to deal with two active generators within the circuit, one at the input and one at the output.

A simplified form of equivalent circuit was developed in Fig. 4.8, which contained only one active generator, which is in the form of a current generator shunting the collector junction diode. Providing that the circuit is working as a linear small-signal amplifier, then the collector junction is reverse biased*, and the emitter junction is forward biased. In this event, the emitter diode can be represented by a resistor r_e with a relatively low value, and the reverse-biased collector diode can be represented by resistor r_d of relatively high value. The equivalent 'tee' circuit is then as shown in Fig. 4.13(a). An alternative T-parameter equivalent circuit is obtained by converting the constant-current generator into a constant-voltage source, shown in Fig. 4.13(b). Typical values for the T-parameter equivalent circuit are $r_e = 20\ \Omega$, $r_b = 500\ \Omega$, $r_d = 50\ \text{k}\Omega$, $k = 50$. The resistor r_b in the figure is representative of the ohmic resistance of the base region.

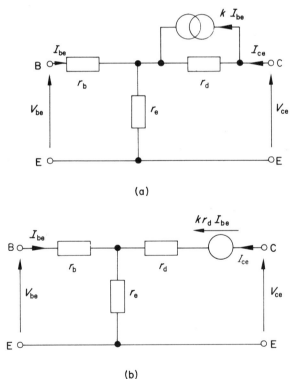

(a)

(b)

Fig. 4.13   Common-emitter T-parameter equivalent circuits.

* When operating as a switch in the ON state, the collector junction is forward biased.

The theoretical relationships existing between the *h*-parameters and the T-parameters are as follows:

$$r_e = h_{re}/h_{oe} \tag{4.9}$$

$$r_b = h_{ie} - h_{re}(1 + h_{fe})/h_{oe} \tag{4.10}$$

$$r_d = (1 - h_{re})/h_{oe} \simeq 1/h_{oe} \tag{4.11}$$

$$k = h_{re} + h_{fe} \simeq h_{fe} \tag{4.12}$$

When the transistor is used in the common-base mode, the equivalent circuit of Fig. 4.13 is modified to that shown in Fig. 4.14(a), merely by interchanging r_b and r_e. The input current to the circuit now becomes I_{eb}, and the disadvantage of Fig. 4.14(a) is that the generator $r_d h_{fe} I_b$ in the collector circuit is not a function of the input current. This can be corrected by noting that $I_b = -I_{eb} - I_{cb}$, to give the equivalent circuit of Fig. 4.14(b).

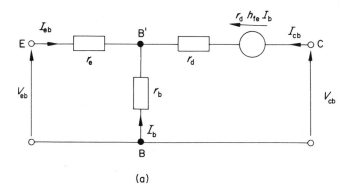

(a)

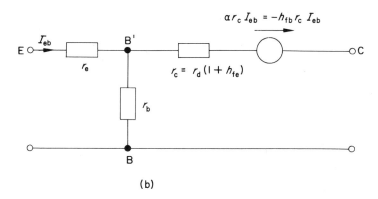

(b)

Fig. 4.14 Common-base *T*-parameter equivalent circuits.

Example 3.1: Evaluate the T-parameters for a transistor having the following common-emitter h-parameters.

$$h_{ie} = 10 \text{ k}\Omega$$
$$h_{re} = 5\cdot5 \times 10^{-4}$$
$$h_{fe} = 350$$
$$h_{oe} = 25 \text{ } \mu\text{S}$$

Solution: Using eqs. (4.9) to (4.12),

$$r_e = 5\cdot5 \times 10^{-4}/25 \times 10^{-6} = 22 \text{ }\Omega$$
$$r_b = 10,000 - 5\cdot5 \times 10^{-4}(1 + 350)/25 \times 10^{-6} = 2280 \text{ }\Omega$$
$$r_d = (1 - 5\cdot5 \times 10^{-4})/25 \times 10^{-6} = 40,000 \text{ }\Omega$$
$$k = 5\cdot5 \times 10^{-4} + 350 = 350 = h_{fe}$$

The equivalent values for the common-base circuit are

$$r_e = 22 \text{ }\Omega$$
$$r_b = 2280 \text{ }\Omega$$
$$\alpha = 0\cdot9972$$
$$r_c = 14\cdot3 \text{ M}\Omega$$

4.11   Transistor parameter variation and its working region

The numerical value of transistor parameters depends upon many factors, including the bias conditions, the collector current, the collector voltage, and the operating temperature. Bias conditions under which acceptance tests should be carried out are specified in data sheets, although it must be borne in mind that parameters vary considerably between transistors of the same type. Figure 4.15 shows the way in which the *h*-parameters for the common-emitter configuration vary with collector current and junction temperature.

4.11.1   Noise in transistors

The ability of a transistor to amplify both small signals and high frequency signals is limited by the electrical noise generated in the device itself. At low frequencies (below about 1 kHz), the noise increases with decreasing frequency, and is known as *1/f noise* due to this inverse relationship. The *1/f* noise has been attributed to surface conduction, and its magnitude varies with working conditions; low noise operation is achieved by working with a low emitter current and a low collector voltage. The amplifier configuration (common-

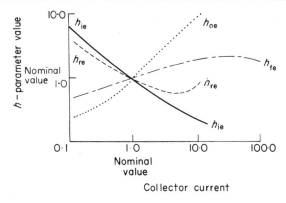

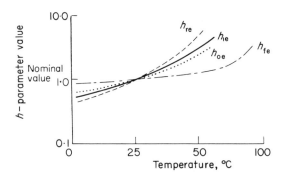

Fig. 4.15 Variation of the common-emitter parameters with collector current and temperature.

emitter, common-base, or common-collector) does not appear to have much influence on the minimum noise level.

Between 1 kHz and the cut-off frequency of the transistor, the noise level remains approximately constant at a fairly low level, and is largely due to shot noise and thermal noise. Above the cut-off frequency, the noise figure again increases, due primarily to the reduction in signal power gain of the transistor.

4.12 Choice of semiconductor materials

Transistors are often described as being suitable for entertainment, industrial or professional applications. Broadly speaking, transistors suitable for *entertainment* applications are produced by the most economical process available, and the parameter tolerances are normally very wide. The circuits in which they are used must be designed to accept the wide spread of parameters.

Devices intended for the *industrial* market, e.g., communications equipment and high quality instrumentation, are subject to more careful checks, including reliability tests. Transistors for the *professional* market are subject to the most exacting tests, and must meet stringent requirements. The price of an individual device reflects, to some extent, the degree of testing required in its selection.

So far as the bulk material of the semiconductor is concerned, manufacturing technology has reached a stage where both silicon and germanium devices are equally reliable. The choice then depends on other factors, such as leakage current, breakdown voltage, etc.

4.13   Semiconductor numbering and outline shapes

Semiconductor devices are registered under many numbering systems, some giving more useful information than others. Most American devices have numbers allocated to them in the order in which they were registered. These numbers are given by the Joint Electronic Device Engineering Council, the numbers being known as JEDEC numbers. In this system, the first numeral gives the number of p-n junctions, '1' for a diode, '2' for a triode-type transistor, '3' for a tetrode device or a thyristor, etc. This is followed by the letter N, and the registration number. Thus, a device numbered 2N2927 is a two-junction semiconductor device, which was the 2927th to be registered.

In the old European system, the first group of numerals gave the heater voltage (on the assumption that it is a thermionic device). When semiconductors came along, this numeral was '0'. The type of device was indicated by a letter, 'A' for a diode, 'C' for a triode, etc. Additional letters, e.g., 'P' for a photoeffect or radiation sensitive device, 'R' for a photoresistive semiconductor material, etc., are also used. Thus, the 0CP71 is a phototransistor.

With the advent of the *PRO ELECTRON* European system, it is expected that the old European system will fall into disuse. The PRO ELECTRON type number has five alpha-numeric symbols, comprising either two letters and three numbers, or three letters and two numbers. The first letter indicates the semiconductor material used, 'A' for germanium, 'B' for silicon, 'C' for gallium arsenide, etc. The second letter indicates the most common application of the device. A list corresponding to this letter is given in Table 4.2. Where two numbers are included in the type symbol, e.g., BFX63, the device is intended for industrial and professional equipment, and when it contains three numerals, e.g., BF194, it is intended for entertainment or consumer equipment. The PRO ELECTRON system also provides for sub-classes by the addition of an alpha-numeric group separated from the basic number by a hyphen, e.g., BTY79-600R. This device is a silicon thyristor, registration number Y79, which has a rated maximum repetitive peak reverse voltage of 600 V. The 'R' denotes reverse connection, i.e., the stud is the anode.

When transistor construction was first commenced, manufacturers devised their own methods of encapsulations or *outlines*. In more recent years, there has

Table 4.2

PRO ELECTRON application type numbers

Letter Application

A	Signal diode
B	Variable capacitance diode
C	A.F. low power transistor
D	A.F. power transistor
E	Tunnel diode
F	H.F. low power transistor
G	Multiple device
H	Field probe
L	H.F. power transistor
M	Hall effect modulator
P	Radiation sensitive device
Q	Radiation generating device
R	Specialized breakdown device
S	Low power switching transistor
T	Power switching device (thyristors, etc.)
U	Power switching transistor
X	Multiple diode
Y	Power diode
Z	Zener diode

been a great deal of standardization, based mainly on JEDEC outlines. The most common metal-can forms for low power devices are cylindrical in shape, having the dimensions given in the first part of Table 4.3. Many outlines have been discontinued, and those listed are in current use. Where the collector power dissipation exceeds about 1 W, the 'diamond' shaped outline of the T03 and T066 is used. In these, the canister has an integral diamond shaped support, which is bolted to the heat sink. In power transistors, the collector is connected internally

Table 4.3

Principal dimensions of JEDEC outlines

JEDEC type	Cylindrical canister	
	Length (in)	Diameter (in)
TO1	0·4	0·23
TO5	0·25	0·34
TO8	0·25	0·5
TO18	0·24	0·2
TO46	0·075	0·19
TO72	0·25	0·19
	Diamond shaped base	
	Length (in)	Width (in)
TO3	1·5	1·0
TO66	1·25	0·7

to the canister to minimize thermal resistance to flow of heat. High power transistors are generally mounted in the T03 canister, while medium power devices use either the T066 or the T08 construction. In the latter case, a clamp is used to fasten the canister to the heat sink, making it into a version of the diamond shaped construction.

In Britain and Europe, other standards have been adopted which are related to the JEDEC outlines. These are the VASCA (Valve and Semiconductor Manufacturer's Association) and IEC (International Electrotechnical Commission) outlines. In addition to these, there are several military and specialized outlines.

In recent years, low priced encapsulated silicon transistors have become available which are designed for the domestic and entertainment market. They possess advantages over metal encased transistors, which include the reduction of parasitic capacitance and elimination of the problem of isolating the canister from contact with other parts of the circuit. However, there are disadvantages. Plastic encapsulated transistors have a lower permissible power dissipation than equivalent transistors in metal canisters, and their long-term reliability is as yet unproven. So far, there is no great uniformity of outline shapes between metal canisters and plastic encapsulations, although some manufacturers do produce plastic devices which have the same general outlines as the T05 and T018 cans.

4.14   Unipolar transistors: FETs

Unipolar transistors are so named because only majority· charge carriers are used to convey current through them. They are also known as *field effect transistors* (FETs), since they depend for their action upon the effect of an electric field. There are two main types of FET,

(a)   the junction-gate FET (JUGFET).
(b)   the insulated-gate FET (IGFET or MOSFET).

Both types are manufactured in epitaxial forms, an n-channel junction-gate FET being shown in Fig. 4.16(a), together with its output characteristics, Fig. 4.16(b).

Junction FETs have two shallow junctions diffused into the main channel, shown in Fig. 4.16(a), the diffused regions being heavily doped to give sharply graded junctions. In an n-channel FET, the current carriers are electrons, and the end of the channel connected to the negative pole of the supply source is known as the *source.* The current carriers are then collected at the *drain,* and the diffused regions are known as the *gate regions.* The gate regions may be used independently, but they are usually connected together. If the FET has a symmetrical structure, either end of the device may be used as the source, and the names 'source' and 'drain' are interchangeable. In p-channel FETs, the current carriers are holes, and the source is connected to the positive pole of v_{DS}. In this case, n$^+$-regions are diffused into the channel, the gate regions being connected to a positive potential with respect to the source. The circuit symbols for p-channel FETs are generally similar to those for n-channel devices, but the directions of the arrows on the diagrams are reversed.

When the gate bias is zero, i.e., the gate and source are connected together, and the drain-source voltage is low, the FET has the characteristic of a resistor with a value between 100 Ω and 1 kΩ. The application of a longitudinal electric field to the channel results in a potential gradient *across* the faces of the two gate regions, resulting in a depletion layer of varying depth. The depletion layer

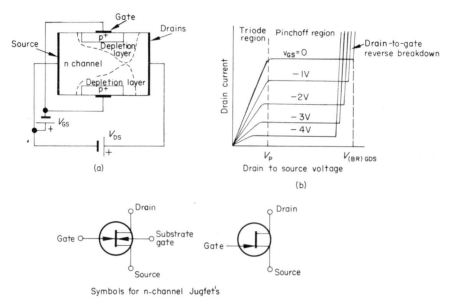

Symbols for n-channel Jugfet's

Fig. 4.16 (a) The principle of operation of an n-channel JUGFET, and (b) the output (drain) characteristics. Typical circuit symbols are also shown in the figure.

results from the fact that p-n gate-channel junction is reverse biased, and is illustrated in Fig. 4.16(a). An increase in v_{DS} increases the reverse bias, and depletion layer depth, resulting in a reduced area for conduction in the channel, giving the resistive-type characteristic for low values of v_{DS}. This region is sometimes known as the *triode region* of the characteristic. At any given value of v_{DS} in this region, an increase in v_{GS} increases the depletion layer depth, which reduces the drain current. In this way, the junction-gate FET can be used as a voltage-controlled resistor, known as a *pinch-effect resistor*. The name 'pinch-effect' derives from the fact that an increase in reverse bias between the gate and the channel 'pinches' or reduces the area of the channel available for conduction. Pinch-effect resistors are sometimes used in monolithic integrated circuits to replace conventional ohmic resistive elements, due to their smaller size for a given value of resistance.

At some value of drain voltage, known as the *pinch-off voltage* V_P, the two depletion regions practically touch one another, when the current is 'pinched' into a thin sheet flow. Any further increase in drain voltage does not significantly increase the drain current. This region is marked as the *pinch-off region* on the output characteristics, in which region the slope resistance is of the order of 10^9 to 10^{11} Ω. In the pinch-off mode of operation, the drain current is controlled by the gate voltage, giving an equivalent circuit similar to the pentode

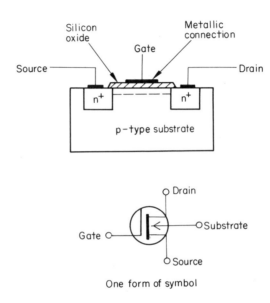

One form of symbol

Fig. 4.17   An n-channel IGFET, together with one form of circuit symbol.

valve. The mutual conductance of JUGFETS lie in the range 0·05 mA/V to about 6 mA/V.

A very large increase in v_{DS} results in avalanche breakdown of the gate-to-channel p-n junction, with consequent sudden increase in collector current. This breakdown voltage is designated $V_{(BR)GDS}$, which is the breakdown voltage when $v_{GS} = 0$. Since the longitudinal voltage v_{DS} reverse biases the p-n junction, an increase in reverse bias due to an increase in v_{GS} results in avalanche breakdown at a lower value of v_{DS}. This is illustrated in the characteristics in Fig. 4.16(b).

A more recent development, the insulated-gate FET or *metal-oxide-silicon FET* (MOSFET), can operate in either of two modes, *depletion* or *enhancement*.

The basic configuration of an n-channel IGFET is shown in Fig. 4.17, the gate electrode being insulated from the semiconductor by an insulating layer. Here, two n⁺-regions are diffused into a silicon p-type layer. In *depletion-mode* IGFETS an *initial channel* of n-type material, which is shown dotted in the figure, connects the source and drain. In p-channel devices, the p-regions and n-regions are interchanged. Upon the application of a drain-source voltage, a steady current flows through the initial channel. The application of a negative potential to the gate terminal causes electrons to be repelled from the semiconductor-insulator interface, resulting in a depletion layer being formed in the channel. This is known as depletion-mode operation, in which an increased negative voltage (in the case of an n-channel IGFET) reduces the drain current. The application of a positive potential to the gate electrode attracts minority charge carriers from the substrate, in the manner of the induced inversion channel described in section 4.4. This has the effect of enhancing conductivity in the initial n-channel, and is known as *enhancement-mode operation.* Many modern IGFETS work in enhancement-mode only, in which case the initial channel is omitted.

The input impedance of both types of FET is very high, being of the order of several hundred megohms, the value in the case of IGFETS being higher than JUGFETS, due to the insulating layer in the former. IGFETS also possess the advantage that either polarity input signal may be applied to the gate without drawing current from the signal source. The input capacitance of the FET is, typically, 5 pF to 20 pF, which is rather less than that of the bipolar transistor (50 pF typically). However, the input resistance of the bipolar transistor is much lower than that of the FET, being of the order of 1 kΩ, with the result that the input time constant of bipolar devices is much smaller than that of FETs. As a direct consequence of this, the high frequency performance and switching performance of FETs are not as good as modern bipolar devices.

Recently, a *tetrode* IGFET has been introduced, which has two independent gates, both of which can be used as control electrodes. This device is now finding increasing use in the radio and telecommunications fields.

4.14.1 Thin-film transistors

A *film circuit* is one in which passive elements, e.g., resistors and capacitors, are manufactured by depositing films of conducting and non-conducting materials on a passive (insulating) substrate. Film circuits are arbitrarily subdivided into *thick-film* and *thin-film* circuits, depending on the conductor depth, which is in turn dependent on the manufacturing process involved. The conductor depth, depending on the material used and the process, varies between 10^{-5} and 10^{-1} mm; the performance of both type of film circuits is generally similar.

During the early years of the development of *film integrated circuits,* it was found to be very difficult to manufacture transistors in the form of a film. To overcome this problem, planar transistors in the form of a 'chip' of semiconductor material were connected externally to the film circuit. The chip is 'flipped' or turned over to enable the connections to be made, the contacts being established by ultrasonic or thermo-compression bonding. This is known as the *flip-chip* method of mounting transistors.

Advances in film circuit technology have resulted in the development of the *thin-film field effect transistor* (TFT). A variety of construction techniques may be adopted, and either depletion-mode or enhancement-mode operation can be achieved. Inexpensive substrate materials, even flexible ones, can be used. These include glass, plastics, paper, and anodized aliminium. The semiconductor is then deposited by conventional techniques, followed by the insulator and the metallic gate. Alternatively, they may be deposited in the reverse order.

4.15   The thyristor

The *thyristor,* formerly known by the trade name of the *silicon controlled rectifier* (SCR), is a p-n-p-n semiconductor switching device with a bistable action, which depends for its operation on internal regenerative feedback. The thyristor has three terminals, an anode, a cathode, and a gate or control terminal, shown in Fig. 4.18. Connection G2 is available on some devices as an additional control region. A simple explanation of the operation of the thyristor, known as the two-transistor analogy, follows.

If the two centre regions of the thyristor are regarded as being split diagonally as shown, the structure reduces to two interconnected transistors T1 and T2. T1 is a p-n-p transistor, and T2 is an n-p-n transistor. When the anode is raised to a positive potential (with no signal applied to the gate), the collector junctions of both transistors are reverse biased, and the only current flowing is the leakage current through the two transistors.

Breakdown can be made to occur in one of two ways. Firstly, if the anode voltage exceeds the breakdown voltage of either collector junction, collector current flows in that transistor. Since this current must also flow through the base region of the other transistor, it also turns that transistor on. This action is regenerative, one transistor holding the other in the ON state, since each provides

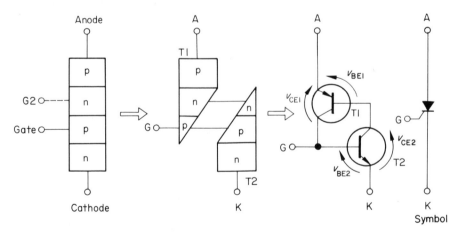

Fig. 4.18   The two-transistor analogy of the thyristor.

the base current of the other. Once this has happened, the thyristor continues to conduct so long as the anode remains positive with respect to the cathode. The characteristic of the thyristor is shown by the full line in Fig. 4.19.

The second method of turning the thyristor ON is to inject a pulse of current into the gate terminal. This turns transistor T2 ON, providing a conduction path for the base current of T1. Once this has occurred each transistor holds the other in the ON state, as described above, and the signal applied to the gate terminal can be removed. This is by far the most common method of triggering a thyristor into its forward conduction mode, since an impulsive current of a fraction of an ampere of a few microseconds duration applied to the gate is adequate for the purpose. Generally speaking, a large gate current causes the thyristor to turn ON more rapidly than a smaller gate current, subject to the limitation that the peak

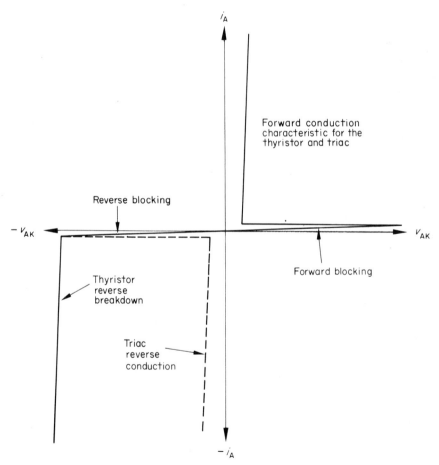

Fig. 4.19   The static characteristic of the thyristor is shown by the full line, and the reverse characteristic of the triac is shown by the broken line.

gate current is not exceeded. Typical values of peak gate current are 1 A at
v_{GK} = 5 V, and 0·5 A at v_{GK} = 10 V.

When the anode becomes negative, the emitter junctions of T1 and T2 are
reverse biased. In this event, the thyristor assumes a blocking state, and it passes
only leakage current. At some value of reverse voltage, reverse breakdown occurs,
which may result in catastrophic failure of the device.

Ratings of devices currently available provide a forward conduction capability
of upwards of 1000 A, with inverse voltage ratings of the order of 2 kV.
A feature of the conduction characteristic is that the p.d. across the thyristor is
substantially constant at all values of current, up to its rated output. This may be
explained by the two-transistor analogy as follows. When a transistor is working
in its saturated mode, i.e., when it is turned ON, V_{BEsat} is of the order of 0·4 V
to 0·7 V, and V_{CEsat} is approximately 0·3 V to 0·5 V. From Fig. 4.17, it is seen
that the anode-to-cathode voltage in the ON state is the sum of these two values,
i.e., approximately 1 V. This voltage is found to increase slightly as the load
current increases, but it may be regarded as substantially constant over the
normal working range.

A device known as a *triac* has been available for some time, which has a low
voltage conduction characteristic in the first and third quadrants, shown in Fig.
4.18. Recent improvements in manufacturing technology have allowed triacs
with current ratings of the order of 200 A, at blocking voltages of 1 kV, to be
developed. A feature of the triac is that it can be fired, or turned ON, with either
polarity anode voltage by either polarity gate signal. Unfortunately, at the
present state of development, the sensitivity to gate signals is much less than that
of an equivalent rated thyristor. This problem is overcome if adequate gate drive
is applied.

4.16   Thermistors

A *thermistor* is a temperature sensitive resistor made of semiconductor material,
usually in the form of metallic oxides of cobalt, manganese, or nickel. The
material is an intrinsic semiconductor with a negative resistance-temperature
coefficient in the region of 3 per cent per C deg. to 4·5 per cent per C deg. at
room temperature.

Thermistors are manufactured in rod, bead, and disc form, with diameters as
small as 0·015 cm, and their working temperatures range from about −100°C to
400°C. Typical ohmic values at room temperature lie between 500 Ω and 100 kΩ.
They are commonly used in one arm of a Wheatstone bridge circuit, the potential
between opposite corners of the bridge being a measure of the thermistor temperature. Thermistors are also used in transistor bias circuits to provide compensation for the effects of temperature on the leakage current of transistors. In
such circuits, the current flowing through the thermistor must be small, otherwise it will cause the temperature of the bulk of the thermistor to rise, resulting
in a further change in resistance.

4.17 The Hall effect

When a current carrying conductor is placed in a transverse magnetic field, an
e.m.f. can be detected between the sides of the conductor which are mutually
perpendicular to the direction of flow of current and the magnetic field, as
shown in Fig. 4.20. This is known as the *Hall effect*. The conditions are
analogous to those existing inside an electric motor. The Hall effect is very small
in metallic conductors, but is of significance in semiconductors.

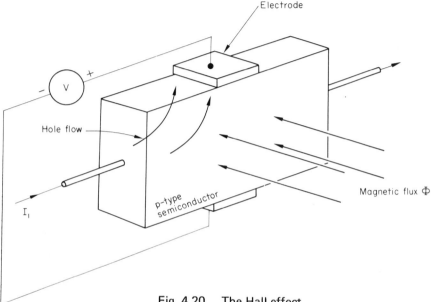

Fig. 4.20   The Hall effect.

In a p-type semiconductor, the current is conveyed by holes. With the conditions shown in Fig. 4.20, the holes are swept towards the upper surface of the
semiconductor, causing it to assume a positive potential with respect to the
lower surface. In an n-type material, under the same conditions, the *Hall e.m.f.*
is of the opposite polarity. Hall devices usually employ indium antimonide or
indium arsenide semiconductors.

The Hall e.m.f. is found to be proportional to the product $I_1 \Phi$, and if the flux
is produced by a current I_2 flowing in a coil, then the Hall e.m.f. is proportional
to the product $I_1 I_2$. Applications of the Hall effect include multipliers, modula-
tors, wattmeters, and fluxmeters.

Problems

4.1 Explain the principle of operation of a junction transistor.

4.2 Why are junction transistors sensitive to variations in temperature? Describe a circuit
which minimizes the effects of temperature variation on the collector current.

Note: see chapter 7 for the final part of the question.

4.3 Give a brief account of the following types of transistor: (a) grown-junction, (b) alloy-junction, (c) alloy-diffused, and (d) planar epitaxial.

4.4 Give an explanation why h-parameters are the most commonly quoted parameters of transistors.

4.5 With the aid of a circuit diagram briefly explain how you would obtain experimentally the output and transfer characteristics of an n-p-n transistor in the common-emitter mode.

Sketch on suitably scaled axes the curves you would expect to obtain for a low power general-purpose transistor. Give typical values of the h-parameters that would be obtained from this test.

Sketch also the characteristics which would be obtained from a p-n-p transistor in the common-emitter mode, clearly indicating the polarities of the voltages and the directions of the currents.

4.6 A certain transistor has a current gain of 0·99 in the common-base configuration; determine its current gain in the common-emitter configuration. Another transistor has $h_{FE} = 80$; determine h_{FB}.

4.7 The input characteristics of a low power transistor are as follows:

i_{BE} (μA)	v_{BE} (mV) at v_{CE} of	
	5 V	10 V
−5	68	75
0	103	108
5	125	128
10	140	142
15	152	155
20	162	165
25	171	174
30	180	183

The output characteristics are linear between the following points:

v_{CE} (V)	2	10	
i_C (mA)	1·9	2·9	when i_B = 20 μA
i_C (mA)	2·7	3·75	when i_B = 30 μA

Estimate the values of the hybrid parameters of this transistor if the quiescent base current and voltage are 25 μA and 170 mV, respectively, and the quiescent collector current is 2·7 mA.

4.8 Convert the h-parameters calculated in question 4.5 into the 'tee' equivalent values.

4.9 Sketch the output characteristics relating the collector current and collector voltage of a bipolar junction transistor when connected in (a) the common-base configuration, and (b) the common-emitter configuration. State the type of transistor. e.g., n-p-n or p-n-p, for which the characteristics are drawn, clearly indicating the polarities of the quantities concerned on the characteristics.

Write down the hybrid parameter equations for the common-emitter configuration, and state typical values for a low power transistor. Sketch the circuit diagram of a suitable test circuit to evaluate the common-emitter parameters.

4.10 Describe, with the aid of diagrams, the essential differences between enhancement mode FETs and depletion mode FETs.

4.11 Using the two-transistor analogy, describe the operation of the thyristor.

4.12 Give a brief account of the Hall effect in semiconductors. Draw a diagram showing the polarity of the Hall voltage for both p-type and n-type semiconductors.

4.13 What restrictions are placed on the use of equivalent circuits for valves and transistors? Draw the *constant voltage* equivalent circuit for a valve and show that

$$i_a = g_m v_g - \frac{v_a}{r_a}$$

where v_g is the signal voltage applied between the grid and the cathode, and v_a and i_a are the corresponding changes produced in the anode-cathode voltage and the anode current.

Draw the equivalent T-circuit for a transistor and show that if the base resistance r_b is negligible:

$$i_c = a i_e - \frac{v_c}{r_c}$$

where i_e is the signal current in the emitter lead, and v_c and i_c are the corresponding changes produced in the collector-base voltage and the collector current.

(C & G)

4.14 Sketch a test circuit for obtaining the collector current-collector voltage characteristics of a transistor. Give a brief description, with typical resulting curves, of an experiment determining these characteristics for a transistor connected in the following configurations:

(a) common-emitter, and
(b) common base.

Sketch a simple transistor amplifier circuit for *either* (a) *or* (b), showing the polarities of the battery connections. (C & G)

5. Photoelectric devices

5.1 Light

Light is electromagnetic radiation which has wavelengths between about $0.02 \, \mu m$
and $100 \, \mu m$. This range of frequencies is commonly given in units known as
Ångstrom units, where $1 \, \text{Å} = 10^{-10} \, m$. Ultraviolet radiation is at the lower end
of the wavelength spectrum (the highest frequency) and infrared is at the
upper end. The *visible band* of the spectrum lies between wavelengths of
$0.4 \, \mu m$ (violet) and $0.7 \, \mu m$ (red). As with any frequency-sensitive detector,
the eye is not equally sensitive to all frequencies, the blue-green colour of
wavelength of $0.55 \, \mu m$ being the most readily detectable.

The output from a tungsten-filament lamp at its normal operating temperature ranges from ultraviolet to infrared, with its peak energy output at about a
wavelength of $1 \, \mu m$ in the infrared region, i.e., above the visible band. Since
the eye cannot detect the longer wavelength signals generated by the tungsten-
filament lamp, the colour we 'see' depends on the greatest energy that is output
in the visible band. In the case considered here, it is in the yellow-red range of
frequencies, and this is the *apparent* colour of the lamp. For this reason, the
eye cannot be used for precise measurement of intensities, particularly where a
band of frequencies is involved.

The S.I. unit of illumination is the *lux (lx)*, or lumen per square metre.
The former British unit of illumination was the foot-candle, which is equivalent
to $10.764 \, lx$.

Light energy can be thought of as travelling in bundles or packages, each
bundle being known as a *photon,* and the energy it carries is a *quantum.* When
light strikes a surface it gives up some of its energy to the atomic structure.
The intensity of light is related to the number of photons arriving at the surface,
and an increase in light intensity results in the surface absorbing a greater amount
of energy.

5.2   Photoemissive cells

Photoemissive cells or *phototubes* employ low work-function cathodes so that they readily emit electrons when radiant energy of the correct frequency strikes them. The cathode, which is enclosed in a glass or quartz envelope which is either evacuated or filled with a gas, is usually made either of caesium on a silver base or of caesium antimony. The former material has a peak spectral

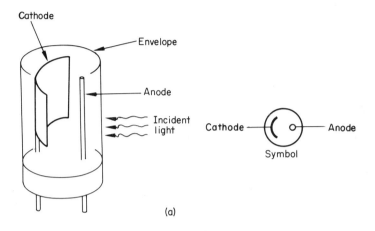

(a)

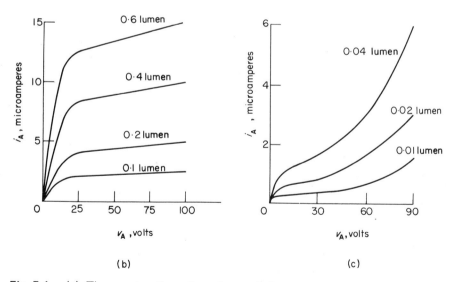

(b)   (c)

Fig. 5.1   (a) The construction of one form of photoemissive cell. The static characteristics in (b) are those of a 20 CV vacuum tube, while those in (c) are of a 20 CG gas-filled tube.

response at about 0·4 μm, and has a response curve generally similar to the human eye. The latter type of cathode material has a peak spectral response of 0·8 μm, which is at the upper end of the visible spectrum. The cathode material should be chosen to suit the particular application in mind.

A diagram of one form of photoemissive cell is shown in Fig. 5.1(a), together with its circuit symbol. The tube is normally operated with a positive anode voltage, so that the emitted electrons are collected by the anode. The characteristics of a vacuum phototube are shown in Fig. 5.1(b), and they are seen to be pentode-like in nature, with the very high slope resistance on the working parts of the curves. The *sensitivity* of the tube illustrated is of the order of 25 μA per lumen, at an anode voltage of 100 V. Vacuum tubes can operate with high frequency modulated light, since the transit time of electrons between the cathode and anode is very small.

By introducing a small quantity of gas into the envelope, the sensitivity is increased by virtue of ionization of the gas molecules. The sensitivity of the gas-filled tube with the characteristic in Fig. 5.1(c) is about 150 μA per lumen; this is the gas-filled equivalent of the tube corresponding to Fig. 5.1(b). The effective increase in sensitivity due to the introduction of gas is known as the *gas amplification factor,* which generally has a value between 7 and 10. Owing to ionization, the characteristics of the gas-filled tube are non-linear, and there is a maximum limit of the anode voltage due to the onset of glow discharge (due to ionic bombardment). Also, since gas has a finite deionization time, the upper frequency limit to which incident light may be modulated is restricted to about 2 kHz.

When phototubes are in complete darkness they pass a small current known as the *dark current,* which is due to a number of factors including electrical leakage and a small amount of thermionic emission. Generally speaking, the dark current is less than 0·1 μA, and may be as small as 1 pA (10^{-12} A).

5.3   Photoconductive cells

When light falls upon a semiconductor, the energy taken up by the atoms causes the spontaneous generation of electron-hole pairs. The net effect is an increase in bulk conductivity, or a decrease in electrical resistance. Materials used as the semiconductor material include lead sulphide, lead selenide, indium anti-monide, and cadmium sulphide (CdS). The last is the most commonly used material since its spectral response curve closely matches the human eye. Cadmium sulphide cells can, therefore, be used in applications where humans could normally estimate illumination levels, e.g., camera exposure settings, street lighting control circuits, smoke detectors, etc.

Other types of photoconductive cells generally have spectral response curves with peak values at wavelengths of 2 μm to 3 μm. These cells are used to detect radiation with a high infrared content.

One form of construction is shown in Fig. 5.2(a), in which a film of semi-conducting material is laid down on an insulating substrate, and electrodes are evaporated on to its surface through a mask. The electrodes have the shape shown in order to increase the contact area and decrease the resistance of the cell. The 'dark' resistance of the cell generally has a value between 10^4 and

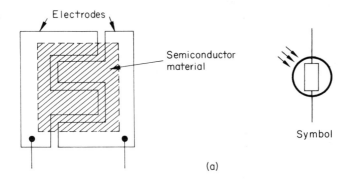

Electrodes

Semiconductor material

Symbol

(a)

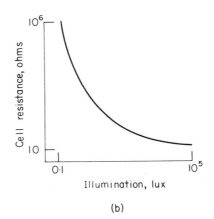

(b)

Fig. 5.2   (a) A typical photoconductive cell, and (b) its characteristic curve.

10^6 Ω, the characteristic of a typical cell being illustrated in Fig. 5.2(b). Photo-conductive cells are also known as *light-dependent resistors* (L.D.R.).

Photoconductor arrays can be manufactured in the form of a film integrated circuit, and densities of about 200 photoconductors per square inch have been achieved. It is possible to connect active devices (e.g. transistors) to the array in order to use it in encoding networks. An array of photoconductors can also be used in punched-card and punched-tape readers and positioning systems.

5.4   Light-activated p-n junction devices

Radiant energy incident upon a semiconductor creates hole-electron pairs, giving rise to a drift current in the presence of an electric field.

The simplest *photojunction device* is the *photodiode,* Fig. 5.3. Photodiodes are generally similar to conventional p-n junction diodes, the difference being that there is a 'window' in the housing of the photodiode to allow light to fall upon it. Photodiodes are operated under reverse bias conditions so that only

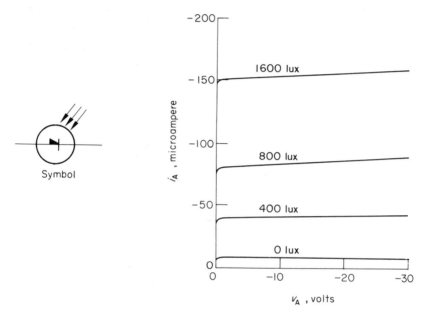

Fig. 5.3   A light-activated p-n junction diode.

leakage current flows through them. The 'dark' current is typically 10 μA in germanium diodes, and 1 μA in silicon diodes. As the illumination intensity increases the leakage current increases, as shown in the characteristics; the sensitivity of photodiodes lies between 10 mA/lm and 50 mA/lm, and the spectral response covers the visible and near infrared frequencies. Incident light can be modulated at very high frequencies with this type of device, and some photodiodes designed for use with laser systems can operate at frequencies of several hundred megahertz.

As with all semiconductor devices, the leakage current increases with an increase in ambient temperature, and the characteristics of Fig. 5.3 are correct at only one temperature.

Bipolar *phototransistors* operate by exposing the base region to light, the radiant energy being equivalent to additional base current. The base connection is brought out for biasing purposes; a circuit symbol and a typical characteristic of a germanium device are shown in Fig. 5.4. The sensitivity of a general-purpose phototransistor is typically 500 mA/lm, and the collector current is usually a few milliamperes, which is adequate to energize the coil of a relay.

Four-layer p-n-p-n *light-activated thyristors* are triggered by incident light, the light providing the energy to the gate region which is normally supplied by the gate circuit. Light-activated thyristors are very sensitive, and to prevent

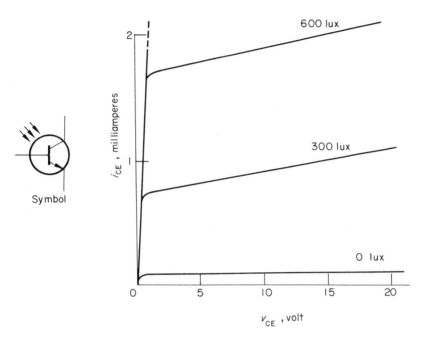

Fig. 5.4   The characteristics of an OCP 71 phototransistor.

inadvertent triggering of the thyristor by induced voltage pulses, the gate and cathode region are usually coupled by a resistor. High power circuits can be controlled directly by photothyristors, many present-day devices being capable of handling several tens of amperes at supply voltages of several hundred volts. It is expected that the ratings will be progressively improved.

5.5   Photovoltaic cells

A photovoltaic cell is a device which generates an e.m.f. between its terminals when light falls upon it, and is made in the form of a silicon p-n crystal. One of the regions is a very thin diffused layer (about 1 μm or 0·04 thou. deep) through which light can pass without much loss of energy. When the light reaches the p-n junction its energy is released into the crystal lattice, causing electrons and holes to be generated in the junction. The normal process of diffusion then takes place. Since one of the regions is very thin, it rapidly saturates with charge carriers, and a differential potential appears between the two regions. The characteristics of a typical photovoltaic cell are shown in Fig. 5.5.

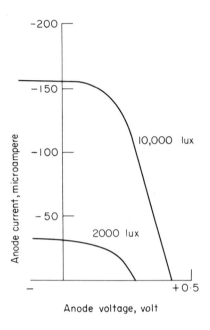

Fig. 5.5   The characteristics of a BPY 10 photovoltaic cell.

Earlier types of photovoltaic cell operated on the barrier-layer principle, the semiconductors being selenium or cuprous oxide.

Applications of photovoltaic cells include exposure meters, punched-tape and card readers, and aerospace projects.

5.6   Light-emitting diodes (L.E.D.)

One form of light-emitting gallium arsenide phosphide (GaAsP) diode is shown in Fig. 5.6(a). The diode is shown in sectional view, and consists of an epitaxial n-type layer of gallium arsenide phosphide which is grown upon a gallium arsenide substrate. A very thin p-region is diffused into the epitaxial layer, and an anode in the shape of a comb or similar configuration is laid down on the p-region. The shape of the anode is selected so as to minimize its masking effect on the emitted light.

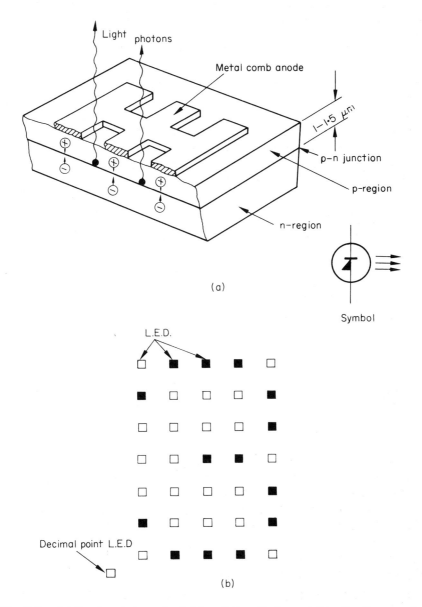

Fig. 5.6 (a) Sectional view of a light-emitting diode, and (b) a 7 x 5 array of light-emitting diodes which can be used to generate any alpha-numeric symbol. In the figure, decimal number 3 is illuminated.

When the diode is forward biased, electrons are injected into the anode region and holes into the cathode region. The charge carriers recombine when they reach the anode and cathode regions, respectively. In some of the recombinations, energy is given off in the form of light, most of which is generated in a region within 0·5 μm of the p-side of the junction. Since the junction is very close to the anode, most of the light reaches the surface, but only a small proportion escapes due to refraction at the surface.

The electroluminous efficiency of the diodes varies with diode current, and in the microampere range the light output is very small. It is normal to operate these diodes with forward current of some tens of milliamperes.

By arranging 35 diodes in a 7 x 5 matrix form, see Fig. 5.6(b), a decimal number or alphabetical character can be generated. In the illustration, the dark squares represent illuminated diodes, the figure giving a decimal 3 readout. A decimal point can be included by having a diode which is offset from the matrix; this is shown at the lower left of the figure. Typical dimensions of the array in Fig. 5.6(b) are 0·4 in x 0·3 in (1 cm x 0·75 cm), each lamp being energized by a current of 50 mA to 100 mA at about 1·7 V to give either a bright red or green indication. Each diode in the matrix can be driven either by a suitable code convertor circuit* or a single integrated circuit code convertor.

Applications of L.E.D.'s include digital read-out devices, tuning indicators, overload indicators, line-of-sight communication systems, and communication systems using fibre optics.

5.7   Photomultipliers

The emission current from a conventional photoemissive cell is relatively small, and it is necessary to amplify it before a useful value of current can be obtained. Devices which amplify the photoemission directly are known as *photomultipliers* or *electron multipliers,* and use the principle of secondary emission.

When light is incident upon the cathode of the photomultiplier, electrons are emitted and are accelerated by the positive charge on an *accelerating anode.* The electrons then enter the first *dynode,* shown in Fig. 5.7(a). There are a series of dynodes in the electron multiplier, each succeeding dynode being at a higher potential than the earlier one. The secondary emission at each dynode results in a multiplication of the space current by a factor having a value between 2 and 8.

In order to obtain a high sensitivity, the potential difference between each pair of dynodes may be 200 V, and in a multi-stage device the final voltage may be 2 kV. Such photomultipliers have sensitivities of more than 200 A/lm, with a 'dark' current of about 1 μA. Each dynode gives rise to a transit-time delay of 2–3 ns, which may be important in some applications. The cathode materials

---

* See *Logic Circuits* by N. M. Morris (McGraw-Hill, 1969).

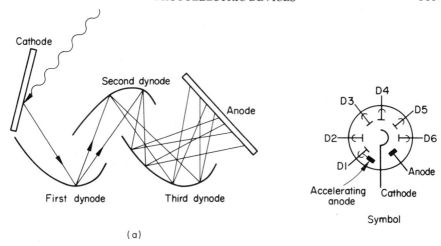

Fig. 5.7   (a) One form of linear dynode construction used in a photomultiplier, and (b) a circuit symbol.

used in photomultipliers are generally similar to those used in photoemissive cells, having similar spectral response curves.

Applications include the measurement of weak light radiation, radiation dosage instruments, and nuclear particle detection in association with an optical scintillator crystal.

Problems

5.1   Write a short essay on the nature of light.

5.2   Sketch the characteristics of a high-vacuum photo-electric cell and illustrate with a second sketch the way in which they would be modified by a low-pressure gas filling. Describe how to test a photo-cell in order to establish that it is in good order.

(C & G)

5.3   Sketch, approximately to scale, typical characteristic curves for (a) vacuum and (b) gas-filled photo-emissive cells. Indicate the magnitude of the voltage and current scales.

Draw a load line on each set of characteristics, and indicate the change in current for typical operating conditions. What special problems are involved in amplifying the voltage resulting from the change in current?

Draw the connection diagram for an industrial application of a photo-electric cell. Description of the operation is not required but state the application.

(C & G)

5.4   A vacuum photocell with the characteristic in Fig. 5.1(b) is connected in series with a 6·67 MΩ resistor to a 100 V d.c. supply. Estimate the change in voltage across the resistor when the luminous flux changes from 0·4 lumen to 0·1 lumen.

5.5   In a second photocell circuit, a 3 MΩ resistor and a gas-filled photoemissive cell with the characteristic in Fig. 5.1(c) are connected in series with a 90 V supply. What is the voltage change across the photocell when the luminous flux changes from 0·01 lumen to 0·04 lumen?

5.6 The anode current—anode voltage characteristics of a vacuum photoemissive cell are given in the following table:

v_A (V)	i_A (μA) at luminous flux of	
	0·1 lumen	0·5 lumen
0	0	0
2	1·0	4·6
4	1·4	6·4
6	1·6	7·4
8	1·7	7·9
10	1·80	8·2
12	1·85	8·4
16	1·85	8·7
20	1·85	8·9
30	1·85	9·2
40	1·85	9·4
50	1·85	9·5
100	1·9	9·6
150	1·95	9·6
200	2·0	9·65

(i) Plot the characteristics, and (ii) plot load lines on the characteristics for a d.c. supply of 200 V and load resistors of (a) 10 MΩ, (b) 20 MΩ, and (c) 100 MΩ. In part (ii), evaluate the anode current and voltage for each load line at each value of luminous flux.

5.7 A substance is found to change its electrical resistance when it is illuminated by light; which of the following names is given to it? (a) Photoelectric, (b) photomultiplier, (c) photoconductive, (d) photovoltaic, (e) photoemissive.

5.8 Sketch the characteristics of a photodiode and a phototransistor, and explain how they operate.

5.9 Sketch the characteristics of a photovoltaic cell, and deduce one form of equivalent circuit for it.

5.10 Describe the principle of operation of the light-emitting diode. Illustrate your answer with suitable diagrams.

6. Power convertors and filter circuits

6.1 Requirements of a power supply

The power supply is the most basic section of any electronic circuit. Practically all electronic systems require one or more power supplies, the simplest and most effective being either an accumulator or a dry battery. While these have the advantages of being simple and having a low output impedance, they suffer from their limited useful life and poor long-term voltage stability.

The most common form of power supply comprises a *rectifier circuit,* which provides a unidirectional but unstabilized output, followed by a filter circuit to smooth out ripples in the output voltage and current.

In some cases, an a.c. power supply must be derived from a d.c. supply, in which case an *invertor circuit* is necessary.

6.2. Single-phase half-wave circuit

The simplest rectifier circuit is the *single-phase half-wave circuit* of Fig. 6.1(a). With a sinusoidal supply voltage $v_s = V_{SM} \sin \omega t$, the current in the conducting half-cycle is

$$i_l = \frac{V_{SM} \sin \omega t}{R_s + r_a + R_L} = I_M \sin \omega t \qquad (6.1)$$

where R_s is the source resistance, r_a is the slope resistance of the diode, and R_L is the resistance of the load. Equation (6.1) is not strictly correct since the true resistance of the diode is slightly greater than r_a, but the equation is accurate enough for most practical purposes.

The instantaneous voltage across the load is always slightly less than instantaneous supply voltage due to the p.d. in the source resistance and in the diode. The peak voltage across the load can be evaluated by proportioning the

117

supply voltage between the resistive components of the circuit. The maximum voltage V_{LM} across the load is

$$V_{LM} = \frac{R_L}{R_s + r_a + R_L} V_{SM} \tag{6.2}$$

In the negative half-cycle, the diode is reverse biased, and only leakage current flows in the circuit. The diode has then to sustain its *peak inverse anode voltage* (p.i.v.), which is $- V_{SM}$ in Fig. 6.1(b).

The *average value** or *d.c. value* of the load current is given by the equation

$$I_L = \frac{1}{2\pi} \int_0^{2\pi} i_l \, d(\omega t)$$

$$= \frac{1}{2\pi} \left[\int_0^{\pi} I_M \sin(\omega t) \, d(\omega t) + \int_{\pi}^{2\pi} 0 \, . \, d(\omega t) \right]$$

$$= (2I_M + 0)/2\pi = 0 \cdot 318 \, I_M \tag{6.3}$$

and the average value of the load voltage is

$$V_L = I_L R_L = 0 \cdot 318 I_M R_L = 0 \cdot 318 V_{LM} \tag{6.4}$$

For a sinewave supply of r.m.s. voltage V_s, then $V_{SM} = \sqrt{2} \, V_s$. Substituting this value into eqs. (6.2) and (6.4) gives

$$V_L = 0 \cdot 45 V_s \frac{R_L}{R_s + r_a + R_L} \tag{6.5}$$

In the case where the sum of R_s and r_a is small compared with R_L, eq. (6.5) reduces to

$$V_L \simeq 0 \cdot 45 V_s \tag{6.6}$$

Example 6.1: A single-phase half-wave circuit with a 475 Ω load is energized by a 200 V supply. If the source resistance is $0 \cdot 5$ Ω and the diode slope resistance is $24 \cdot 5$ Ω, calculate the peak load voltage, the average load voltage and current, and the power dissipated in the load.

Solution: The total resistance in the circuit is

$$R_s + r_a + R_L = 0 \cdot 5 + 24 \cdot 5 + 475 = 500 \, \Omega$$

---

* A convenient fact to remember is that the *area* of *one half-cycle of a sinusoidal wave* is twice the peak value. Thus, if the peak value is 10 A, then the area of one half-cycle is 10 x 2 = 20 ampere-radians.

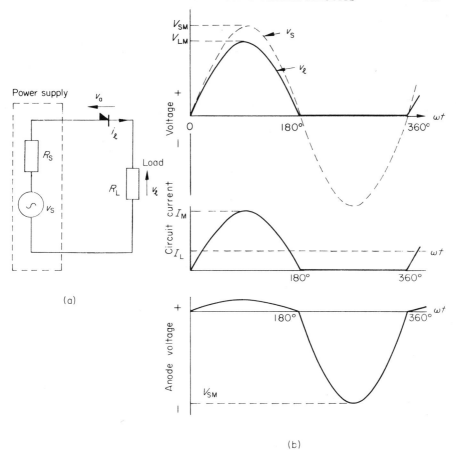

Fig. 6.1   (a) A single-phase half-wave rectifier circuit, and (b) typical waveforms.

and since

$$V_{SM} = \sqrt{2} \times 200 \text{ V, then from eq. (6.2)}$$

$$V_{LM} = \sqrt{2} \times 200 \times 475/500 = 269 \text{ V}$$

From eq. (6.5)

$$V_L = 0.45 \times 200 \times 475/500 = 85.5 \text{ V}$$

(**Note:** the approximate eq. (6.6) gives 90 V)

$$I_L = V_L/R_L = 85.5/475 = 0.18 \text{ A}$$

$$P_L = V_L I_L = 85.5 \times 0.18 = 15.4 \text{ W}$$

6.3 Single-phase full-wave circuits

The full-wave *centre-tap circuit* or *bi-phase circuit* is shown in Fig. 6.2(a). In this circuit a centre-tapped transformer is used to provide a bi-phase supply. The dots on the windings in Fig. 6.2(a) indicate the ends of the windings which are at the same instantaneous potential. Thus, when the upper end of the primary winding is at a positive potential, the corresponding ends of the two secondary windings are also positive. In this operating state, the upper diode is

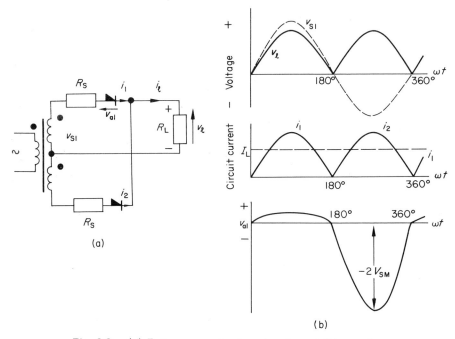

(a)

(b)

Fig. 6.2   (a) Full-wave centre-tap circuit, and (b) waveforms.

forward biased, and the lower diode is reverse biased. In the following half-cycle the operating states are reversed, and the lower diode is forward biased and passes load current.

    Since current flows in both half-cycles, the average load current and voltage are twice the values of the half-wave circuit. That is

$$I_L = 0.636 I_M \tag{6.7}$$

$$V_L = 0.9 V_s \frac{R_L}{R_s + r_a + R_L} \tag{6.8}$$

If $(R_s + r_a) \ll R_L$, eq. (6.8) reduces to $V_L = 0.9 V_s$.

    In the second half-cycle, the upper diode is reverse biased while the lower diode is forward biased. Under peak voltage conditions, the potential of the anode of the upper diode is $- V_{SM}$ while its cathode is $+ V_{SM}$. The peak inverse voltage applied under this condition is, therefore, twice the peak supply voltage, or

$$\text{p.i.v.} = 2V_{SM} = 2\sqrt{2}V_s = 2{\cdot}828V_s$$

This is the value of the repetitive inverse voltage applied to both diodes, and the p.i.v. rating of any diodes used in this type of circuit should be greater than this value. As a general rule of thumb, the p.i.v. rating should be at least four times the r.m.s. value of the supply voltage.

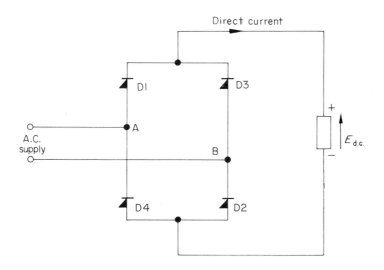

Fig. 6.3   Single-phase bridge rectifier circuit.

A *bridge* configuration of four diodes is shown in Fig. 6.3. In this arrangement, diagonally opposite pairs of diodes conduct simultaneously, so that when point A is positive with respect to point B, diodes D1 and D2 conduct, D3 and D4 being reverse biased. When point B is positive with respect to point A, diodes D3 and D4 conduct, while D1 and D2 are reverse biased.

This circuit possesses two advantages over the centre-tap circuit. Firstly, a centre-tapped transformer is not required and secondly, the repetitive peak inverse voltage is one-half that in the centre-tap circuit. A disadvantage of this circuit is that it is not possible to simultaneously earth one side of the power supply and one side of the output, otherwise part of the bridge circuit will be short-circuited.

The bridge circuit is commonly used in measuring instruments to convert alternating voltage signals into unidirectional signals, the load resistor being replaced by a microammeter.

6.4   Harmonic generation

The output voltage from a simple rectifier circuit consists of a direct potential (or current), with a superimposed alternating ripple component. These are shown in Fig. 6.4 for a single-phase full-wave rectifier. The unidirectional output V_L is the time-average of the waveform taken over one cycle, and has a constant value. The ripple voltage waveform is usually non-sinusoidal, and consists of a large number of *harmonics* of the supply frequency. A harmonic is a frequency which is an integral multiple of the supply frequency.

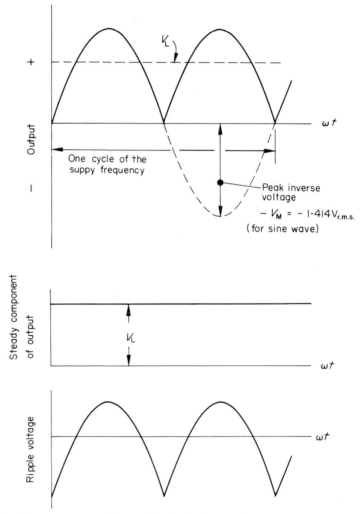

Fig. 6.4   The output waveform of a single phase full-wave circuit showing its constituent unidirectional and harmonic components.

The number and magnitude of the harmonics are dependent on the supply frequency and the connection used. Both the number and the magnitude can either be calculated from a knowledge of the waveshape by *Fourier analysis,* or be measured in the laboratory by harmonic analyzers, which are basically highly selective tuned circuits. In the waveform in Fig. 6.4, the second harmonic (100 Hz with a 50 Hz supply) predominates. Other harmonics present include the fourth (200 Hz), the sixth (300 Hz), the eighth (400 Hz), etc. Fortunately, the magnitude of the harmonics diminishes rapidly as their order increases. It is the function of *filter circuits* to reduce or even eliminate the harmonic content of the output from a rectifier circuit.

6.5   Three-phase half-wave circuit

For outputs greater than about 2 kW the ripple content in the output from a single-phase rectifier is very high. If an output power greater than this is required, or a low ripple output is a desirable feature, then polyphase rectifier circuits must be employed. The simplest three-phase circuit is the half-wave circuit in Fig. 6.5(a). Here, a three-phase delta-star transformer is used to provide power to the circuit. When the potential of phase r is greater than that of either y or b, diode D1 conducts since it is forward biased and diodes D2 and D3 are reverse biased. This occurs between the two points marked 30 degrees and 150 degrees in Fig. 6.5(b). After the 150 degrees point, v_y has a greater potential than v_r, forcing the current to commutate to diode D2. The instant this occurs, the cathode of D1 is raised to a higher potential than that of its anode, and the current through it falls to zero. When v_b rises to a higher potential than v_y, the current commutates to diode D3. Later D1 takes the current up again, and the cycle is recommenced. The use of a three-phase supply increases the average output voltage to

$$V_L = 3 \times \frac{1}{2\pi} \left[\int_0^{30°} 0 \, d(\omega t) + \int_{30°}^{150°} V_{SM} \sin \omega t \, d(\omega t) \right.$$

$$\left. + \int_{150°}^{360°} 0 \, d(\omega t) \right]$$

$$= 0 \cdot 827 V_{SM} = 1 \cdot 17 V_s$$

The ripple magnitude at the output of a three-phase rectifier is much smaller than in either of the single-phase circuits, and the ripple frequency is higher (three times the supply frequency in Fig. 6.5). The inverse voltage waveform for diode D1 is plotted in Fig. 6.5(b), and it is found that there are two peaks, each having a magnitude of $\sqrt{3} \, V_{SM}$ or $2 \cdot 449 V_s$.

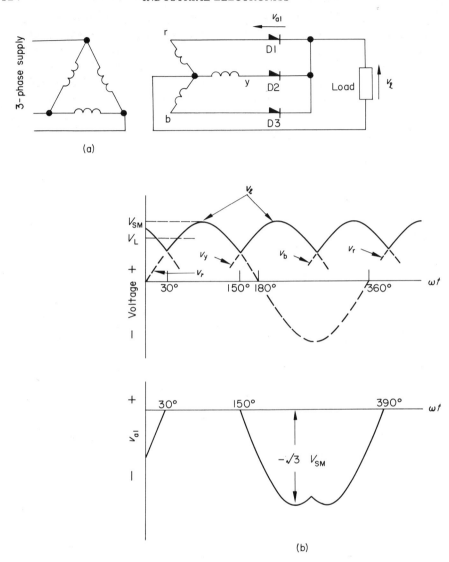

Fig. 6.5   (a) A three-phase half-wave rectifier circuit. and (b) typical waveforms.

6.6   Three-phase full-wave circuits

The *three-phase centre-tap circuit,* Fig. 6.6(a) is basically similar to the single-phase equivalent in that each secondary winding has a centre-tap and two diodes per phase are employed. The unidirectional output voltage is

$$V_L = 1 \cdot 35 V_s$$

and the diode p.i.v. is $2V_s$. This circuit is also described as a *six-phase half-wave circuit* since the centre-tap connection converts each single-phase supply into a bi-phase supply.

The three-phase bridge circuit of Fig. 6.6(b) is very frequently used since it is simple and does not require a centre-tap transformer. In the bridge circuit, when any one supply phase is more positive that the others, the upper connection of the load is connected to that supply line by the appropriate diode. Thus, when phase *a* is positive with respect to phases *b* and *c*, diode D1

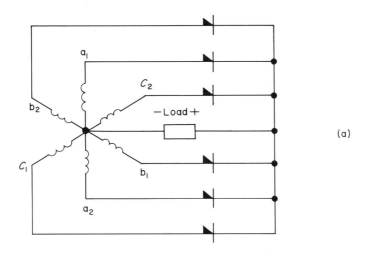

(a)

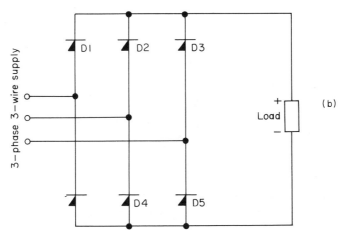

(b)

Fig. 6.6   Three-phase full-wave centre-tap circuit (a), and (b) a three-phase bridge circuit.

is forward biased, and diodes D2 and D3 are reverse biased. The current in line *a* is returned to the supply via diodes D4 and D5.

This circuit gives the highest output voltage of those discussed so far, and is given by the expression

$$V_L = 2 \cdot 3 V_s$$

and each diode is subject to a peak inverse voltage of V_{SM}, which is not as great as in other polyphase rectifier systems.

The principal advantage of polyphase rectifier systems is that they are capable of supplying a large amount of power while generating only a low ripple signal. Six-, twelve-, and twenty-four phase rectifier circuits have been used in order to capitalize on this advantage.

6.7 Transformer utility factor

The secondary windings of rectifier circuit transformers have to carry harmonic currents in addition to the useful average current. The harmonic components do not make any useful contribution to the output and, in addition, contribute to raising the temperature of the transformer. Since transformers are rated on a volt-ampere basis, harmonic currents materially contribute to a reduction in efficiency. Transformer utilization is indicated by the *utility factor,* where

$$\text{Utility factor} = \frac{\text{D.C. power output from the winding}}{\text{Total volt–amperes carried by the winding}}$$

Clearly, the utility factor is a function of the form-factor of the waveform, and the numerical values of this factor for the circuits described earlier are given in Table 6.1.

Table 6.1

A list of transformer utility factors

	Single-phase			Three-phase	
	Half-wave	Centre-tap	Bridge	Half-wave	Bridge
Secondary U.F.	0·287	0·574	0·813	0·666	0·95
Primary U.F.	0·287	0·813	0·813	0·666	0·95

The reason for the difference between the two utility factors in the case of the single-phase centre-tap is as follows. The transformer secondary configuration is such that each diode circuit is equivalent to a half-wave circuit, and the utility factor for each half winding is equal to the half-wave case, giving a utility factor for the secondary of 2 x 0·287 = 0·574. Owing to the bi-phase nature of the secondary winding, current flows in the primary winding in both half-cycles, giving a sinusoidal primary current. The primary utility factor is, therefore,

$$\frac{\text{Average power}}{\text{r.m.s. power}} = \left(\frac{1}{\text{Form factor}}\right)^2 = \frac{1}{1 \cdot 11^2} = 0 \cdot 813$$

6.8   Voltage multiplying circuits

There is a need in some instruments for a d.c. power supply which has a very high voltage as, for example, in the oscilloscope where a voltage of the order of several kilovolts is required. In these applications, voltage multiplying circuits can be employed. The circuits in Fig. 6.7 develop an unloaded output voltage of approximately twice the peak value of the a.c. input signal. Figure 6.7(a) shows

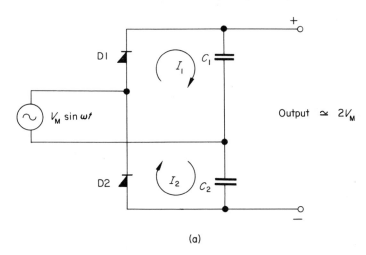

(a)

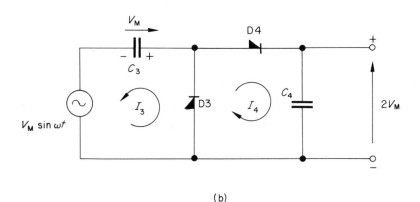

(b)

Fig. 6.7   Voltage doubling circuits.

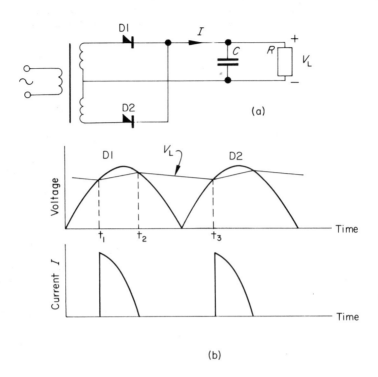

(a)

(b)

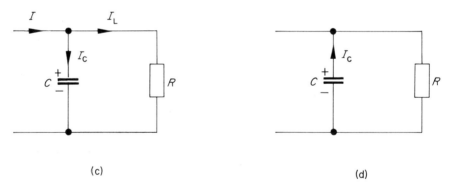

(c)

(d)

Fig. 6.8   (a) A simple capacitor filter, and (b) the waveform diagram of the load voltage. The circuit conditions between t_1 and t_2 are shown at (c), and those between t_2 and t_3 are shown in (d).

a bridge configuration in which diode D1 charges capacitor C_1 to V_M volts in one-half of the input cycle. In the second half-cycle, diode D2 charges C_2 to V_M volts. Since the two capacitors are series connected, the output voltage is approximately $2V_M$ volts.

In the circuit in Fig. 6.7(b), C_3 is charged to V_M during the *negative* half-cycle of the supply voltage. The potential between the terminals of C_3 now acts as a battery in series with the supply. In the positive half-cycle of the supply voltage, C_4 is charged to a voltage equal to the sum of the peak supply voltage and the voltage across C_3.

When a current is drawn from either of these circuits, a large ripple voltage is generated; this factor limits the use of these circuits to low current applications. By cascading voltage multiplying circuits, d.c. voltages greater than four times the peak alternating supply voltage can be generated.

6.9   Capacitor filter

The simplest form of harmonic filter circuit is the shunt capacitor circuit shown in Fig. 6.8(a). To provide adequate filtering with this circuit, the reactance of the capacitor at the ripple frequency should be much lower than the resistance of the load. In this case, the load presented to the rectifier is no longer completely resistive, and a large proportion of the diode current flows into the capacitor. The voltage and current waveforms associated with the circuit are shown in Fig. 6.8(b). Between t_1 and t_2 (Fig. 6.8(b)), the supply potential is greater than the voltage across the capacitor, and current flows from the supply into the capacitor and load. This circuit condition is illustrated in Fig. 6.8(c). Between t_2 and t_3, the supply potential is lower than that on the capacitor, causing both diodes to be reverse biased. During this period, the capacitor discharges and maintains the load potential. The capacitor voltage during the discharge period is a function of the resistance of the connected load, and with a small load resistance the capacitor voltage drops rapidly between charging pulses.

To reduce the ripple voltage to a low level, a large value of capacitance must be used, resulting in a larger charging current flowing for a shorter period than is the case with a small value of capacitance. Figure 6.9 illustrates the change in charging current waveform when the capacitance is increased by a factor of 100. Values of capacitance used vary between a few microfarads and several thousand microfarads, electrolytic capacitors being used to achieve the high value required. It is permissible to connect high values of capacitance to the terminals of thermionic diode circuits, since the maximum current that may be drawn is limited by the temperature of the cathode. In the case of semiconductor diodes, excessive peak current can be drawn during the charging periods, resulting in overheating of the p-n junction and possibly leading to the diode being damaged. It is common practice to include a current limiting resistance R in series with the rectifier, shown in Fig. 6.10.

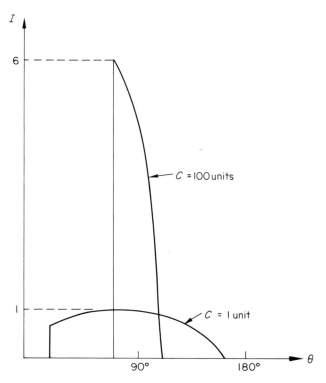

Fig. 6.9   Illustrating the effect of the capacitor value on the diode current waveform.

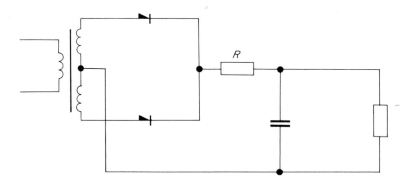

Fig. 6.10   Resistor R is frequently necessary in semiconductor circuits to avoid damage to the diodes.

In the case of gas-filled valves, any attempt to draw an excessive cathode current results in positive-ion bombardment of the cathode, with consequent disintegration of the cathode surface. For this reason, it is inadvisable to connect large values of capacitance directly to gas diode rectifiers.

If the discharge time constant of the capacitor circuit is long compared with the periodic time of the ripple waveform ($1/f$ in half-wave circuits and $1/2f$ in full-wave circuits), i.e., if $CR_L \gg T$, then the ripple waveform approximates to the triangular waveshape in Fig. 6.11. A basic assumption of this waveform is that the capacitor charges in zero time to the peak supply voltage. If the

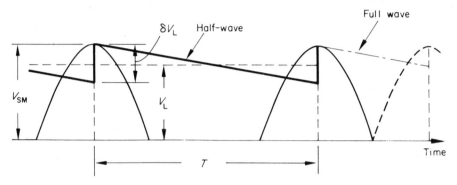

Fig. 6.11   A method of estimating the average output voltage from a rectifier circuit.

assumption is valid, the capacitor voltage falls linearly at the rate of $V_{SM}/R_L C$ volts/second, which continues for T seconds. Thus, the total change in potential δV_L is given by

$$\delta V_L = V_{SM} T / R_L C$$

and the average output voltage of a half-wave circuit is

$$V_L \cong V_{SM} - \frac{\delta V_L}{2} = V_{SM} - \frac{V_{SM} T}{2 R_L C} = V_{SM} \left(1 - \frac{T}{2 R_L C} \right) \qquad (6.9)$$

For example, if a unidirectional output voltage of 200 V is required from a 50 Hz half-wave rectifier circuit with a simple capacitor filter, the load resistance being 1 kΩ and the shunt capacitance 100 μF, then the peak supply voltage must be

$$V_{SM} = 200/(1 - 0 \cdot 02/2 \times 10^3 \times 100 \times 10^{-6}) = 222 \text{ V}$$

that is to say, the r.m.s. voltage provided by the transformer secondary must be $222/\sqrt{2} = 157$ V. In selecting a suitable transformer, the secondary no-load voltage should be greater than 157 V by about 10 to 15 per cent to allow for the voltage drop in the diode and the transformer windings, and also to allow for the errors introduced by the simplifying assumptions made in the theory.

In the full-wave rectifier circuit, the discharge period is $T/2$ s, and eq. (6.9) is modified as follows:

$$V_L \simeq V_{SM}(1 - T/4R_L C)$$

If a full-wave rectifier is used in the above example, the r.m.s. value of the transformer secondary voltage needs to be 150 V (i.e., a 150-0-150 V bi-phase secondary).

6.10   Inductor filter

This filter depends for its operation on the property of an inductance to oppose any change of current in its winding. A simple inductor filter is shown in Fig 6.12. The inductor or *choke* fulfils the function of providing a high series impedance to the alternating ripple frequencies, while presenting only a low resistance to the flow of unidirectional current. For example, a typical 10 henry choke has a resistance of about 225 Ω. If the supply frequency is 50 Hz, then the reactance of the inductance to the lowest ripple frequency (100 Hz) in the output of a full-wave rectifier is 6300 Ω. The rejection of the higher harmonics of the supply frequency is further improved, since the reactance of the inductor increases with frequency.

The design of an inductor filter is often a compromise, since the weight, inductance, and resistance are interdependent. An increase in inductance improves the performance of the filter, but it causes an increase in weight, resistance, and cost of the choke. The resistance can be reduced by the use of larger diameter wire, but this increases the weight and cost further.

Inductor filters are seldom used with half-wave rectifier circuits, since the output voltage in such cases is very low when compared with the average value of the rectified wave. This type of filter circuit, when used with a full-wave circuit, provides a steady output voltage even with a large load current variation. Also, the inductor has the effect of 'smoothing out' current impulses, and the peak current drawn from the diodes is low when compared with the capacitor filter circuit.

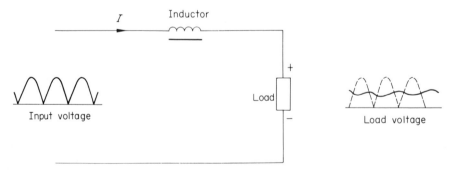

Fig. 6.12   A simple inductor filter.

6.11 Choke input filter

The *choke input filter* or *inverted-L filter* is shown in Fig. 6.13(a). In this
circuit, the inductor provides a high impedance to harmonic currents, while
the capacitor shunts the remaining ripple current from the load. For low values
of load current, the circuit tends to operate as a simple capacitor filter; under
this condition, the rectifier current is discontinuous during the capacitor discharge periods. This operating condition is associated with a rapid reduction
in terminal voltage with load current, shown in Fig. 6.13(b). When the average
value of the load current reaches the value shown by I_K in the figure, the slope
of the characteristic is reduced and the output voltage does not fall as rapidly
with increase in load current as in the earlier part of the characteristic. This
is the point at which the inductor causes the load current to become continuous,

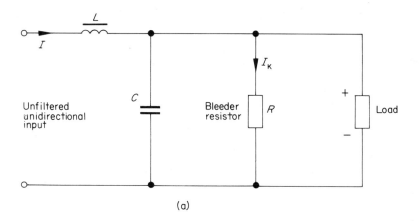

(a)

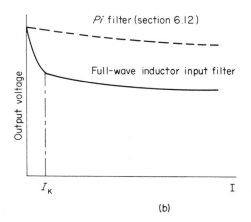

(b)

Fig. 6.13 (a) A choke input filter, and (b) its load characteristic.

that is the rectifier current does not fall to zero. Beyond this point, the characteristic is generally similar to that of a simple inductor filter.

To overcome the problem of the rapid reduction in terminal voltage with load current at low values of load, the output is shunted by resistor R (see Fig. 6.13(a)), known as a *bleeder resistor,* which carries a current equal to or greater than I_K. Owing to the inductor-input nature of the circuit, the peak rectifier current is small when compared with capacitor-input circuits. Values of capacitance used lie between about 5 μF and 100 μF, and inductors between about 3 H and 30 H are used. The drop in terminal voltage with circuits incorporating a bleeder resistor is largely due to the effects of the resistance and leakage reactance of the transformer and choke.

6.12   The π filter

The performance of the choke input filter is improved by adding a capacitor in parallel with its input terminals, as shown in Fig. 6.14. Since the configuration of the reactive components is not unlike the mathematical π symbol, it gets its name from this.

In this circuit, the rectifier delivers pulses of current to C_1 in much the same manner as in the shunt capacitor filter. The filter section comprising L and C_2 smoothes out the ripple voltage appearing across the terminals of C_1. This circuit provides a higher output voltage than the choke input filter described above, but the rectifiers must provide a higher peak current to the π filter than to the choke input filter because of the capacitor input nature of the circuit.

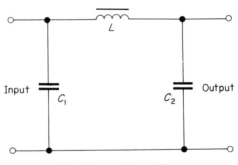

Fig. 6.14   The π filter.

6.13   Ripple filters in a.c. circuits

The function of filters on the d.c. side of the circuit is to reduce the harmonic currents in the load, and reduce inductive interference with nearby circuits. In most rectifier circuits, harmonic currents also flow in the a.c. side, the order and magnitude of the harmonic currents depending on the rectifier connection.

The most satisfactory method of reducing harmonics on the a.c. side is to fit resonant filters to the circuit. These comprise tuned *L-C* circuits, one for each of the principal harmonics present. For higher frequency harmonics, broadband filters may be used to reduce groups of two or three harmonics.

6.14   Parallel operation of diodes

Where the current supplied by one rectifier element is insufficient to supply the load, it is possible to connect two or more diodes in parallel. The way in which the total current divides between the diodes is then dependent upon their

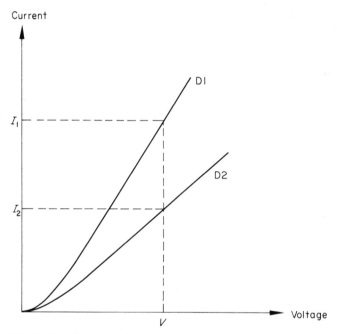

Fig. 6.15   Current sharing between two diodes in parallel.

characteristics. The case of the parallel connection of two thermionic diodes or semiconductor diodes is shown in Fig. 6.15. Since the two diodes are in parallel, the same potential difference appears between their anodes and cathodes; if this is V, then diode D1 carries current I_1, and D2 carries I_2, the total circuit current being $I_1 + I_2$. A more even current distribution results if a resistor is connected in series with diode D1. This reduces both the current through D1 and the total current. To illustrate the magnitude of the problems involved in current sharing in rectifier circuits, in a large computer it is necessary to provide a total current of approximately 1500 A at about 15 V.

It is often inconvenient to operate gas discharge tubes in parallel without some form of load sharing circuit. The idealized anode characteristics of two

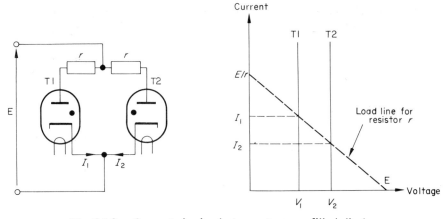

Fig. 6.16   Current sharing between two gas-filled diodes.

gas discharge tubes T1 and T2 are shown in Fig. 6.16. The p.d. across these tubes may differ by as much as 10 V between old and new tubes of the same type. Consequently, when T1 begins to conduct, the p.d. across T2 is V_1. This is insufficient to cause it to conduct, and all the load current is then carried by T1. Load sharing between the two tubes is forced by including a low value of resistance r in series with each anode. As a result of this, with a supply voltage of E, tube T1 carries current I_1 and T2 carries I_2. Improved performance is obtained if the centre-tapped resistor is replaced by a centre-tapped choke.

6.15   Invertors

Inversion is the name given to the process of converting a unidirectional power supply into an alternating supply. Simple relay invertors have been used for many years in motor vehicles to provide power supplies to radios. More recently, transistorized invertors have been used in standby power supplies from batteries. The basic principle of the electronic d.c.-a.c. invertor is shown in Fig. 6.17. The electronic switch (usually consisting of two transistors) connects the

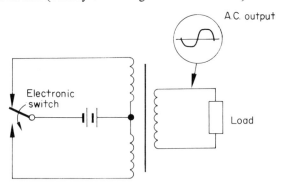

Fig. 6.17   The basic principle of an a.c. to d.c. convertor.

negative pole of the battery alternately to one half of the primary winding of a transformer, and then to the other half. When this changeover occurs the core flux is reversed, causing a reversal of the induced e.m.f. in the secondary winding. The process of inverted operation of controlled rectifiers has already been described in section 2.8.

6.16   Inductive loads

The inherent property of an inductance to try to maintain the current flowing in its windings even if the supply voltage is disconnected, or even reversed, can present problems in some circuits. It tends to maintain the current by generating

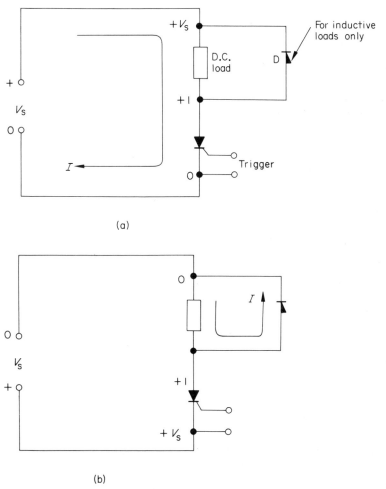

(a)

(b)

Fig. 6.18   The use of a commutating diode or flywheel diode with an inductive load.

a 'back' e.m.f. within its windings whenever the current changes. A method of overcoming this difficulty with an inductive load is illustrated in Fig. 6.18, which shows a simple thyristor circuit. In the positive half-cycle, the potentials at the various points in the circuit are as shown in Fig. 6.18(a), the forward p.d. across the thyristor being about 1 V. In this half-cycle, diode D is reverse biased, and does not carry current.

At the commencement of the negative half-cycle, the inductive effect of the load tends to maintain the flow of current through the thyristor. In so doing, the e.m.f. of self-inductance in the inductive load causes its lower terminal to become positive with respect to its upper terminal, so forward biasing diode D. The voltage across the load is then restricted to the forward conducting voltage drop across the diode. In the meantime, the cathode of the thyristor is raised to a more positive potential than its anode, and the current through it falls to zero. In the absence of the diode, which is known as a *commutating diode* or *flywheel diode,* the inductance of the load forces the thyristor to continue to conduct until the energy stored in the load is dissipated.

Commutating diodes are also used in transistor circuits in which the load is inductive, e.g., when the transistor has to drive a relay. The diode is connected across the load so that it is reverse biased when the transistor is conducting (see Fig. 6.18). When the base signal applied to the transistor forces the collector current to be rapidly reduced, the diode provides an alternative path for the flow of current during the period that the inductive energy is discharged. This artifice limits the collector potential to a safe value during this period of time.

Commutating diodes may, alternatively, be replaced by resistors, the only problem here being that the current through the resistor adds to the normal circuit current.

Problems

6.1   Draw circuit diagrams of single-phase (a) half-wave, (b) centre-tap, and (c) bridge circuits, and explain their principles of operation.

6.2   (a) Sketch the anode current–anode voltage characteristic of a vacuum diode. Comment on the shape of the various sections of the curve.

(b) A vacuum diode with a slope resistance of 150 Ω supplies an 850 Ω resistive load. A moving-iron ammeter and a moving-coil ammeter are connected in series with the load. If the reading of the moving-iron ammeter is 50 mA, calculate (i) the reading of the moving-coil ammeter, (ii) the rectifier efficiency, and (iii) the ripple factor.

6.3   Which of the following is the frequency of the principal ripple voltage at the output of a full-wave rectifier circuit operating from a 50 Hz power supply? (a) 25 Hz, (b) 50 Hz, (c) 100 Hz, (d) 150 Hz.

6.4   For a full-wave centre-tap circuit of the type in Fig. 6.2(a), calculate the average load voltage, the average load current, and the average power consumed by the load if R_L = 900 Ω, the slope resistance of the diodes is 100 Ω, and the r.m.s. voltage induced in *each half* of the transformer secondary is 70·7 V.

State the value of the peak inverse voltage applied to each of the diodes in the circuit. If the output voltage is smoothed by a capacitor of very large value, what would then be the peak inverse voltage applied to the diodes? Give reasons for your answer.

6.5   Sketch the characteristic curve relating grid voltage and anode voltage for a small gas-filled triode. Show how the curve is modified by the action of an extra grid in a tetrode. State the advantages of the tetrode.

A gas-filled triode having a control ratio of 20, is used as a rectifier with an anode load of 10,000 Ω. Sketch, approximately to scale, one cycle of the waveform of applied voltage and current in the load if an alternating voltage of 300 V peak is applied between the anode and cathode and the grid bias is -10 V d.c. The arc and extinction voltage are both 20 V.

(C & G)

6.6   Draw a diagram to show the essential constructional features of a low-voltage ignitron. Indicate the materials used. Explain the operation of the device and give sketches of anode current waveform to illustrate the effect of retarding the ignition.

Compare the ignitron with the thyratron and the mercury arc rectifier and state a typical application of each of these devices.

(C & G)

6.7   Explain, with sketches where appropriate, the operation of a mercury arc rectifier and state why starting electrodes and auxiliary anodes are required. Describe one method of starting.

A six-phase rectifier is to be supplied from a three-phase system. Draw a connection diagram, including the transformer. Number the transformer/rectifier anode connections to show the correct firing order.

(C & G)

6.8   Explain the operation of one form of voltage multiplying circuit, and state one application.

6.9   State which of the following is a typical value for a capacitor for a filter circuit on a 50 Hz supply: (a) 10 pF, (b) 2000 pF, (c) 20 μF, (d) 10 F.

6.10   A rectifier circuit, operating at 50 Hz, supplies an output of 50 mA at 250 V. The output voltage is smoothed by a capacitor of 10 μF. Calculate the value of the load resistance.

Estimate the r.m.s. value of the transformer secondary voltage if the rectifier circuit is single-phase (a) half-wave, and (b) full-wave centre-tap.

6.11   Draw a circuit diagram of a choke input filter, and describe its operation.

6.12   When compared with a simple capacitor, does the choke input filter provide: (a) more filtering action and a smaller direct voltage drop, (b) more filtering action and a larger direct voltage drop, (c) less filtering action and a smaller direct voltage drop, (d) less filtering action and a larger direct voltage drop.

6.13   Using a thyristor as the regulating element, describe the principle of power inversion.

6.14   Explain why it is sometimes necessary to shunt an inductive load in an electronic circuit with a diode.

7. Amplifiers I

7.1 Amplifier classification

Amplifiers are classified in many ways according to such factors as their frequency range, the method of interstage coupling used, the bias point at which the transistors or valves operate, and the aspect of the output signal which is of particular interest, e.g., the voltage, current, or power.

Circuits which amplify a wide band of frequencies are known as *untuned amplifiers* or *broadband amplifiers,* and those which are tuned to amplify a narrow band of frequencies are known as *tuned amplifiers* or *narrowband amplifiers.* The method of coupling between stages of amplification modifies the performance to some extent, the most common method being *a.c. coupling* or *alternating current coupling.* In this type of interconnection, low frequency components (including unidirectional or d.c. signals) are not transmitted to the following stage. Some amplifiers are *direct coupled,* so that they transmit every frequency down to unidirectional signals to the following stage. The latter type of amplifier is one form of *d.c. amplifier* or *direct current amplifier,* which also includes a range of circuits known as *chopper amplifiers.* In chopper amplifiers, the input signal is 'chopped' into a series of pulses by a device known as a 'chopper', which is either a semiconductor circuit or a synchronous relay, so that the signal is converted into an alternating signal. This is amplified by a.c. coupled amplifiers, and is reconverted into d.c. at the output by a process which is the inverse of 'chopping'.

The point on the characteristics to which electronic devices are biased is related to the amplifying function carried out by that stage. For the moment, we merely classify the operation in terms of the relationship between the input signal and the current flowing in the load circuit.

CLASS A   Current flows in the load during the whole period of the input signal cycle.

CLASS AB   Load current flows for more than one-half cycle, but less than the full cycle of the input signal waveform.

CLASS B   Load current flows for one-half cycle of the input signal waveform.

CLASS C   Current flows for less than one-half cycle of the input signal waveform.

Tuned and untuned voltage amplifiers, and low power audio-frequency amplifiers generally work in class A, while audio-frequency power amplifiers work in class B, as do some tuned radio-frequency amplifiers. Oscillators and radio-frequency amplifiers usually operate in class C.

Switching amplifiers and *pulse amplifiers* often form another grouping. These amplifiers are largely concerned with transmitting signals which rapidly change between two voltage or current levels, without delaying the signal or distorting the waveform.

The *gain* of an amplifier is the ratio of the magnitude of the output signal to that of the input signal. *Voltage amplifiers* increase the magnitude of the input voltage signal, but are generally incapable of providing a significant amount of output power. *Current amplifiers* and voltage amplifiers are generally bracketed together so far as amplifier classification goes (it has already been shown that current generator models and voltage generator models are interchangeable), the principal difference between them being the relative values of their output impedance. The operating efficiency of voltage amplifiers is not generally a major consideration, and greater attention is paid to keeping distortion to a minimum.

Power amplifiers are designed to provide an adequate signal power to drive output devices. The output power may range from a few watts to many mega-watts, and the choice of amplifying device is very wide. Operating efficiency is a major consideration in such amplifiers, whereas distortion may not be an important factor. In high power control systems, the input signal to the power amplifier may be in the form of a continuous voltage, and the output may appear as a succession of pulses. The amount of distortion that is acceptable depends upon the application.

7.2   The common-emitter class A voltage amplifier

The basic circuit of this amplifier is shown in Fig. 7.1. In this arrangement, two separate voltage sources are necessary, one for the collector supply and one for the base circuit. The base circuit supply V_{BB} is used to bias the transistor to its *operating point* or *quiescent point*; for class A operation, the steady value of base current I_B must bias the transistor to about the mid-point of its characteristics.

Since the incoming signal (the information signal) may be in the form of an alternating voltage which is superimposed upon an unwanted steady voltage, it is necessary to separate the alternating information signal before amplification.

A simple method of doing this is shown in Fig. 7.1. By connecting capacitor C in series with the input lead, it effectively blocks the steady voltage from the transistor input. If the reactance of the capacitor is small enough at the signal frequency, it provides an unimpeded path to the flow of information signals. This is one form of *a. c. interstage coupling.*

A change in input signal of v_s causes a current i_b to flow through the capacitor, giving a total instantaneous base current of $i_B = I_B + i_b$. The change

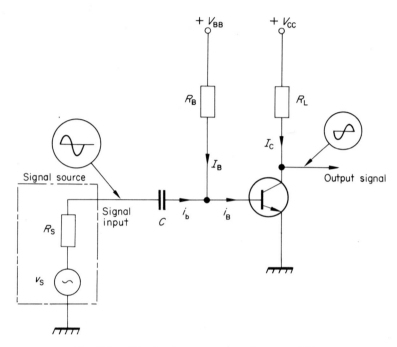

Fig. 7.1   The basic common-emitter amplifier.

i_b in base current causes the collector current to increase by $h_{fe}i_b$. Since the collector current is increased, the p.d. across load resistor R_L increases, causing the collector potential to fall in accordance with the collector circuit equation which is given below:

or

$$\left. \begin{aligned} V_{CC} &= v_{CE} + i_C R_L \\ v_{CE} &= V_{CC} - i_C R_L \end{aligned} \right\} \tag{7.1}$$

That is, an increase in signal voltage applied to the common-emitter amplifier causes a reduction in collector voltage, and a reduction in signal voltage causes the collector voltage to rise. For this reason, the common emitter amplifier is described as a *phase-inverting amplifier,* since the *change* in input and output signals are anti-phase to one another.

7.3   Dynamic transfer characteristics

The dynamic transfer characteristic of a common-emitter amplifier gives the relationship between the collector current and base current, and is deduced from a knowledge of the static output characteristics and the load line. There is one such characteristic for each and every load line; for a given amplifier with a fixed value of load resistance, there is one dynamic transfer characteristic.

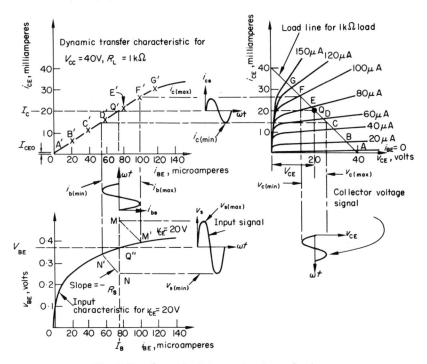

Fig. 7.2   Graphical determination of gain.

The construction of the characteristic is illustrated in Fig. 7.2. The load line is first drawn upon the static output characteristics; in the figure, V_{CC} is 40 V and $R_L = 1$ kΩ. The load line has a slope of $-1/1000$ ampere per volt, and terminates at $v_{CE} = 40$ V, $i_{CE} = 0$ at the lower end, and

$$v_{CE} = 0, \qquad i_{CE} = 40 \text{ V}/1000 \ \Omega = 40 \text{ mA}$$

at the upper end. Point A is the intersection of the $i_{BE} = 0$ characteristic and the load line, and gives the collector current when $i_{BE} = 0$. This is the collector leakage current I_{CEO}, and is plotted on the dynamic transfer characteristic as point A'. Point B on the output characteristics gives the collector current corresponding to a base current of 20 μA; this gives point B' on the dynamic transfer characteristic. Points C', D', E', F', and G' are located in the same way.

The dynamic transfer characteristic is seen to be fairly linear at the lower end due to the equidistant spacing between the static characteristics, but becomes non-linear at the upper end due to the reduction in h_{FE} as saturation conditions are approached. The dynamic transfer characteristic allows the current gain K_i of the *amplifier* (as distinct from the current gain of the transistor) to be calculated as follows. For a maximum value $i_{b(max)}$ of base current there is a maximum value $i_{c(max)}$ of collector current, and corresponding to $i_{b(min)}$ there is a minimum current $i_{c(min)}$ in the collector circuit.

Hence,

$$\text{Current gain } K_i = \frac{i_{c(max)} - i_{c(min)}}{i_{b(max)} - i_{b(min)}}$$

$$= \frac{(26 - 14 \cdot 5)\,\text{mA}}{(100 - 55)\,\mu\text{A}} = 255$$

The current gain can also be evaluated if the r.m.s. value of the collector current I_c and the r.m.s. value of the base current I_b are known, when $K_i = I_c/I_b$. An approximate value for the current gain can also be computed from the slope of the linear part of the dynamic transfer characteristic.

The r.m.s. value of the current in the load, assuming sinusoidal collector current, is

$$I_c = \frac{i_{c(max)} - i_{c(min)}}{2\sqrt{2}}$$

$$= \frac{11 \cdot 5}{2 \cdot 828} = 4 \cdot 06\,\text{mA}$$

and the a.c. power dissipated in the load is

$$I_c^2 R_L = (4 \cdot 06 \times 10^{-3})^2 \times 10^3\,\text{W} \qquad \text{or} \qquad 16 \cdot 5\,\text{mW}$$

The efficiency of the collector circuit is given by the equation

$$\eta = \frac{\text{Signal power developed in the load}}{\text{Average power delivered by the collector supply}}$$

$$= \frac{I_c^2 R_L}{V_{CE} I_C} = \frac{16 \cdot 5 \times 10^{-3}}{20 \times 20 \times 10^{-3}}$$

$$= 0 \cdot 0412 \text{ per unit or } 4 \cdot 12 \text{ per cent}$$

The reason for the low efficiency in this case is due to the relatively large values of quiescent collector voltage and current. A maximum theoretical efficiency for this type of amplifier is 50 per cent (see chapter 8). The efficiency of this amplifier can be improved by simply working at a lower value of V_{CC}; a supply voltage of 20 V would give a collector efficiency of about 16 per cent, but there is some risk of signal distortion with the magnitude of input signal considered here.

Since the circuit is to work as a class A amplifier, the quiescent voltage should be about mid-way along the part of the load line that lies in the active region of the characteristics. In the case of the transistor, this is approximately $V_{CC}/2$. This allows a large base current swing, while still permitting collector current to flow continuously. Here, we select a quiescent collector voltage V_{CE} of 20 V, with its associated quiescent collector current I_C of 20 mA. As stated in section 7.1, the quiescent point Q is fixed by the bias circuit, the design of this circuit being dealt with in section 7.5. The *voltage gain* of the stage can now be estimated by relating the input characteristic corresponding to the quiescent value of v_{CE} to the other characteristics, shown in Fig. 7.2.

The total instantaneous base-emitter voltage is given by

$$v_{BE} = V_{BE} + v_s - R_s i_b \tag{7.2}$$

where V_{BE} is the steady bias voltage, v_s is the instantaneous value of the signal voltage, R_s is the output resistance of the signal source, and i_b is the instantaneous value of the current drawn from the signal source. To implement eq. (7.2) we consider a particular value of v_{BE}, say the maximum value. This is projected horizontally to point M, where it meets the quiescent base current I_B. A line of slope $-1/R_s$ is then drawn from this point until it intersects the input characteristic at point M'. This gives the base current corresponding to the maximum signal voltage. Projecting this point upwards onto the dynamic transfer characteristic gives the maximum collector current. The corresponding minimum collector voltage is determined by further horizontal projection on to the output characteristics. This process is repeated for as many points as considered necessary. Once completed, the output voltage swing can be determined, and the voltage gain computed as follows

$$m = -\left\{ \frac{\text{Output voltage swing}}{\text{Input voltage swing}} \right\}$$

$$= -\left\{ \frac{25{\cdot}5 - 14}{0{\cdot}49 - 0{\cdot}25} \right\} = -48$$

The negative sign in the calculation implies phase inversion, which can be seen from the waveforms of v_{ce} and v_s in Fig. 7.2.

An advantage of the graphical construction is that the collector voltage waveshape can be developed. This permits visual inspection for distortion, which may be introduced by the curvature of the input and dynamic characteristics of the circuit. Fortunately, the curvature on the two characteristics tends to produce opposite effects which cancel each other out, keeping distortion within reasonable limits.

7.4 The small-signal equivalent circuit of the amplifier

The small-signal a.c. equivalent circuit of the complete amplifier is drawn on the assumption that all d.c. sources either have zero resistance or that their resistance can be lumped with the value of some other resistance connected in series with them. That is, in Fig. 7.1, the internal resistance of V_{CC} can be lumped with R_L and that of V_{BB} with R_B. At normal operating frequencies the reactance of capacitor C (Fig. 7.3(a)) is very small when compared with the values of other components in the input circuit; capacitor C is replaced in the a.c. equivalent circuit by a short-circuit. At very low frequencies, when the capacitive reactance of C has a large value, it must be accounted for in calculations (see section 7.5). The basic a.c. equivalent circuit is then as shown in Fig. 7.3(b).

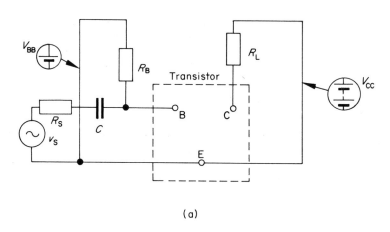

(a)

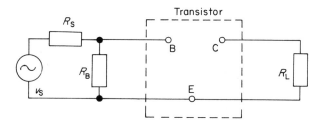

(b)

Fig. 7.3   Small-signal equivalent circuit of the basic common-emitter amplifier.

7.5 Calculation of component values

When working in class A, the peak collector voltage approaches V_{CC} when $i_{BE} = 0$, and the minimum collector voltage is V_{CEsat} when the transistor enters the saturated region of the characteristics. The maximum permissible collector swing is, therefore, slightly less than $V_{CC} - V_{CEsat}$. Since V_{CEsat} has a value between 0·15 V and 0·8 V (depending on the transistor and its base current), the maximum collector voltage swing can be taken to be approximately equal to V_{CC} without significant loss of accuracy. Thus, if $V_{CC} = 40$ V, and the maximum allowable collector current I_{CM} is of the order of 40 mA, then

$$R_L = \frac{V_{CC}}{I_{CM}} = \frac{40 \text{ V}}{40 \text{ mA}} = 1 \text{ k}\Omega$$

The load line corresponding to a 1 kΩ load is drawn on the output characteristics as shown in Fig. 7.2. If a quiescent collector voltage of 20 V is chosen, then the base current required is approximately 75 μA.

When working in class A, the steady value of base-emitter voltage is small (0·37 V in Fig. 7.2), and can often be neglected when compared with the base bias voltage V_{BB}. Since the bias current has to be supplied through resistor R_B in Fig. 7.1, its value is calculated as follows

$$R_B = (V_{BB} - V_{BE})/I_B \simeq V_{BB}/I_B$$

For convenience, both V_{BB} and V_{CC} are usually supplied from a common voltage source, giving

$$R_B = 40 \text{ V}/75 \text{ μA} = 533 \text{ k}\Omega$$

At the lowest frequency to be amplified, the reactance of input capacitor C (Fig. 7.1) must not cause an excessive voltage drop when input current flows through it. At this frequency, a base-emitter voltage which is 0·707 of that at higher frequencies is generally regarded as acceptable (see also section 8.2). That is, the effective input impedance of the circuit at this frequency is $1/0·707 = 1·414$ times greater than its value at higher frequencies.

The a.c. equivalent circuit of the amplifier input circuit is shown in Fig. 7.4(a). Here R_{in} is the small-signal input resistance of the transistor, which is computed from the slope of its input characteristic at the quiescent point. In the case of the characteristic in Fig. 7.2, R_{in} has a value of approximately 1·3 kΩ. Where the input characteristic is not available, R_{in} may be assumed to have the value of the parameter h_{ie}. The effective resistance to small signals between the base and emitter is, therefore,

$$R'_{in} = R_B R_{in}/(R_B + R_{in})$$

$$= 533{,}000 \times 1300/(533{,}000 + 1300)$$

$$\simeq 1300 \text{ }\Omega$$

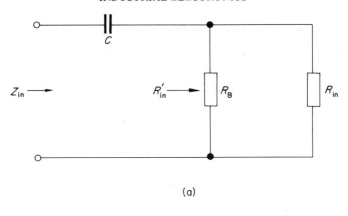

(a)

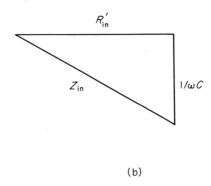

(b)

Fig. 7.4   The linearized input circuit of a common-emitter amplifier.

At high frequencies, when the reactance of capacitor C can be neglected, the
input impedance Z_{in} of the amplifier is equal to R'_{in}. At the lowest frequency to
be amplified, we must account for the reactance of capacitor C by using the
approximation given above, when

$$Z_{in} = 1 \cdot 414 R'_{in} = \sqrt{[R'_{in}{}^2 + 1/(\omega C)^2]}$$

Squaring both sides gives

$$2R'_{in}{}^2 = R'_{in}{}^2 + 1/(\omega C)^2$$

or

$$R'_{in} = 1/\omega C$$

hence,

$$C = 1/\omega R'_{in}$$

If the lowest frequency of interest is 31·8 Hz, then

$$C = 1/2\pi \times 31 \cdot 8 \times 1300 \qquad F = 3 \cdot 85 \ \mu F$$

7.6 Stabilization of the operating point

Variations in such factors as leakage current and transistor parameters can cause a shift in the quiescent point. The most common cause of short-term changes in these values is temperature variation.

The relative importance of these factors depends on the type of semi-conductor material used in the construction of the transistor. For instance, the variation in leakage current with temperature in a germanium transistor is far greater than in an equivalent silicon device, while the variation in h_{FE} with temperature in a silicon device is much greater than in a germanium transistor. In both types of semiconductor, v_{BE} reduces with increasing temperature. The effects are related to the thermal generation of electron-hole pairs, as follows. An increase in temperature results in additional electron-hole pairs appearing in the emitter junction, leading to increased conductivity in that region, and a reduction in v_{BE}. In the collector junction, the additional charge carriers result in an increase in leakage current.

The basic common-emitter circuit in Fig. 7.1 is found to be unsatisfactory from a *thermal stability* viewpoint for the following reasons. Under normal working conditions at constant temperature, the steady current in the base circuit is approximately V_{BB}/R_B. A rise in ambient temperature results in the thermal generation of electron-hole pairs within the transistor, which has an equivalent effect to a rise in external base drive. This results in an increased collector

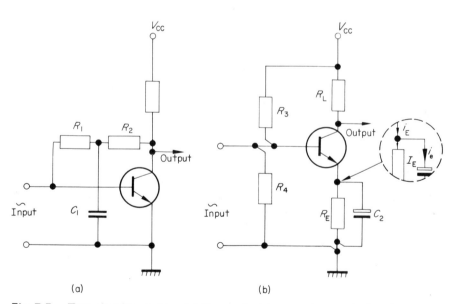

(a)   (b)

Fig. 7.5   Two circuits used to stabilize against temperature variation. Insert shows how the unidirectional component I_E and the varying component i_e of the total emitter current i_E divide in the emitter circuit of (b).

current and a reduced collector voltage. The quiescent point thus moves along the load line with change in temperature. In the circuit in Fig. 7.1 no attempt is made to prevent this happening.

In a power amplifier circuit, this condition leads to an increase in internal power dissipation which can result in a further increase in junction temperature. If the heat generated rises at a greater rate than it can be dissipated, the effect becomes cumulative, leading to *thermal runaway*. As a result, the transistor may be damaged. Thermal runaway is of primary importance in power amplifier stages.

Some improvement of thermal stability is achieved by the circuit in Fig. 7.5(a). For the moment, we will ignore capacitor C_1. Making the assumption that v_{BE} is small compared with V_{CE}, the bias current is approximately $V_{CE}/(R_1 + R_2)$. As the ambient temperature rises it causes the quiescent collector voltage to fall in value, and with it the bias current. By selecting suitable values for R_1 and R_2, the base bias current is reduced by the correct amount to bring the quiescent collector voltage back to its original value. Capacitor C_1 has a low reactance at the signal frequencies, and its function is to effectively earth the junction of R_1 and R_2 to the signal frequencies, so preventing alternating signals from being fed back to the base from the collector.

An even better circuit is shown in Fig. 7.5(b); a load line analysis is shown in Fig. 7.6(a). For the moment, we will ignore capacitor C_2. In this circuit, the base bias is provided by the resistor chain R_3R_4; an additional resistor R_E is included in the emitter circuit. Since part of the supply voltage is dropped across R_E, its value should be as small as possible so that the largest possible voltage can be developed across R_L. The value of R_E cannot be too small, however, otherwise the temperature stability of the circuit will suffer. A p.d. of about 1 V at the quiescent current is usually adequate, so that $R_E \cong 1/I_C$.

Under normal operating conditions, the average potential at the base connection is maintained at a constant value over a wide temperature range by the resistive potential divider chain. As the ambient temperature rises the collector current rises, and with it the p.d. across R_E. Now,

$$v_{BE} = V_B - I_E R_E$$

hence, the increase in $I_E R_E$ with temperature causes v_{BE} to fall in value. This causes a reduction in base current, which compensates the collector current for the rise in temperature.

Variations in h_{FE} with temperature produce generally similar effects to those of variations in leakage current, and the circuits so far described reduce the effects of both. The change in v_{BE} with temperature is generally much smaller than the magnitude of the steady bias voltage applied to the circuit, and changes in v_{BE} are effectively swamped.

It was stated above that the value of R_E must be sufficiently large to provide adequate temperature compensation; however, R_E carries the steady emitter current and dissipates power. To keep the power loss to a minimum, R_E should have a small value. The design of a circuit of this type is, at best, a compromise between thermal stability and other factors such as voltage gain and efficiency.

The total resistance in series with the transistor in Fig. 7.5(b) is $(R_L + R_E)$, and the so-called *d.c. load line* has a slope of $-1/(R_L + R_E)$. The operating point Q, Fig. 7.6(a), lies on this line. In order to prevent variation of the emitter potential when an input signal is applied, R_E is shunted by capacitor C_2 which has a reactance which is small at the lowest frequency of interest when compared

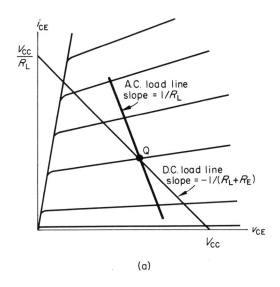

(a)

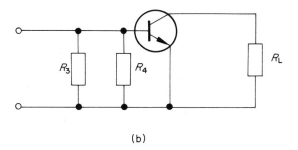

(b)

Fig. 7.6   (a) The Q-point lies on the d.c. load line, and the voltage gain is estimated from the a.c. load line. (b) A method of accounting for the bias network in circuit analysis.

with the resistance of R_E. The emitter capacitor effectively causes the varying component i_e of the emitter current i_E to bypass resistor R_E. So far as the varying component of the circuit current is concerned, the effective resistance in series with the transistor is R_L, since R_E is short-circuited by C_2. This is illustrated in the inset in Fig. 7.5(b). The *a.c. load line* or *dynamic load line* thus has a slope $-1/R_L$, and passes through the quiescent point Q in Fig. 7.6(a). The stage gain is then computed from the a.c. load line.

A simplified a.c. circuit of the amplifier is shown in Fig. 7.6(b), R_E being replaced by a short-circuit. It is seen in Fig. 7.6(b) that both R_3 and R_4 effectively shunt the input circuit, so reducing the input impedance of the amplifier below that of the transistor. This is one of the less desirable features of this method of stabilization.

The temperature stabilization provided by Fig. 7.5(b) can be improved by incorporating temperature-sensitive resistors in the circuit. If R_4 is shunted by a resistor with a negative resistance-temperature coefficient, or R_E is replaced by one with a small positive resistance-temperature coefficient, then the base-emitter voltage changes rapidly with temperature to reduce the effects of temperature change on the circuit. Alternatively a diode could be included in series with R_4; as the ambient temperature rises the forward voltage drop across the diode falls, so reducing the base-emitter voltage to restore thermal stability. A silicon diode is normally used with a silicon transistor, and a germanium diode with a germanium transistor.

The effectiveness of the bias circuit in controlling the change in collector current with temperature is given either by the *stability factor S* or by the *thermal stability factor k,* where

$$S = \frac{\partial I_C}{\partial I_{CBO}} \quad \text{and} \quad k = \frac{\partial I_C}{\partial I_{CEO}} \tag{7.3}$$

hence,

$$S = h_{FE}k$$

The stability factor of the circuit in Fig. 7.5(b) can be evaluated from the equations of the circuit, and is given by

$$S = \frac{1 + R_E\left(\dfrac{1}{R_3} + \dfrac{1}{R_4}\right)}{\dfrac{1}{1 + h_{FE}} + R_E\left(\dfrac{1}{R_3} + \dfrac{1}{R_4}\right)}$$

For example, if R_E = 1 kΩ, R_3 = 50 kΩ, R_4 = 5 kΩ, and h_{FE} = 99, then S = 5·3.

In circuits with poor thermal stability, S has a high value, the worst circuit being that of Fig. 7.1, in which $S = 1 + h_{FE}$. Maximum thermal stability is achieved when S has unity value. Generally speaking, more care must be taken with circuits employing germanium transistors since the leakage current is significant at room temperature, and it doubles for each 8°C to 10°C temperature rise. In low power epitaxial silicon transistors, the leakage current is of the order of only 0·01 μA to 0·1 μA at room temperature. Even though the leakage current in silicon transistors increases at the same relative rate as in germanium transistors, it is of little importance in many applications. In silicon transistors, it is the variation of h_{FE} which is important and its value at 50°C is typically twice the value at 25°C.

Example 7.1: A single-stage audio-frequency voltage amplifier uses an n-p-n transistor which works in class A. The base voltage of the transistor is + 1 V which is derived from a potential divider chain which is connected between the + 10 V supply line and earth. The bias network comprises a 10 kΩ resistor between the base and the earth line, and a resistor R_1 between the base and the 10 V line. If the quiescent collector current is 2 mA, and h_{FE} = 50, determine a suitable value for the resistor R_1.

If the quiescent value of the base-emitter voltage is 0·3 V, estimate the ohmic value of a resistor R_E which must be connected in series with the emitter. This resistor is shunted by a capacitor C_E; determine its value if the lowest signal to be amplified has a frequency of 31·8 Hz

Solution: Since I_{CE} = 2 mA, and h_{FE} = 50, then

$$I_{BE} = 2/50 = 0.04 \text{ mA} \quad \text{or} \quad 40 \text{ } \mu\text{A}$$

The current flowing in the 10 kΩ resistor is 1 V/10 kΩ = 0·1 mA, hence, the current in R_1 is 0·04 + 0·1 = 0·14 mA. The p.d. across R_1 is

$$V_{CC} - V_{BE} = 10 - 1 = 9 \text{ V},$$

hence,

$$R_1 = 9/0.14 = 64.3 \text{ k}\Omega$$

Since V_{BE} = 0·3 V, the p.d. across the emitter resistor is 1 − 0·3 = 0·7 V, and the quiescent emitter current is $I_{CE} + I_{BE}$ = 2 + 0·04 = 2·04 mA.
Hence,

$$R_E = 0.7/2.04 = 0.343 \text{ k}\Omega \quad \text{or} \quad 343 \text{ } \Omega$$

At the lowest frequency of interest, the reactance of C_E should be about one-tenth the value of the resistance of R_E, therefore,

$$C_E \geqslant 10/\omega R_E = 10/2\pi \times 31.8 \times 343 = 146 \text{ } \mu\text{F}$$

7.7   Class A valve amplifiers

Two class A amplifier circuits are shown in Fig. 7.7. In both circuits, the grid bias voltage is developed across R_K.

Hence,

$$V_{GK} = I_K R_K \qquad \text{or} \qquad R_K = V_{GK}/I_K$$

where V_{GK} is the quiescent grid bias voltage and I_K is the quiescent cathode current. As with the transistor, the valve is biased to the middle of the working range of the characteristics. The quiescent point lies on the d.c. load line of slope $-1/(R_L + R_K)$, the voltage gain being computed from information provided by the a.c. load line of slope $-1/R_L$ which passes through the quiescent point. The graphic construction is generally similar to that shown in Fig. 7.6(a).

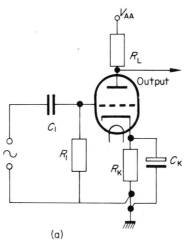

(a)

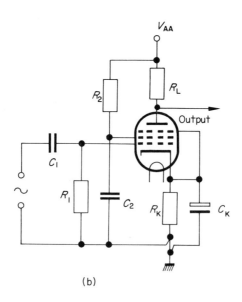

(b)

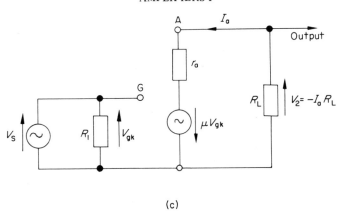

(c)

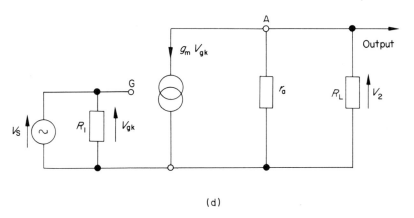

(d)

Fig. 7.7   Class A (a) triode amplifier, and (b) pentode amplifier. Equivalent circuits for (c) the triode amplifier, and (d) the pentode amplifier.

Ambient temperature changes have little effect upon the operating point of valve circuits, and temperature stabilizing circuits are not required. Since the control grid and the cathode are not electrically connected, an external resistor R_1 is necessary to apply the bias voltage to the grid. Since the grid is always at a negative potential with respect to the cathode when the valve works in class A, the grid does not draw current and R_1 can have a high value. A 1 MΩ resistor is commonly used for this purpose. Owing to the high input impedance of the circuit, the capacitance of C_1 is usually only a fraction of a microfarad.

To ensure that a steady voltage is developed across R_K, it is shunted by capacitor C_K which has a reactance at the lowest frequency of interest which is much less than the resistance of R_K. As a rule-of-thumb guide, the reactance of C_K should be about one-tenth of the resistance of R_K at the lowest operating frequency. If this frequency is ω_L, then

$$C_K \geqslant \frac{10}{\omega_L R_K} \, \text{F}$$

In the circuit of Fig. 7.7(b), the screen grid potential is fixed by the values of V_{AA} and R_2; capacitor C_2 effectively short-circuits the screen-grid to ground so far as the signal frequency is concerned. One significant difference between the pentode and the triode is that in the pentode the steady cathode current is $(I_A + I_{G2})$, while it is I_A in the triode. In the pentode, the screen current may represent a significant proportion of the cathode current. This is often not the case in the beam tetrode, in which the screen grid and control grid wires are optically aligned, and the screen grid current may be very small when compared with the anode current.

The most frequently used a.c. equivalent circuit of the *common cathode* triode amplifier is the voltage-source circuit in Fig. 7.7(c), and that of the common cathode pentode amplifier is the current-source circuit in Fig. 7.7(d).

Example 7.2: A triode in an amplifier similar to the one shown in Fig. 7.7(a) passes a quiescent anode current of 8 mA when the grid bias is −4 V. Estimate suitable values for R_K and C_K if the lowest frequency of interest is 15·92 Hz. If the grid coupling resistor R_1 has a value of 1 MΩ, and the input capacitor C_1 reduces the gain at 15·92 Hz to 0·707 of its value at higher frequencies, estimate a suitable value for C_1.

Solution:

$$R_K = |V_{GK}|/I_{AK} = 4/8 = 0\cdot5 \text{ k}\Omega \quad \text{or} \quad 500 \ \Omega$$
$$C_K \geqslant 10/\omega R_K = 10/2\pi \times 15\cdot92 \times 500 \text{ F} \quad \text{or} \quad 20 \ \mu\text{F}$$

For C_1 to produce an attenuation of 0·707 at 15·92 Hz, then

$$R_1 = 1/\omega C_1$$

or

$$C_1 = 1/\omega R_1 = 1/2\pi \times 15\cdot92 \times 10^6 \text{ F} \quad \text{or} \quad 0\cdot01 \ \mu\text{F}$$

7.7.1 Analysis of valve circuits

Voltage-source equivalent circuit (Fig. 7.7(c))
The current flowing in the anode circuit is

$$I_a = \mu V_{gk}/(r_a + R_L)$$

Assuming the polarities shown in Fig. 7.7(c), current flows upwards through R_L so that the output (upper) terminal is negative with respect to terminal K, hence

$$V_2 = -I_a R_L = -\mu R_L V_{gk}/(r_a + R_L)$$

and the voltage gain of the amplifier is given by

$$m = \frac{V_2}{V_{gk}} = \frac{-\mu R_L}{r_a + R_L}$$

the negative sign in the equation implying overall phase inversion of the signal. The r.m.s. power dissipated in the load resistor is

$$P_l = I^2_a R_L = \left(\frac{\mu V_{gk}}{r_a + R_L}\right)^2 R_L$$

It can be shown either by plotting values of P_l to a base of R_L or by the calculus that maximum power is developed in the load when $r_a = R_L$. In this event, the voltage gain is limited to a value of $-\mu/2$.

The input impedance of the amplifier is equal to the value of the grid resistor R_1, since the resistance of the grid-cathode path of the valve when it is operating in class A may be taken to be infinity. The output resistance of the amplifier is equal to the value of r_a.

Current-source equivalent circuit (Fig. 7.7(d))
The current generator in the anode circuit develops a current of $g_m V_{gk}$, which flows through the parallel circuit comprising r_a and R_L. The output voltage developed across the combination is

$$V_2 = -g_m V_{gk} \frac{r_a R_L}{r_a + R_L}$$

hence,

$$m = \frac{V_2}{V_{gk}} = -\frac{g_m r_a R_L}{r_a + R_L}$$

In pentode valves and beam tetrode valves, r_a generally has a much greater value than R_L, and the equation for gain then reduces to the approximate equation

$$m \simeq -g_m R_L$$

As in the case of the triode, the input resistance of the pentode amplifier is equal to the value of the grid resistor R_1, and the output resistance of the amplifier is equal to the value of r_a (which is normally very high).

Example 7.3: A triode valve amplifier has an amplification factor of 20 and a slope resistance of 10 kΩ. It is used in a class A voltage amplifier, and has an anode load of 50 kΩ. Calculate the stage gain, and the a.c. power dissipated in the load if the input signal is 1 V r.m.s.

Solution:

$$m = -20 \times 50/(10 + 50) = -16 \cdot 66$$

$$P_l = V_a{}^2/R_L = (1 \times 16 \cdot 66)^2/50 = 55 \cdot 5 \text{ mW}$$

Example 7.4: Calculate the voltage gain of an amplifier which employs a pentode valve with $g_m = 5$ mA/V, $r_a = 200$ kΩ, and $R_L = 20$ kΩ.

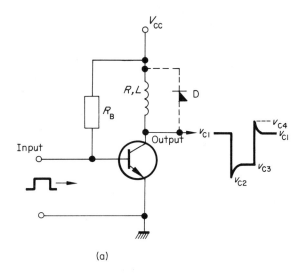

(a)

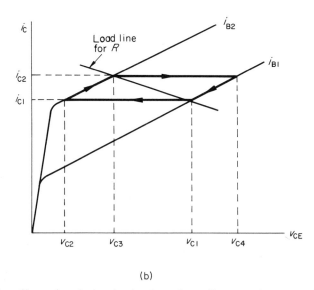

(b)

Fig. 7.8 The effect of an inductive load on the collector voltage waveform.

Solution:

$$m = -5 \times 200 \times 20 / (200 + 20) = -91$$

Note: the approximate equation gives $m \simeq -5 \times 20 = -100$.

7.8   Effect of reactive loads

Imagine that a signal with a square waveform is applied to the input of the common emitter amplifier in Fig. 7.8(a), which has an inductive load. The initial base current and collector voltage are i_{B1} and v_{C1}, respectively.

When the base current is suddenly increased to i_{B2}, the inductance of the load causes the collector current to remain initially constant at i_{C1}, causing the collector voltage to fall to v_{C2}. This transition on the characteristics is shown by the lower horizontal line in Fig. 7.8(b). When the inductance has dissipated its excess stored energy, the collector voltage rises to v_{C3} along a constant base current characteristic. A reduction in input current to i_{B1} causes the collector voltage to suddenly rise to v_{C4} along a line of constant collector current, following which the collector voltage slowly decays to v_{C1} along a line of constant base current.

When the input signal changes in a negative direction, a positive-going voltage spike is generated at the collector. The peak value of this spike can exceed V_{CC}, and may even exceed the voltage rating of the transistor. In circuits of this type, transistors are often protected against overvoltage by shunting the load by diode D which functions in the manner described in section 6.16. It is not usually necessary to protect valves in this way, since they can sustain large overvoltages without damage.

If the input signal has a sinusoidal waveshape, the excursion traced out on the characteristics is approximately elliptical in shape, the movement round the ellipse being in a clockwise direction.

With a capacitive-type load there is a tendency for current spikes to be generated, since the capacitor initially maintains a constant voltage between its terminals, resulting in a rush of charging or discharging current.

7.9   Transistor analysis in terms of the h-parameters

The following analysis applies to any transistor configuration, e.g., common-emitter, common-base, or common-collector, since the general h-parameters h_i, h_r, h_f, and h_o are used. The gain, input impedance, output impedance, etc., of any specific configuration are then obtained by inserting the appropriate h-parameters into the general equations. For instance, if the transistor is used in the common-emitter mode the parameters h_{ie}, h_{re}, h_{fe}, and h_{oe} are inserted in place of h_i, h_r, h_f, and h_o, respectively.

A block diagram which represents the amplifier is shown in Fig. 7.9(a), where V_s is the r.m.s. value of the source voltage, and R_s is the output imped-

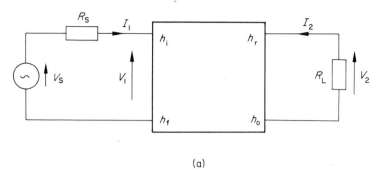

(a)

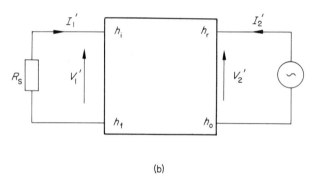

(b)

Fig. 7.9   Amplifier analysis using the h-parameters.

ance of the signal source. The r.m.s. values of the input voltage to the transistor and the output voltage from the transistor are V_1 and V_2, respectively, while I_1 and I_2 are the respective values of the input and output current. In the common-emitter configuration, the quantities V_1, V_2, I_1, and I_2 correspond to V_{be}, V_{ce}, I_{be}, and I_{ce}, respectively.

The general hybrid equations of Fig. 7.9(a) are

$$V_1 = h_i I_1 + h_r V_2 \tag{7.4}$$

$$I_2 = h_f I_1 + h_o V_2 \tag{7.5}$$

Also, in Fig. 7.9(a)

$$V_2 = - I_2 R_L \tag{7.6}$$

Substituting eq. (7.6) into eq. (7.5) yields

$$I_2 = h_f I_1 + h_o(-I_2 R_L)$$

from which the current gain is

$$K_i = \frac{I_2}{I_1} = \frac{h_f}{1 + h_o R_L} \tag{7.7}$$

The input resistance of the stage is computed by substituting eq. (7.6) into eq. (7.4)

$$V_1 = h_i I_1 + h_r(-I_2 R_L) = h_i I_1 - h_r R_L I_2$$

$$= h_i I_1 - h_r R_L K_i I_1 = I_1(h_i - h_r R_L K_i)$$

hence, the input resistance of the amplifier is

$$R_{in} = V_1/I_1 = h_i - h_r R_L K_i \tag{7.8}$$

The overall voltage gain of the amplifier is given by the ratio V_2/V_1. Now, $V_1 = I_1 R_{in}$, and $V_2 = -I_2 R_L$, hence,

$$m = \frac{V_2}{V_1} = -\frac{I_2}{I_1}\frac{R_L}{R_{in}} = -K_i \frac{R_L}{R_{in}} \tag{7.9}$$

The output resistance is computed from a consideration of Fig. 7.9(b). Here the input signal voltage is reduced to zero, and a generator is connected to the output terminals. The output resistance is then given by the equation

$$R_{out} = V_2'/I_2'$$

The voltage at the input terminals is given by $V_1' = -I_1' R_s$; substituting this expression into eq. (7.4) gives

$$-I_1' R_s = h_i I_1' + h_r V_2'$$

or

$$I_1' = -h_r V_2'/(h_i + R_s)$$

Thus, from eq. (7.5)

$$I_2' = -\frac{h_f h_r V_2'}{h_i + R_s} + h_o V_2' = V_2'\left(h_o - \frac{h_f h_r}{h_i + R_s}\right)$$

therefore,

$$R_{out} = \frac{V_2'}{I_2'} = \frac{1}{h_o - \dfrac{h_f h_r}{(h_i + R_s)}} \tag{7.10}$$

The power gain is expressed as the ratio of the a.c. power delivered into the load to that delivered to the input of the transistor.

$$\text{Power gain } K_p = \frac{I_2^2 R_L}{I_1^2 R_{in}} = \left(\frac{I_2}{I_1}\right)^2 \frac{R_L}{R_{in}} = K_i^2 \frac{R_L}{R_{in}} \tag{7.11}$$

The equations for the common-emitter and the common-base configurations are listed in Table 7.1. The approximate equations for the common-emitter configuration are obtained by noting that the feedback voltage $h_{re}V_{ce}$ is small compared with V_{be}, and that the shunting effect of h_{oe} on the collector current is very small. That is to say that h_{re} and h_{oe} are assumed to have zero value; this concept was first developed in chapter 4, and is dealt with more fully in section 7.9.1. Equations for the common-collector configuration are omitted, since transistor parameters for this mode of operation are rarely quoted

Table 7.1

Equations for the common-emitter and common-base amplifier configurations

	Common-emitter		Common-base
	Exact	Approximate	
K_i	$\dfrac{h_{fe}}{1 + h_{oe}R_L}$	h_{fe}	$\dfrac{h_{fb}}{1 + h_{ob}R_L}$
R_{in}	$h_{ie} - h_{re}R_L K_i$	h_{ie}	$h_{ib} - h_{rb}R_L K_i$
m	$-K_i \dfrac{R_L}{R_{in}}$	$-h_{fe}\dfrac{R_L}{h_{ie}}$	$-K_i \dfrac{R_L}{R_{in}}$
R_{out}	$1 \Big/ \left(h_{oe} - \dfrac{h_{fe}h_{re}}{(h_{ie} + R_s)} \right)$	$\dfrac{1}{h_{oe}}$	$1 \Big/ \left(h_{ob} - \dfrac{h_{fb}h_{rb}}{(h_{ib} + R_s)} \right)$
K_p	$K_i{}^2 \dfrac{R_L}{R_{in}}$	$\dfrac{h_{fe}{}^2 R_L}{h_{ie}}$	$K_i{}^2 \dfrac{R_L}{R_{in}}$

Table 7.2

Limiting values of the common-emitter amplifier constants

	Maximum value	Minimum value	Condition
K_i	h_{fe}		$R_L \to 0$
		0	$R_L \to \infty$
R_{in}	h_{ie}		$R_L \to 0$
		$h_{ie} - \dfrac{h_{fe}h_{re}}{h_{oe}}$	$R_L \to \infty$
m	$-\dfrac{h_{fe}}{h_{oe}R_{in}}$		$R_L \to \infty$
		0	$R_L \to 0$
R_{out}	$1 \Big/ \left(h_{oe} - \dfrac{h_{fe}h_{re}}{h_{ie}} \right)$		$R_s \to 0$
		$1/h_{oe}$	$R_s \to \infty$

in manufacturers' literature; the common-collector circuit (the emitter follower) is dealt with in chapter 10.

The limiting values of the common-emitter amplifier constants are related to the maximum permissible values of R_L and R_S; these values are listed in Table 7.2, together with the conditions under which they are obtained.

Example 7.5: A transistor having the following parameters is used in a common-emitter amplifier.

$$h_{ie} = 1500 \ \Omega \quad h_{re} = 0 \cdot 0002$$

$$h_{fe} = 50 \qquad h_{oe} = 10^{-5} \ S$$

The output resistance of the signal source is 1 kΩ, and the collector load resistance is 10 kΩ. Determine the current gain, the input resistance, the voltage gain, the output resistance, and the power gain of the amplifier.

Solution: The current gain is evaluated from eq. (7.7)

$$K_i = 50/(1 + 10^{-5} \times 10^4) = 45 \cdot 5$$

and from eq. (7.8)

$$R_{in} = 1500 - 0 \cdot 0002 \times 10^4 \times 45 \cdot 5 = 1409 \ \Omega$$

The voltage gain, from eq. (7.9), is computed as

$$m = -45 \cdot 5 \times 10^4/1409 = -323$$

The output resistance, from eq. (7.10), is

$$R_{out} = 1/(10^{-5} - 50 \times 0 \cdot 0002/(1500 + 1000)) = 167 \ k\Omega$$

and the power gain, from eq. (7.11)

$$K_p = 45 \cdot 5^2 \times 10/1 \cdot 409 = 14,700$$

To illustrate the variation of the amplifier constants with load resistance and source resistance, curves showing their values are given in Fig. 7.10.

7.9.1   Simplified analysis of transistor amplifiers

In many applications, the foregoing analysis can be simplified by making a few approximations. Firstly, the parameter h_r is generally very small and may be neglected. The equation for the input impedance of the amplifier becomes

$$R_{in} \simeq h_i$$

hence, the input voltage equation is $V_1 \simeq h_i I_1$. Secondly, h_o is usually small, and the current flowing in the load is approximately $h_f I_1$. Hence,

$$I_2 \simeq h_f I_1$$

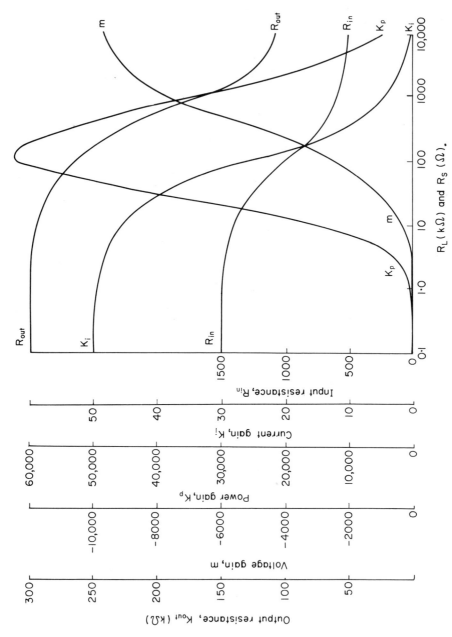

Fig. 7.10   Variation of the common-emitter amplifier properties with R_L and R_S.

therefore,

$$K_i \simeq h_f$$

and the voltage across the load is

$$V_2 = -I_2 R_L \simeq -h_f I_1 R_L$$

but

$$I_1 \simeq V_1/h_i$$

hence,

$$V_2 \simeq -\frac{h_f}{h_i} V_1 R_L$$

Note: h_f/h_i is the effective mutual conductance or transconductance of the transistor.

The voltage gain of the amplifier is

$$m = \frac{V_2}{V_1} \simeq -\frac{h_f}{h_i} R_L$$

The approximate value of the output impedance of the amplifier is $1/h_o$, and the expression for power gain is

$$K_p = K_i^2 \frac{R_L}{R_{in}} \simeq h_f^2 \frac{R_L}{h_i}$$

Example 7.6: Compute the input impedance, the current gain, the voltage gain, the output impedance, and the power gain of a common-emitter amplifier which uses a transistor with $h_{ie} = 1500 \ \Omega$, $h_{fe} = 50$, and $h_{oe} = 10^{-5}$ S. The load resistance is 10 kΩ.

Solution:

$$R_{in} \simeq h_{ie} = 1500 \ \Omega$$

$$K_i \simeq h_{fe} = 50$$

$$m \simeq -\frac{h_{fe}}{h_{ie}} R_L = -\frac{50}{1500} \times 10{,}000 = -333$$

$$R_{out} \simeq \frac{1}{h_{oe}} = \frac{1}{10^{-5}} = 100 \ \text{k}\Omega$$

$$K_p \simeq h_{fe}^2 \frac{R_L}{h_{ie}} = 50^2 \times \frac{10^4}{1500} = 16{,}700$$

These results should be compared with the values of 1409 Ω, 45·5, −323, 167 kΩ, and 14,700, respectively, obtained for the same amplifier by exact methods in example 7.5. Bearing in mind the spread of parameters in transistors of the same type, the results of the approximate analysis suffice for many practical purposes

7.10   Tuned voltage amplifiers

Where it is necessary to amplify either a single frequency or a narrow band of frequencies, amplifiers with tuned L-C loads are used. These circuits effectively reject frequencies other than those within the pass-band of the tuned circuit.

The simplest form of load is the parallel L-C circuit. If the inductor in the parallel circuit has resistance $R,$ then the effective impedance at resonance (which is a pure resistance) is L/CR $\Omega,$ and is known as the *dynamic impedance* of the circuit. The dynamic impedance of the circuit is also given by $\omega_0 LQ,$ where ω_0 is the resonant frequency of the circuit, and Q is the Q-factor of the coil. This arrangement ensures that a high impedance load is presented at resonance, and a low impedance at other frequencies. For example, if a 250 μH coil of resistance 8 Ω is tuned to a frequency of 0·5 MHz by a shunt capacitor, its dynamic impedance is 77 kΩ. The impedance of the tuned circuit to a frequency of 0·45 MHz is only 3·7 kΩ.

To utilize the full gain offered by the high impedance load, either a transistor or a pentode should be used as the active device in the amplifier. In the case of transistor circuits, a step-down inductive coupling is necessary between the L-C circuit and the following stage to avoid loading effects due to the low input impedance.

7.11   Common-base and common-grid amplifiers

One form of common-base amplifier, together with its h-parameter equivalent circuit, is shown in Fig. 7.11. The potential divider and emitter resistor temperature stabilization circuit developed in section 7.6 is used in the circuit described here, the base being effectively connected to the common line by capacitor C_2 so far as signal frequencies are concerned. Capacitor C_1 acts as a blocking capacitor, which prevents external unidirectional voltages from being applied to the amplifier. Resistor R_1 does not appear in the a.c. equivalent circuit since, so far as signal frequencies are concerned, both ends are coupled to the common line (one by C_2 and the other by the supply source).

Under class A operating conditions the emitter diode is forward biased, and the input impedance of the common-base approximates to that of a forward biased diode. This is normally of the order of 20 Ω to 100 Ω, although its value varies widely with load resistance. The output impedance approaches that of the reverse biased collector diode and has a high value, 50 kΩ to 2 MΩ typically. Since the alternating current flowing in the emitter circuit has a value which is slightly greater than that in the collector, the current gain of the amplifier is less than unity. The voltage gain for a given value of load resistance is equal to that obtained from a common-emitter amplifier, while the power gain is lower than that of a common-emitter stage.

When a positive-going voltage signal is applied to the input of the amplifier, it raises the emitter potential and reduces the current flowing in the transistor.

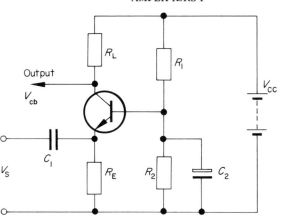

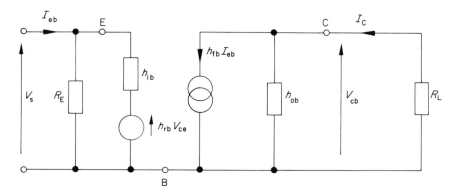

Fig. 7.11   The common-base amplifier.

This has the effect of reducing the p.d. across R_L and increasing the collector potential. Thus, the common-base amplifier is a *non-phase-inverting amplifier* so far as voltage gain is concerned. When signal currents are considered, the amplifier is phase-inverting since an increase in signal current flowing *into* the emitter causes an increase in the signal current flowing *out of* the collector.

The mathematics relating the h_e and h_b parameters is tedious, and is not included here. For convenience, the relationships between the parameters is given in Table 7.3. Amplifier constants can then be computed from the equations in Table 7.1.

Owing to the low input impedance and high output impedance of the common-base amplifier, it is not usual to cascade amplifier stages since the low

Table 7.3

Relationships between the h_e and h_b parameters

$h_{ib} = \dfrac{h_{ie}}{1 + h_{fe}}$	$h_{ie} = \dfrac{h_{ib}}{1 + h_{fb}}$
$h_{rb} = \dfrac{h_{ie}h_{oe}}{1 + h_{fe}} - h_{re}$	$h_{re} = \dfrac{h_{ib}h_{ob}}{1 + h_{fb}} - h_{rb}$
$h_{fb} = -\dfrac{h_{fe}}{1 + h_{fe}}$	$h_{fe} = -\dfrac{h_{fb}}{1 + h_{fb}}$
$h_{ob} = \dfrac{h_{oe}}{1 + h_{fe}}$	$h_{oe} = \dfrac{h_{ob}}{1 + h_{fb}}$

input impedance of the driven stage causes a massive reduction in output voltage. Audio-frequency applications of the common-base amplifier are therefore restricted to cases in which the output impedance of the signal source is low and the load impedance is high. A typical application is where the signal source is a 30 Ω microphone, and the load is the input circuit of a common-emitter amplifier stage. Owing to the high cut-off frequency of the common-base amplifier, extensive use is made of this type of circuit at V.H.F.

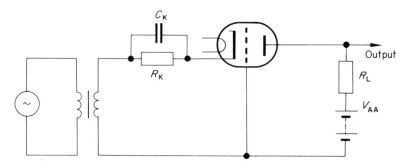

Fig. 7.12   The common-grid amplifier.

A common-grid valve amplifier is shown in Fig. 7.12 This type of amplifier is most commonly used with a tuned anode load, but an untuned circuit is considered here for simplicity. In the figure, the signal source is transformer coupled to isolate the signal source from the quiescent cathode current. The bias voltage is developed across the parallel combination R_K and C_K. As with the common-base circuit, the input impedance is low (when compared with the common-cathode amplifier) and the output impedance is quite high. Again, the voltage amplification offered is non-phase-inverting.

Example 7.7: The parameters of a transistor in the common-base configuration are $h_{fb} = -0\cdot98$, $h_{ob} = 5\ \mu S$, $h_{ib} = 30\ \Omega$, and $h_{rb} = 0\cdot0005$. The transistor is used in a common-base amplifier circuit with a collector load of 20 kΩ, the signal source resistance being 1 kΩ. Calculate the current gain, the input resistance, the voltage gain, the output resistance, and the power gain of the amplifier.

Solution: From eq. (7.7)

$$K_i = h_{fb}/(1 + h_{ob}R_L) = -0.98/(1 + 5 \times 10^{-6} \times 20 \times 10^3)$$

$$= -0.89$$

The negative sign implies that the current flows *out* of the collector when current flows *into* the emitter, or vice versa. From eq. (7.8),

$$R_{in} = h_{ib} - h_{rb}R_L K_i$$

$$= 30 - 0.0005 \times 20 \times 10^3 \times (-0.89) = 38.9 \ \Omega$$

The voltage gain is obtained from eq. (7.9):

$$m = -K_i R_L/R_{in} = -(-0.89) \times 20 \times 10^3/38.9$$

$$= +458$$

It should be noted that the common-base amplifier voltage gain is non-phase-inverting, indicated by the positive sign associated with the numerical value of *m*. From eq. (7.10),

$$R_{out} = 1 \bigg/ \left(h_{ob} - \frac{h_{fb}h_{rb}}{h_{ib} + R_s} \right)$$

$$= 1 \bigg/ \left(5 \times 10^{-6} - \frac{(-0.98) \times 0.0005}{30 + 1000} \right) = 182.5 \ k\Omega$$

and the power gain is given by

$$K_p = K_i^2 R_L/R_{in} = (-0.89)^2 \times 20 \times 10^3/38.9 = 408$$

Note: The approximate analysis gives $K_i \simeq h_{fb} = -0.98$; $R_{in} \simeq h_{ib} = 30 \ \Omega$; $m \simeq -h_{fb}R_L/h_{ib} = 654$; $R_{out} \simeq 1/h_{ob} = 200 \ k\Omega$; and $K_p = h_{fb}^2 R_L/h_{ib} = 640$.

7.12 Field-effect transistor circuits

It was shown in section 4.14 that a depletion-mode FET operated like a valve, in that with zero gate voltage the drain current was high. To bias a depletion-mode FET to its correct operating point for class A operation, it must be *biased-off*. That is, a reverse bias voltage must be applied between the gate and source. One bias circuit for an N-channel junction-gate depletion-mode FET is shown in Fig. 7.13(a). Here, the bias voltage is developed across R, while C effectively shunts the varying component of the source current from

R. Resistor R_1 merely provides an ohmic connection between the common line and the gate to allow the bias voltage to be applied to the gate.

In an enhancement-mode FET, the drain current is zero when V_{GS} is zero. In this case, the transistor must be *biased-on* to its characteristics; one circuit is shown in Fig. 7.13(b). Here, the bias voltage developed at the junction of R_2 and R_3 is applied to the gate via R_4. Since the input resistance of the FET is high the value of R_4 is usually large, typically 10 MΩ. So far as the signal

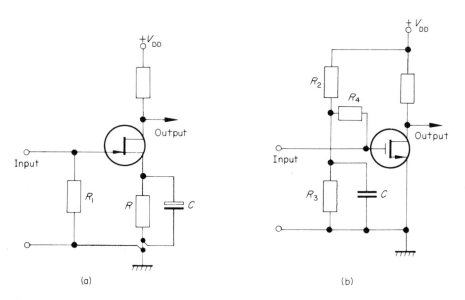

Fig. 7.13   Bias circuits for (a) an N-channel JUGFET, and (b) an N-channel IGFET.

frequency is concerned, capacitor C connects the junction of R_2 and R_3 to the FET source electrode.

The main advantage of FETs over bipolar devices is their high input impedance, and it is in applications where this fact is of paramount importance that they are gaining ground. At the present stage of their development, FETs are predominantly associated with voltage amplifier stages, but there is no reason to suppose that high power FETs will not be developed in the future. Field-effect transistors have an inherent high frequency capability; present restrictions on frequency range are imposed by parasitic capacitances between the gate, the source, and the drain. There is no reason why manufacturing technology will not allow them to catch up and even overtake conventional bipolar transistors in high frequency applications.

Another major advantage of FETs over bipolar transistors and conventional resistors is that they can be manufactured more easily in large numbers in the form of monolithic integrated circuits. Interconnections between the elements are then made by depositing conducting films on the microcircuit. In these applications, resistors can be replaced by FETs which operate as pinch-effect resistors.

7.13   Photoelectric circuits

A number of circuits employing photoelectric devices are shown in Fig. 7.14. In Fig. 7.14(a) a photoemissive cell is used to control the base bias current flowing in the transistor. Under dark conditions, practically no base current

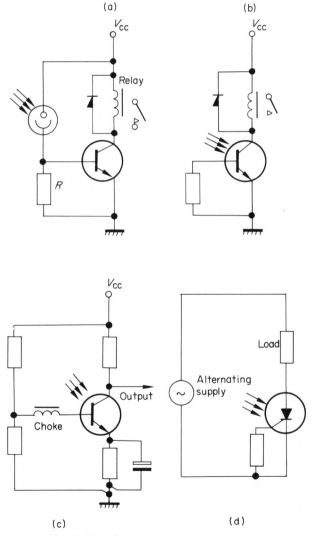

Fig. 7.14   Some photoelectric circuits.

flows, and the relay is de-energized. Upon the incidence of light at a suitable illumination level, the base current becomes sufficiently great to cause the collector current to pull the relay armature in. If resistor R and the photocell are interchanged, the circuit energizes the relay under dark conditions, and de-energizes it under conditions of high illumination. It is possible to use a photoconductive cell to replace the photoemissive cell in these applications.

Figure 7.14(b) illustrates the application of a phototransistor to control a relay. The transistor is protected from voltage spikes by the commutating diode connected across the relay. Phototransistors may be operated with the base open-circuited, or with a resistor of about 5 kΩ connected between the base and the emitter. The resistor reduces the dark current and improves thermal stability. Both circuits in Figs. 7.14(a) and (b) employ unmodulated light sources, but some applications require the use of a modulated (a rapidly varying) light source. One such circuit is shown in Fig. 7.14(c). The base bias current in this circuit is provided by the resistive potential-divider chain and the emitter resistor. The choke effectively open-circuits the base to variations in current at the modulation frequency, but provides a low resistance path to the flow of bias current. The circuit works as a class A amplifier.

A photothyristor circuit is shown in Fig. 7.14(d). When light is incident upon the photothyristor, it switches into its forward conducting mode (if the anode is positive), and current flows in the circuit. Photothyristors can also be triggered by fibre-optic means. The gate electrode is normally coupled to the cathode by a resistor to prevent inadvertent firing of the thyristor by induced voltages in the gate circuit.

Photoelectric devices are also used in conjunction with hard valves and gas-filled devices. An example of the use of a photodiode and a thyratron is given below.

Example 7.8: The thyratron in Fig. 7.15 has a control ratio of 25, and its striking voltage with zero grid bias is 15 V. The characteristic may be assumed to be linear. If the sensitivity of the photodiode is 25 mA/lm, and it has negligible dark current, estimate the minimum illumination level required to trigger the thyratron into conduction.

Solution: The grid-cathode voltage of the thyratron is

$$v_{GK} = 2I - 7$$

where I is in mA; the equation for the anode voltage at which the thyratron fires is

$$V_A = 15 - 25v_{GK} = 15 - 25(2I - 7) = 190 - 50I$$

In this case, the anode voltage at which the thyratron fires is 100 V, hence,

$$I = (190 - 100)/50 = 1·8 \text{ mA}$$

and the minimum illumination level to cause conduction is

$$1\cdot 8 \text{ mA}/25 \text{ mA/lm} = 0\cdot 072 \text{ lm}$$

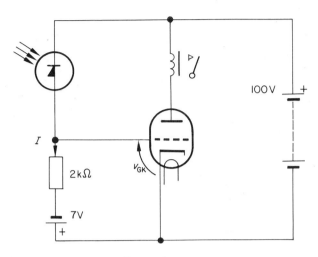

Fig. 7.15   Figure for example 7.8.

Problems

7.1   Discuss the methods used to classify amplifiers; are other methods of classification possible?

7.2   Explain the essential differences between: (a) class A and class B operation, (b) d.c. coupled and a.c. coupled amplifiers, (c) voltage amplifiers and power amplifiers.

7.3   The output characteristics of a transistor connected in the common-emitter mode are linear between the following points:

v_{CE} (V)	2	25	
i_C (mA)	4·5	5·5	when $i_B = 50 \ \mu A$
i_C (mA)	14·0	16·0	when $i_B = 150 \ \mu A$
i_C (mA)	23·5	26·5	when $i_B = 250 \ \mu A$

Draw the characteristics and construct the load line for a collector load of 1 kΩ and a collector supply voltage of 30 V.

If the peak-to-peak sinusoidal current swing in the base circuit is 200 μA, about a mean value of 150 μA, estimate: -

(a) the r.m.s. value of the a.c. component of collector current.
(b) the r.m.s. value of the voltage developed across the load.
(c) the r.m.s. value of the a.c. power developed in the load, and
(d) the values of h_{fe} and h_{oe} at the quiescent point.

7.4 The output characteristics of an n-p-n transistor connected in the common-emitter configuration are linear between the following points:

v_{CE} (V)	10	40	
i_C (mA)	4·5	5·0	when i_B = 20 μA
i_C (mA)	9·4	11·6	when i_B = 40 μA
i_C (mA)	14·5	17·0	when i_B = 60 μA
i_C (mA)	19·4	26·5	when i_B = 80 μA
i_C (mA)	25·0	40·0	when i_B = 100 μA
i_C (mA)	31·0	52·0	when i_B = 120 μA

The leakage current is negligible. Plot the static transfer characteristics of the transistor relating the collector current to the base current for collector voltages of (a) 10 V, (b) 20 V, (c) 40 V. Estimate the current gain at each value of v_{CE}.

The transistor is to be used with a load resistor of 1 kΩ and a collector supply voltage of 40 V. Draw the dynamic transfer characteristic of the amplifier and estimate its current gain.

7.5 Estimate the component values to be used in a common-emitter amplifier of the type shown in Fig. 7.1 if the collector supply voltage is 6 V and the maximum allowable collector current is 2 mA. The quiescent collector current is to be 1 mA and the value of h_{FE} is 50.

7.6 If the small-signal current gain of the amplifier in question 7.5 is 45, compute the voltage gain and the power gain of the stage if the input impedance is 1 kΩ.

7.7 If a 600 Ω load is connected to the amplifier in question 7.6, calculate the new overall voltage gain.

Note: The effective value of load resistance is now R_L in parallel with 600 Ω.

7.8 Give three basic configurations of the transistor amplifier. What precautions are necessary to reduce the effect of temperature variation on the d.c. working point of each circuit?

Explain in detail the operation of a typical stabilizing circuit and show how it modifies the input and output impedance of the amplifier.

How can the effect of the feedback be restricted to direct current and very low frequencies?

(C & G)

7.9 Define the thermal stability factor of a transistor amplifier in terms of (a) I_{CBO} and (b) I_{CEO}. Show how the two are related.

7.10 Given the following details, estimate the values of circuit components in an amplifier of the type in Fig. 7.5(b). The collector supply voltage is 20 V, the quiescent collector current is 2 mA, and the quiescent collector-emitter voltage and base-emitter voltage are 5·5 V and 0·25 V, respectively. The value of h_{FE} at the operating point is 100. In order to provide adequate thermal stability, R_E = 1 kΩ. The lowest frequency of interest is 31·8 Hz.

7.11 Explain why temperature stabilization is less important in the design considerations of common-base amplifiers than it is in grounded-emitter amplifiers.

7.12 Describe the techniques necessary to achieve the correct operating point in valves and transistors (including FETs). Comment on whether or not it is necessary to stabilize against temperature change in each case.

7.13 Explain the meaning of the terms p-n-p and n-p-n as applied to transistors. What is the essential difference in the circuits appropriate to these two types of transistors?

The characteristics of a transistor in the grounded base connections are, for small signals, represented by the following equations:

$$V_e = h_{ib}I_e + h_{rb}V_c$$
$$I_c = h_{fb}I_e + h_{ob}V_c$$

In a particular transistor, $h_{fb} = 0.97$, $h_{ob} = 5\mu S$, $h_{ip} = 20\ \Omega$, $h_{rb} = 0.0005$. If this transistor is used in the grounded base connection with a load resistor of 50 kΩ in the collector circuit, calculate (a) the voltage amplification, (b) the input resistance.

(C & G)

7.14 The output characteristics for a transistor connected in the common-base configuration are as follows:

$v_{CB}(V)$	0	-10	-20	
i_C (mA)	-57	-57.5	-58	when $i_E = -60$ mA
i_C (mA)	-38	-38.5	-39	when $i_E = -40$ mA
i_C (mA)	-19	-19.5	-19.75	when $i_E = -20$ mA

Draw the characteristics, and construct a load line for a load of 330 Ω and a collector supply voltage of 20 V.

If the quiescent emitter current is 40 mA and a sinusoidal current of 20 sin ωt is injected into the emitter, estimate (i) the r.m.s. output voltage, and (ii) the r.m.s. output power.

7.15 A triode, used in a class A amplifier, has the following constants: $r_a = 10$ kΩ; $g_m = 2.5$ mA/V. This valve is coupled to the grid of the following stage by means of a capacitor of negligible reactance at the operating frequency. The grid resistor of the following stage has a value of 100 kΩ.

If the quiescent anode voltage and current are 120 V and 4 mA, respectively, determine the value of the supply voltage and the voltage gain when the anode load is (a) a resistance of 40 kΩ, (b) a choke of negligible resistance and extremely high reactance at the signal frequency.

7.16 The following table contains information relating to a triode:

Anode current	Anode voltage when bias is		
	Zero	-4 volts	-8 volts
10 mA	100	180	260
3 mA	40	120	200

Plot the anode characteristics (assumed to be linear).

The triode is to be used in class A, with a resistive load of 25 kΩ with a supply voltage of 300 V. A fixed bias of -4 V is provided.

Draw, on the static characteristics, a load line representing these conditions and hence determine:

   (i) the mean anode current and voltage.
   (ii) the a.c. output voltage if the magnitude of the sinusoidal input signal is 2.83 volts r.m.s.
   (iii) the stage gain.
   (iv) the output power.

(C & G)

7.17 Draw the constant voltage equivalent circuit of the amplifier in Fig. 7.7(a).

A load of 100 kΩ is coupled to the amplifier via a capacitor of 0·05 μF; draw the constant voltage equivalent circuit of the combination. If the input signal is 2 V peak-to-peak, and its frequency is 10 kHz, estimate the r.m.s. voltage across the 100 kΩ load. The valve parameters are μ = 30, g_m = 5 mA/V, and a load resistor of 10 kΩ is used. Justify any assumptions made in the calculations. If an additional load of 100 kΩ is connected in parallel with the output, estimate the new value of output voltage.

7.18 With the aid of a connection diagram, explain the operation of a circuit using a photo-electric cell and an amplifier stage to count objects passing along a conveyor belt.

State, with reasons, which type of cell and amplifier valve would be used.

What factors govern the rate at which counting may proceed?

(C & G)

7.19 With the aid of a diagram explain the operation of a circuit using a photoelectric cell which could be used as an industrial alarm system.

State with reasons what type of cell you would use and explain its action.

(C & G)

7.20 Define the four hybrid parameters of a transistor in the common-emitter mode and state the units, if any, in which they are measured.

If the signal applied between the base and emitter of a transistor is 1 mV and the collector load is 1 kΩ, calculate the output voltage given that

$$v_b = (10^3)i_b + (10^{-3})v_c$$
$$i_c = (80)i_b + (2 \times 10^{-5})v_c$$

(C & G)

7.21 Draw the circuit diagram and explain the action of a grounded-grid triode amplifier. Show the equivalent circuit diagram and deduce the amplification factor when a triode is used in this way.

In what circumstances are grounded-grid triodes normally used, and what advantages do they possess under such conditions?

(C & G)

7.22 Describe briefly an experiment to determine the voltage gain of a single stage amplifier using a triode valve with a resistive load.

Sketch and explain an equivalent circuit for this amplifier, neglecting stray capacitances. Use this circuit to find an expression for the gain of the stage in terms of the amplification factor of the valve.

The slope resistance of the valve is 10 kΩ and its amplification factor is 12. Calculate the input voltage that will give 3 V r.m.s. output across an anode load of 5 kΩ.

(C & G)

7.23 Draw the equivalent circuit of a transistor with the following hybrid parameters: h_{ie} = 1 kΩ, h_{re} = 2 x 10^{-4}, h_{fe} = 50, h_{oe} = 20 μS.

Assuming that the bias conditions remain unchanged throughout, calculate (a) the input resistance and the current gain with the collector load on short-circuit, and (b) the voltage gain and the power gain with a collector load of 50 kΩ.

(Based on a C & G question)

7.24 A pentode amplifier employs a valve with g_m = 6 mA/V, an anode load resistor of 22 kΩ, a cathode resistor of 500 Ω, and a screen grid resistor of 150 kΩ. If the anode and cathode potentials are 140 V and 3 V, respectively, and the h.t. voltage is 250 V, calculate (a) the cathode current, (b) the anode current, (c) the screen-grid current, and (d) the screen-grid voltage.

(C & G)

7.25 With reference to a thermionic triode valve, define the terms:

(a) anode slope resistance (r_a).

(b) amplification factor (μ).

(c) mutual conductance (g_m).

A tuned-anode triode valve amplifier operates at resonance. For the triode, $\mu = 30$, $r_a = 15$ kΩ; the anode load consists of a 500 μH inductor (L) of 30 Ω resistance (R) with a parallel-connected tuning capacitor (C) of 450 pF.

Sketch the simple equivalent circuit of the amplifier and calculate its gain at resonance, given that the dynamic resistance of the resonant circuit is L/CR.

(C & G)

7.26 State the impurity elements used to make p and n types of germanium. Why are these elements used?

A p-n-p junction transistor has $a = 0.95$, an emitter-base resistance of 500 Ω and an output resistance of 1·2 MΩ.

Draw a circuit diagram showing the transistor used as a voltage amplifier with common base and estimate the voltage gain if the load resistance is 4·7 kΩ.

Calculate the collector-base current gain expected when used in the grounded emitter connection. Further compare these two connections on the score of input and output impedance.

(C & G)

8. Amplifiers II

8.1 The decibel

When dealing with electronic and electrical circuits it is often convenient to think in terms of the transmission of sinusoidal signals, rather than non-sinusoidal signals. As a result of the circuit constants, sinusoidal signals are subject either to *gain* or to *attenuation* and to a *phase-shift*.

Gain and attenuation are most conveniently expressed as a logarithmic ratio. Historically, the logarithmic ratio of two values of power is given in *bels* (after the scientist Alexander Graham Bell), as follows

$$\text{Power ratio} = \lg P_2/P_1 \text{ bels (B)}$$

where P_1 and P_2 are two values of power. If the two power levels are produced in a resistance of value R by the application of voltages V_1 and V_2, respectively, then

$$\text{Power ratio} = \lg \frac{V_2{}^2/R}{V_1{}^2/R} = \lg(V_2/V_1)^2 = 2 \lg V_2/V_1 \text{ B}$$

The bel is an inconveniently large unit, and the *decibel* ($1 \text{ dB} = 0 \cdot 1 \text{ B}$) is commonly used in electronics, when

$$\text{Power ratio} = 20 \lg V_2/V_1 \text{ dB} \tag{8.1}$$

The use of the ratio is strictly only correct when V_1 and V_2 are developed across equal values of resistance. The definition of the decibel is distorted in its application to electronic circuits, since the two voltages may be developed across unequal values of resistance. Thus, an amplifier with an input signal of $0 \cdot 1$ V r.m.s., and an output signal of 10 V r.m.s. is said to have a voltage gain of $20 \lg 10/0 \cdot 1 = 40$ dB, even if the input resistance is 1 kΩ and the load resistance is 100 kΩ.

Logarithmic ratios are also very useful in indicating when a *change* in the signal level has occurred. When no change has occurred, then $V_2 = V_1$, and the 'gain' is 20 lg 1 = 0 dB. When the gain in decibels has a positive value, then V_2 is greater than V_1; when the decibel gain has a negative value then V_2 is less than V_1, and the signal is said to be *attenuated*. The simplest method of computing the 'gain' in decibels of an attenuator is as follows:

$$20 \lg V_2/V_1 = 20 \lg (V_1/V_2)^{-1} = -20 \lg V_1/V_2 \qquad (8.2)$$

Thus, if V_1 = 1 V, and V_2 = 0·1 V, the gain in dB is

$$20 \lg \frac{0·1}{1} = -20 \lg \frac{1}{0·1} = -20 \lg 10 = -20 \text{ dB}$$

A feature of logarithmic gain ratios is that the overall gain of cascaded amplifier and attenuator stages can be calculated simply by *adding* the logarithmic gain values together. For instance, if two voltage amplifiers with gains of 20 dB and 40 dB, respectively, are cascaded with a circuit giving 30 dB of attenuation, the overall gain is

$$20 + 40 - 30 = 30 \text{ dB}$$

An alternative method of expressing power ratios is given by the *neper* (after the Scottish farmer-mathematician John Napier). The power ratio in nepers is defined by the relationship

$$\text{Power ratio in neper} = \ln P_2/P_1$$

The neper is equal to 8·686 decibel.

8.1.1 Reference level

The decibel is merely a means of comparing two signal levels and no reference level is necessarily implied in the calculations. However, it is sometimes useful to quote a change in signal level by stating that it is X decibels above or below a reference level. There are two common examples of this; the first is in amplifier applications, in which the gain is substantially constant over a range of frequencies, but is reduced at other frequencies. Without implying the reference value of gain, it is permissible to say that the gain at a certain frequency is, say, 3 dB below the reference gain. The second example is in general telecommunications practice, where a reference level of 1 mW is often used. In this case a power level of 2 mW would be expressed as 10 lg 2/1 = 3·01 dB.

8.2 Bandwidth

The *bandwidth* or *pass-band* of an amplifier is the band of frequencies over which the power gain does not drop to a value less than an assigned fraction— usually one-half—of the 'mid-band' power gain (see Fig. 8.1). In tuned amplifiers, the reduction in power gain is expressed in terms of the gain at resonance.

The bandwidth is derived from the amplifier *frequency response diagram* shown in the figure, in which gain (both power and voltage) is plotted to a base of frequency.

Curve A shows a response curve which is 'flat' down to d.c. or zero frequency, and is typical of a d.c. amplifier. At high frequency the gain begins to fall as a result of some frequency-dependent component in the circuit, e.g., stray capacitance which shunts the output. If the output power at frequency f_2 is one-half

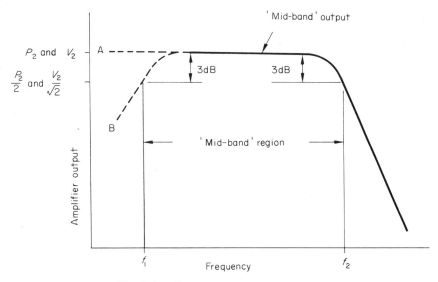

Fig. 8.1   Frequency response curves.

of the mid-band power output, then the comparative reduction in gain at f_2, expressed in logarithmic terms, is

$$\lg \frac{P_2}{0 \cdot 5 P_2} = \lg 2 = 0 \cdot 301 \text{ bel} = 3 \cdot 01 \text{ dB}$$

For practical purposes this is taken to be 3 dB. The equivalent output voltage at f_2 is

$$\text{output voltage} = \sqrt{(\text{Output power} \times \text{Load resistance})}$$

$$= \sqrt{\left[\frac{P_2}{2} \times R \right]} = \sqrt{\left[\frac{1}{2} \left(\frac{V_2{}^2}{R} \right) R \right]}$$

$$= \frac{V_2}{\sqrt{2}} = 0 \cdot 707 V_2 \tag{8.3}$$

Frequency f_2 is known variously as a *corner frequency*, a *cut-off frequency*, a *half-power point*, and a *break-point*.

In the case of characteristic A, the gain is constant down to d.c. The bandwidth of the d.c. amplifier is therefore $(f_2 - 0) = f_2$. A criterion for selecting circuits is the *gain-bandwidth product,* which is the product

Numerical value of gain x Bandwidth

If amplifier A has gain m, then the gain-bandwidth product is mf_2.

The magnitude of the linear slope of the high frequency end of the frequency response curve is related to the number of reactive elements in the circuit, and the way in which they are connected into the circuit. Generally speaking, the slope of the high frequency end of the response curve lies between 6 dB/octave and 12 dB/octave (20 dB/decade and 40 dB/decade), where one octave represents a frequency doubling (one decade is a tenfold increase in frequency). Thus, on the linearly falling part of the high frequency response, if the gain is 30 dB at 20 kHz, a slope of −6 dB/octave (−20 dB/decade) gives gain values of 24 dB, 18 dB, and 10 dB at frequencies of 40 kHz, 80 kHz, and 200 kHz, respectively.

Curve B is typical of an a.c. amplifier, in which the low frequency response falls off in much the same way as the high frequency response. Here, the lower corner frequency is f_1, and the bandwidth of the amplifier is $(f_2 - f_1)$. The gain-bandwidth product is $m(f_2 - f_1)$. In broadband a.c. amplifiers, f_1 may have a value of about 20 Hz and f_2 may be several megahertz, giving a bandwidth approximately equal to f_2. In tuned amplifiers, f_1 and f_2 may be close together, in which case both frequencies must be accounted for in calculations.

In many cases, it is necessary to cascade amplifier stages in order to achieve the desired gain or some other aspect of performance. Unfortunately, in many instances, this has the effect of reducing the bandwidth, as illustrated in the following. Suppose that an amplifier has an upper cut-off frequency of 1 MHz and a mid-band gain of 20 dB; if two such amplifiers are cascaded, the overall mid-band gain is 40 dB (assuming that one does not load the other), but at 1 MHz the gain is 3 + 3 = 6 dB below this value. The new corner frequency (i.e., when the total gain is 39 dB) occurs when each stage contributes a reduction of gain of 1·5 dB. This gives a bandwidth for the cascaded amplifier of about 0·6 MHz. Another example of this effect is given in question 8.6 at the end of this chapter.

8.3   Drift in d.c. amplifiers

The principal feature of interest in direct current amplifiers is their ability to amplify slowly varying signals; otherwise they are generally similar to a.c. amplifiers. They find a wide range of applications in the more specialized fields of electronics, including measuring instruments, automatic control systems, voltage regulators, cathode-ray oscilloscopes, and analogue computers.

Unfortunately, d.c. amplifiers are subject to a phenomenon known as *drift*, which is a slow variation of the output voltage or current of the amplifier when the input signal is maintained at a constant level. This imposes a limitation upon the permissible gain of some designs of amplifier. Drift is caused by a shift in the operating point on the characteristics of the amplifying device, due to several factors including variations in

(a) The parameters of the device used.
(b) The anode (or collector) supply voltage.
(c) The heater voltage (in valves).

In transistor circuits, it is principally the temperature-dependent parameters which are involved. These are the current gain, the base-emitter voltage, and the leakage current. Both (b) and (c) above are due to mains voltage variation, and their effects can be minimized by using stabilized power supplies.

Drift is particularly troublesome when the magnitude of the input signal is small and its frequency is very low; in this case, the drift and input signal effects may be indistinguishable from one another.

The temperature dependence of the gain of semiconductor d.c. amplifiers can be reduced by utilizing the temperature-dependent effects of other semiconductor devices, e.g., thermistors and diodes to compensate for the original changes. Thus, by replacing the resistor in the collector of a common-emitter amplifier by a thermistor, its effective resistance falls with rising temperature. If the thermistor is correctly selected, it compensates for the increase in current gain of the transistor with temperature.

A circuit in which the effects of drift are minimized is the *long-tailed pair amplifier,* described in section 8.3.2 of this chapter.

8.3.1   The Darlington connection

The Darlington connection is a method of cascading two transistors with no additional circuitry, as shown in Fig. 8.2. Two transistors connected in this way are also known as a *super-alpha pair.* The emitter current of Q1 is the sum of its collector and base currents. If the current gain of Q1 is h_{FE1}, then

$$i_{B2} \simeq h_{FE1} i_{B1} + i_{B1} = i_{B1}(1 + h_{FE1})$$

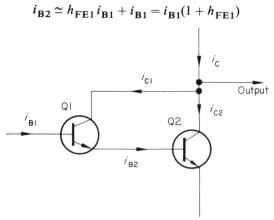

Fig. 8.2   Darlington-connected transistors.

If the current gain of Q2 is h_{FE2}, then the collector current taken by the composite circuit is

$$i_C = i_{C1} + i_{C2} \simeq h_{FE1} i_{B1} + h_{FE2} i_{B2}$$

$$= h_{FE1} i_{B1} + h_{FE2} i_{B1}(1 + h_{FE1})$$

$$= i_{B1}(h_{FE1} + h_{FE2}(1 + h_{FE1}))$$

and the current gain of the combination is

$$\frac{i_C}{i_{B1}} = h_{FE1} + h_{FE2}(1 + h_{FE1})$$

Thus, if h_{FE1} has a value of 49, and h_{FE2} has a value of 50, the current gain is

$$49 + 50(1 + 49) = 2549$$

The circuit is described as a super-alpha pair since the effective value of α (the modulus of the common-base current gain) for the combination approaches unity. In the example above,

$$\alpha = \frac{2549}{1 + 2549} = 0 \cdot 9999$$

When two separate transistors are used in this way, it is advisable to mount them on a common heat sink. This ensures that they work at approximately the same temperature, when variations in temperature-sensitive parameters are similar in both of them. A better alternative is to use two transistors which have been manufactured in the same chip of semiconductor material.

8.3.2   Emitter-coupled and cathode-coupled amplifiers

Emitter-coupled and cathode-coupled circuits are widely used in d.c. amplifiers. In the cathode-coupled amplifier, Fig. 8.3, the cathodes are connected together and taken via a common resistor either to earth potential or a negative potential, as shown. When both input signals are zero, the potential of the cathode connection will adjust itself to become slightly positive to provide the bias voltage. With equal values of anode load resistors and balanced input signals, then $I_1 = I_2$ and $V_o = V_o'$. The output signal from these circuits is usually taken as the differential signal (the signal difference) between the two anodes; its value is zero under the conditions considered. A twin-triode or double-triode valve is normally used in these circuits so that the heater voltage is common to both valves, as is the anode supply voltage. Any variation in either the heater voltage or the anode supply voltage causes the anode currents of both valves to change by equal amounts, and the differential output voltage remains unchanged. Any drift effects due to temperature variation or supply voltage variation thus causes little change in the differential output voltage.

If the two input signals v_1 and v_2 are equal in magnitude and are in phase with one another, then the two anode voltages change by equal amounts, and the differential output voltage is again zero. When the input signals are in phase with one another, they are known as *common-mode signals* since both signals have the same polarity with respect to the common line; since the circuit in Fig. 8.3 (and also that in Fig. 8.4) gives practically zero differential output signal for common-

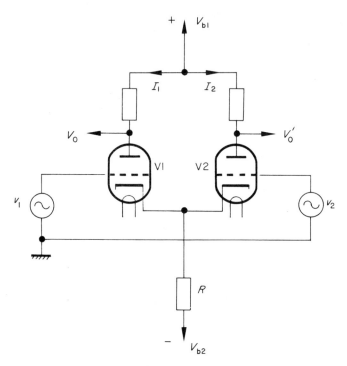

Fig. 8.3   Cathode-coupled amplifier.

mode input signals, it is said to have a high *common-mode rejection* figure. This is very desirable in d.c. amplifiers, where electrical noise often appears as a common-mode signal.

If one of the two input signals is earthed, say v_2, and signal v_1 becomes positive with respect to earth, it has the effect of increasing the current through valve V1. This has two simultaneous effects. Firstly, it causes the anode potential of V1 to fall, and secondly the cathode potential rises. The increase in the cathode potential increases the bias voltage applied to the grid of V2, resulting in a reduction of current through V2. Thus, as V_o is reduced, V_o' increases, giving a differential output between the two anodes. Thus, the circuit amplifies the difference in the two input signals. For this reason, it is known as a *difference amplifier* or *differential amplifier*. The magnitude of the output signal is further

increased if the differential input is increased. One method of achieving this end
is to energize V1 with a signal which is antiphase to the signal applied to V2.

In circuits where a large voltage gain is required, differential amplifier stages
can be cascaded so that outputs V_o and V'_o are used as the inputs to a second
differential amplifier.

The operation of the differential amplifier may, alternatively, be explained
by the following argument. If the magnitude of the negative supply voltage is

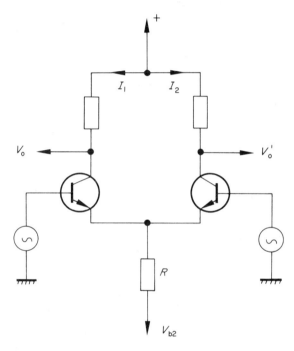

Fig. 8.4   Emitter-coupled amplifier.

large, and resistor R has a large value, then the cathode bias circuit acts as a
constant-current source which limits the sum of the two cathode currents to
V_{b2}/R. In this event, an increase in the differential mode input voltage causes an
increase in current through one valve, and an equal reduction in current in the
other valve since the sum $I_1 + I_2$ remains constant. The common-cathode resistor
R gives rise to an alternative name for the circuit—the *long-tailed pair*.

The bipolar transistor version, the *emitter-coupled amplifier,* is shown in Fig.
8.4. Owing to the versatility of semiconductor devices, other configurations are
possible. For instance, resistor R can be replaced by a transistor which derives its
base current from a reference voltage source, thereby ensuring a constant current
supply to the two transistors. The stage gain of the circuit in Fig. 8.4. can be
improved by replacing each transistor with a super-alpha pair, shown in Fig. 8.2.

Circuits using FETs to replace the bipolar transistors in Fig. 8.4 are popular where a high input impedance is desirable; examples include measuring instruments and oscilloscopes. These amplifiers are known as *source-coupled amplifiers.*

To minimize the effects of temperature variation in semiconductor versions, it is preferable to use dual transistors in the same canister, which can be mounted on a heat sink if necessary.

The small-signal gain of the long-tailed pair amplifier is determined as follows. The instantaneous potential of the common-emitter connection is approximately equal to the average value of the two input signals. That is

$$v_e = \tfrac{1}{2}(v_1 + v_2)$$

Now, the instantaneous base-emitter voltage applied to each transistor is equal to the difference between the input voltage and v_e, hence,

$$v_{be1} = v_1 - v_e = v_1 - \tfrac{1}{2}(v_1 + v_2) = \tfrac{1}{2}(v_1 - v_2)$$

$$v_{be2} = v_2 - v_e = v_2 - \tfrac{1}{2}(v_1 + v_2) = -\tfrac{1}{2}(v_1 - v_2) = -v_{be1}$$

If the mutual conductance of the transistors is $g_m \, (\simeq h_{fe}/h_{ie})$, then,

$$i_{c1} \simeq g_m v_{be1} \quad \text{and} \quad i_{c2} \simeq g_m v_{be2} = -i_{ce1}$$

Assuming balanced loading conditions, with each transistor having a collector load of R_L, then the changes in the collector potentials are

$$v_{ce1} = i_{c1} R_L = \tfrac{1}{2} g_m (v_1 - v_2) R_L$$

$$v_{ce2} = i_{c2} R_L = -\tfrac{1}{2} g_m (v_1 - v_2) R_L = -v_{ce2}$$

and the differential output voltage is

$$v_{ce1} - v_{ce2} = g_m R_L (v_1 - v_2)$$

and the magnitude of the stage gain is

$$g_m R_L = \frac{h_{fe}}{h_{ie}} R_L$$

The theory given above is based on the assumption that the devices used in the amplifier have transistor-like (or pentode-like) characteristics. If triodes are used, the assumption that the change in output current is equal to the product of g_m and the change in grid-cathode voltage is not justified, and the results will be in error.

Example 8.1: An emitter-coupled transistor amplifier using n-p-n transistors has a +10 V collector supply and a −8 V emitter supply. If the emitter resistor has a value of 2 kΩ, and the quiescent collector voltage of the transistor with zero input signal is to be about +5 V, estimate suitable values for the collector resistors. If the transistors used have h_{fe} = 50, h_{ie} = 1 kΩ, estimate the magnitude of the voltage gain.

Solution: With zero input signal, the potential of the emitter connection is approximately zero, and the current in the emitter resistor is approximately

$$I_E = 8 \text{ V}/2 \text{ k}\Omega = 4 \text{ mA}$$

Since this current divides between the two transistors, the collector current of each is 2 mA. This current causes a p.d. of 5 V in the collector resistor, hence

$$R_L = 5 \text{ V}/2 \text{ mA} = 2 \cdot 5 \text{ k}\Omega$$

and the magnitude of the stage gain is

$$g_m R_L \simeq \frac{h_{fe}}{h_{ie}} R_L = \frac{50}{1} \times 2 \cdot 5 = 125$$

8.4   Interstage coupling methods

The essential differences between a.c. coupling and direct coupling are shown in Fig. 8.5. The output from almost any amplifier consists of a quiescent voltage with a superimposed alternating component; it is the alternating component which generally carries the signal information. In a.c. amplifiers, the alternating component is separated from the d.c. component by the coupling network. In Fig. 8.5(a) this is achieved by an *R-C network*. It can equally well be done by a transformer. The capacitor *C* effectively blocks the d.c. component, so that only the a.c. component of the output signal causes a current to flow in resistor *R*. The voltage developed across *R* is then applied to the next stage for further amplification. In the context of Fig. 8.5, resistor *R* is the external load resistance of the amplifier which includes the input resistance of the following stage. For satisfactory operation, the reactance of capacitor *C* at the lowest frequency of interest must be much smaller than the resistance of *R* (see chapter 7).

It has already been shown that when $R = 1/\omega C$, the voltage across resistor *R* is 0·707 of the voltage applied to the *R-C* circuit. At the frequency at which this occurs, the *R-C* network introduces a reduction in voltage gain of 0·707 or 3 dB.

Example 8.2: If the input signal to a broadband amplifier is applied via a blocking capacitor, and the input impedance of the amplifier is 1 kΩ, calculate the capacitance of the capacitor if it introduces a reduction in gain of 3 dB below the mid-band gain at a frequency of 31·8 Hz.

Solution: For a 3 dB reduction,

$$R = 1/\omega C$$

or

$$C = 1/\omega R = 1/31 \cdot 8 \times 2\pi \times 1000 = 5 \text{ } \mu\text{F}$$

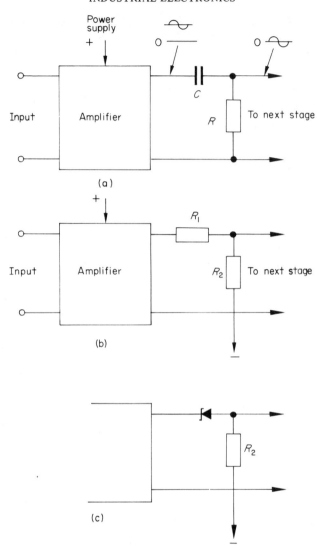

Fig. 8.5   (a) *R-C* interstage coupling for a.c. amplification; (b) and (c) two methods of direct coupling between stages.

8.4.1   Direct coupling circuits

The basis of direct coupling circuits is shown in Fig. 8.5(b). The output of one stage is coupled to the next by the resistor chain $R_1 R_2$, which has its lower end taken to a negative supply rail (positive supply in the case of p-n-p transistor circuits). The function of the negative power supply is to balance out the quiescent output voltage of the amplifier. By this means, any variation in output voltage from the first stage, even a very slow change, is applied directly to the following stage. Thus, undesirable low frequency drift voltages are amplified as well as signal frequencies by subsequent stages. It is common practice in d.c. amplifiers to use low drift stages, e.g., a long-tailed pair amplifier, at the input end to reduce the effects of drift.

A disadvantage of the resistive network in Fig. 8.5(b) is that the varying component of the output voltage is attenuated by the resistor chain. This reduces the overall gain of the stage. Some improvement is effected by replacing R_1 by a Zener diode, as shown in Fig. 8.5(c). The diode acts as a battery which opposes the d.c. component of the output voltage while allowing the a.c. component to pass with very little attenuation.

8.5   Multi-stage class A operation

A three-stage R-C coupled class A transistor amplifier is shown in Fig. 8.6(a). It contains three similar common-emitter stages, which are cascaded by R-C networks. Resistors R_1, R_2, and R_E provide base bias and temperature stabilization. It is common practice to develop about 1 V across the emitter resistor, the quiescent collector potential being approximately mid-way between the emitter potential and V_{CC}. At normal operating frequencies (i.e., mid-band frequencies), the reactance of the coupling capacitor is small enough to be neglected when compared with the input resistance of the driven stages. This being so, they are omitted from the small-signal a.c. equivalent circuit of the stage, shown in Fig. 8.6(b). In the figure, the equivalent circuit of the second stage is shown.

The input to Q2 is effectively shunted by R_1 and R_2, while its output is shunted by R_L, R_3, R_4 and R_{in} (the input resistance of the third stage).

Figure 8.6(c) illustrates the technique of graphical analysis as applied to this circuit. The quiescent point Q is selected at an appropriate point on the d.c. load line of slope $-1/(R_L + R_E)$. The a.c. load line has a slope of $-1/R'_L$, where R'_L is the value of the parallel combination R_L, R_3, R_4, and R_{in}. The output voltage swing can then be computed from a knowledge of the change in base current. A quantitative example is given in example 8.3 below.

Example 8.3: A transistor has the following h-parameters

$$h_{ie} = 1500 \ \Omega$$

$$h_{re} = 0\cdot0002$$

$$h_{fe} = 50$$

$$h_{oe} = 10^{-5} \ \text{S}$$

Estimate the voltage gain of an amplifier using the transistor in a circuit similar to that of Fig. 8.6(a), given that $R_1 = R_3 = 60 \ \text{k}\Omega$, $R_2 = R_4 = 10 \ \text{k}\Omega$, and $R_L = 2 \ \text{k}\Omega$. The input resistance of the driven stage is 1 kΩ.

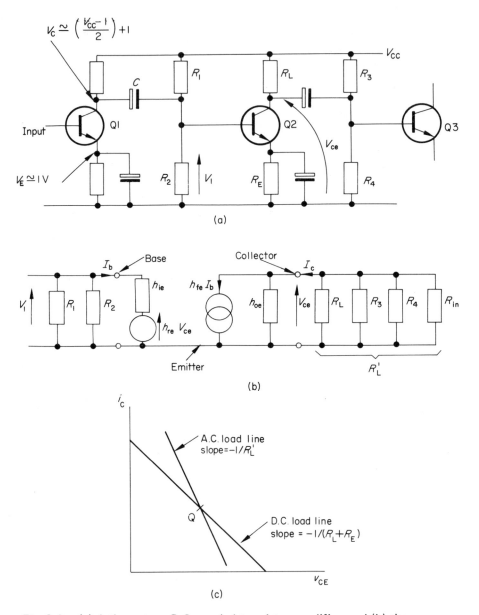

Fig. 8.6   (a) A three-stage R-C coupled transistor amplifier, and (b) the small-signal equivalent circuit of the second stage. (c) The application of the load line technique to determine the gain.

Solution: The effective load resistance connected to the amplifier is R'_L, where

$$R'_L = 1\bigg/\left(\frac{1}{R_L} + \frac{1}{R_3} + \frac{1}{R_4} + \frac{1}{R_{in}}\right) = 1\bigg/\left(\frac{1}{2} + \frac{1}{60} + \frac{1}{10} + \frac{1}{1}\right)k\Omega$$

$$= 618\ \Omega$$

The current gain is computed from eq. (7.7):

$$K_i = h_{fe}/(1 + h_{oe}R'_L) = 50/(1 + 10^{-5} \times 618)$$

$$\simeq 50$$

From eq. (7.8), the input resistance of the transistor is

$$R_{in} = h_{ie} - h_{re}h_{fe}R'_L/(1 + h_{oe}R'_L)$$

$$= 1500 - 0 \cdot 0002 \times 50 \times 618/(1 + 10^{-5} \times 618)$$

$$\simeq 1494\ \Omega$$

Since this is shunted by bias resistors R_1 and R_2 so far as a.c. signals are concerned, the effective input resistance R'_{in} of the amplifier is

$$R'_{in} = 1\bigg/\left(\frac{1}{R_{in}} + \frac{1}{R_1} + \frac{1}{R_2}\right) = 1\bigg/\left(\frac{1}{1 \cdot 494} + \frac{1}{60} + \frac{1}{10}\right)$$

$$= 1 \cdot 27\ k\Omega$$

and the stage gain, from eq. (7.9), is

$$m = -K_i R'_L/R'_{in} = -50 \times 618/1270 = -24 \cdot 3$$

The value obtained from m above is relatively small when compared with the maximum values of voltage gain that can be achieved in common-emitter amplifiers. The reason for the low value of gain here is the low value of R'_L used in the amplifier.

Simplified solution: If the shunting effect of the bias circuit is ignored, the effective load resistance is R_L shunted by R_{in} of the driven stage, giving an effec-load resistance of

$$2 \times 1/(2 + 1)\ k\Omega = 667\ \Omega$$

The current gain of the stage is $K_i \simeq h_{fe} = 50$, and the input resistance is approximately equal to h_{ie}. This gives an approximate voltage gain of

$$m = -50 \times 667/1500 = -22 \cdot 3$$

which is in error (when compared with the earlier result) by $8 \cdot 5$ per cent.

8.6   The transformer

The transformer, in its electronic context, is employed as an impedance trans-
forming device. That is to say, it is employed to modify the value of a load
resistance connected between its secondary terminals to a different value when
viewed from the primary side. Since we are dealing with flow of current in transformer windings, the following theory is applicable to a.c. signals.

In Fig. 8.7, the resistance coupled to the secondary winding of the transformer is given by

$$R_L = V_2/I_2$$

where V_2 and I_2 are the respective r.m.s. values of the secondary voltage and
current. In the figure, there are N times as many turns of wire on the primary

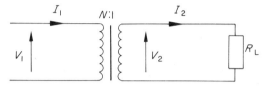

Fig. 8.7   An ideal transformer.

winding as there are on the secondary winding, hence the r.m.s. primary values
are

$$V_1 = NV_2 \qquad I_1 = I_2/N$$

Hence, the input impedance between the primary terminals is

$$R'_L = \frac{V_1}{I_1} = \frac{NV_2}{I_2/N} = N^2 \frac{V_2}{I_2} = N^2 R_L$$

That is to say, the transformer has converted the value of the load resistance R_L
connected between its secondary terminals into an apparent load value of $N^2 R_L$
between its primary terminals. The apparent value is known as the *reflected load
resistance*. Thus, if $N = 5$ and $R_L = 15\ \Omega$, then the a.c. input resistance between
the primary terminals is $5^2 \times 15 = 375\ \Omega$.

8.7   Conditions for maximum power transfer into a load

The output circuit of many electronic amplifiers can be simplified to the form
shown in Fig. 8.8, where E is the no-load output voltage, R_s is the output resistance, and R_L is the load resistance.

When $R_L = 0$, the voltage developed across it is zero, and the power consumed
in it must also be zero. At the other extreme, when $R_L = \infty$, the circuit current is
zero, and the power consumed in the load is again zero. At some point between

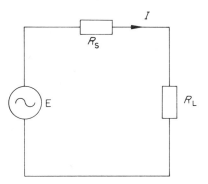

Fig. 8.8   The condition for maximum power transfer into R_L.

the two extremes lies a condition for maximum power consumption in the load. A knowledge of this condition is vital to the design of power amplifier stages.

The circuit current in Fig. 8.8 is

$$I = E/(R_s + R_L)$$

and the power consumed in the load is

$$P_L = I^2 R_L = E^2 R_L/(R_s + R_L)^2 \qquad (8.4)$$

To determine analytically the value of R_L which consumes the maximum power, it is necessary to differentiate eq. (8.4) with respect to R_L, and equate the result to zero. This gives the relationship

$$R_L = R_s \qquad (8.5)$$

That is, maximum transfer to power into the load occurs when the load and source resistances are equal in value. Under this condition, the load is said to be *matched* to the source.

If the load is coupled to the source by a transformer with a turns ratio of $N{:}1$, maximum power transfer occurs when

$$R_s = N^2 R_L \qquad \text{or} \qquad N = \sqrt{(R_s/R_L)}$$

The value of N given above is known as the *optimum turns ratio* for maximum power transfer.

Transformers used in electronic circuits have power transfer efficiencies of the order of 0·5 to 0·8 per unit (50 to 80 per cent). The losses in the transfer must be provided by the primary circuit. To allow for the efficiency of the transformer, the reflected load resistance becomes $N^2 R_L \eta$, where η is the efficiency of the transformer. The optimum turns ratio for an imperfect transformer is, therefore, $\sqrt{(R_s/R_L \eta)}$.

The condition for maximum power transfer should not be confused with the condition for maximum voltage gain. Generally speaking, the voltage gain of an amplifier operating under maximum power transfer conditions is one-half of the maximum possible value.

It is an unfortunate fact of life that transformers used in practical circuits depart from the ideal by a considerable measure. In a practical transformer, there is a *core power loss,* which may be represented by a resistor shunting the primary winding. There are, also, *leakage fluxes* which link with one winding only, and do not contribute to the transmission of power. In addition, there is a shunt *capacitance* associated with each winding, and another capacitance between each pair of windings. The net result of these imperfections is a reduction in operating efficiency, and in the presence of resonant effects at high frequencies. In general, the quality of a transformer is reflected in its cost, and an expensive unit will give a better general performance than a cheaper unit. Even so, ideal conditions can never be achieved in transformers.

8.8   Waveform distortion

Under class A operating conditions the output signal waveform should be a faithful reproduction of the input signal waveform. When we speak of *waveform distortion,* we mean that the output signal waveform is not a true reproduction of the input signal. It is possible to recognize three basic forms of distortion, which may either exist separately or simultaneously. They are:

(a) Non-linear distortion or harmonic distortion.
(b) Amplitude-frequency distortion.
(c) Phase-frequency distortion or delay distortion.

Non-linear distortion or *harmonic distortion* occurs when the amplifier introduces additional harmonic component frequencies of the input signal. Illustrative examples of the effects of harmonic distortion are given in Fig. 8.9.

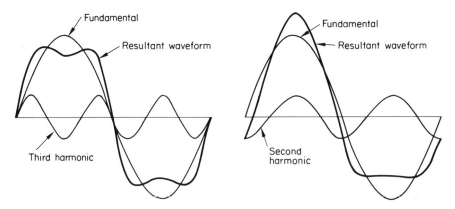

Fig. 8.9   Distortion introduced by (a) a third harmonic, and (b) a second harmonic.

The introduction of a third harmonic term which is in phase with the fundamental signal distorts the waveform in the manner shown in Fig. 8.9(a), the resultant waveform being the sum of the two harmonic components. The harmonics so introduced are generated by the non-linear characteristics of the amplifying device used or of the circuit used.

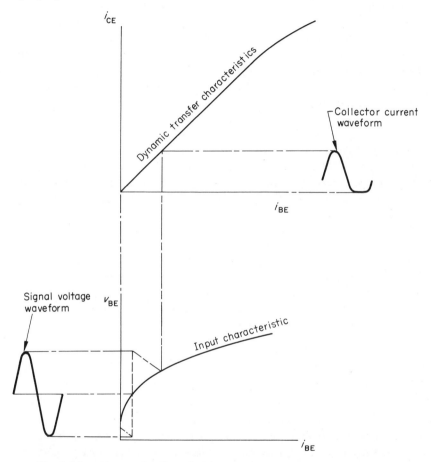

Fig. 8.10   Harmonic distortion is introduced by non-linear effects.

The way in which the phase-shift between the fundamental and the harmonic term can modify the waveform is illustrated in Fig. 8.9(b). Here, a second harmonic signal is introduced, the harmonic signal lagging behind the fundamental by an angle equal to 90 degrees of the harmonic waveform. Since this is a second harmonic, the angle of lag is equivalent to 45 degrees of the fundamental waveform.

One way in which harmonic distortion is introduced is illustrated in Fig. 8.10. Owing to the non-linearity of the input characteristic of the transistor, the collector current falls to zero for part of the cycle, causing the collector current waveform to be non-sinusoidal in the negative half-cycle of the input waveform. The collector current waveform in Fig. 8.10 is seen to be generally similar to the resultant waveform in Fig. 8.9(b), resulting in the introduction of a large second harmonic term in the output.

Any *complex waveform,* such as the resultant waveforms in Fig. 8.9, can be shown to consist of a fundamental sinusoid, together with a number of harmonics. When the input signal to the amplifier is itself a complex waveform, the non-linear characteristic of the amplifier can also result in *intermodulation distortion.* That is, frequencies which are the sum and difference of all the harmonic frequencies present at the input may be generated. This is used to advantage in communications equipment.

Amplitude-frequency distortion or *amplitude distortion* occurs when the component frequencies of a complex waveform are not amplified by equal amounts. For example, an amplifier with the frequency response diagram shown in Fig. 8.1 (page 180) gives greater amplification to a signal frequency in the mid-band range (between f_1 and f_2), than it does to a signal frequency which is either greater than f_2 or less than f_1. Over the mid-band frequency range, the amplitude-frequency distortion is zero.

Phase-frequency distortion or *delay distortion* arises when all the component frequencies of a complex wave are not delayed by an equal time interval. Ideal amplifiers either give zero phase-shift (zero delay) to all frequencies, or give a phase-shift to each component frequency which is proportional to the frequency. In the former case, there is no delay, and the output waveform is undistorted. In the latter case, all harmonic components are delayed by an equal time interval, and the complex wave is delayed as a single unit, giving zero phase distortion.

8.9   Class A power amplifiers

In class A audio frequency amplifiers, the load can either be connected directly in the collector circuit or it can be transformer-coupled to it. The latter method is often preferred, since it enables the load to be matched to the amplifier in order to realize the maximum power gain, while keeping the d.c. power loss small because of the small resistance of the transformer primary winding.

Since we are concerned here with delivering the maximum a.c. power into the load, consistent with consuming the least possible d.c. power, a factor of interest is the *conversion efficiency* η of the amplifier. This is defined as the ratio of the r.m.s. value of the alternating power delivered into the load to the average power drawn from the collector (or anode) supply source. An illustrative example of the conversion efficiency is given at the end of this section.

A circuit diagram of a transistor power amplifier operating in class A is shown in Fig. 8.11. In the figure, the driving stage is represented by a constant current generator. Transformer T1 is a transformer which provides impedance matching between the output of the driver stage and the input of the amplifier to give *minimum distortion,* and *not* maximum power transfer into the amplifier. The reasons for this are fully discussed at a later point in this section (see page 201). Base bias and temperature stability are provided by resistors R_1, R_2, R_E, and thermistor R_3. Capacitor C provides a low impedance path for the flow of signal currents in the base-emitter loop. The ratio of T2 is chosen on a basis of maxi-

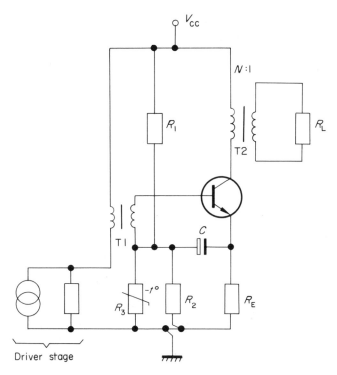

Fig. 8.11   A class A power amplifier.

mum power transfer. Since the quiescent current of a class A power amplifier is high, the base current is proportionally high, and the ohmic values of the bias circuit resistors are much lower than in an equivalent class A voltage amplifier. Accordingly, the value of R_E may only be 0·5 Ω, and that of the parallel combination R_2 and R_3 only 10 Ω. So that capacitor C has a low reactance when compared with these values, the capacitance of C is very high, and a value of 1000 μF is not uncommon. In some cases, C is omitted if the negative feedback effects of R_E do not significantly affect the performance of the amplifier.

The maximum theoretical value of the conversion efficiency is deduced from the output characteristic in Fig. 8.12. Under zero-signal conditions, the effective resistance in the collector circuit is that of the primary winding of the transformer. This has a very small value, and is assumed to be zero in the figure. Thus, the d.c. load line is a vertical line rising from V_{CC}. If the transformer winding resistance is R_p, the d.c. load line has a slope of $-1/R_p$ terminating at V_{CC} on the v_{CE} axis. The quiescent point Q is set by the base bias circuit, and lies very near to the maximum average power dissipation curve $P_{C(max)}$. Since the transistor is worked on its power limit, the collector region of the transistor is normally connected to the case of the transistor, which is secured to a heat sink.

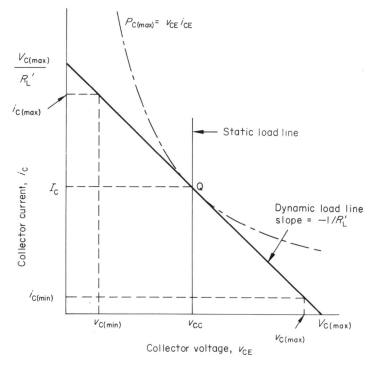

Fig. 8.12   Determination of the efficiency of a class A amplifier.

In valves, the internal heat losses are dissipated by conduction and radiation, and internal temperatures are limited by the point at which occluded gases are released.

When the collector current is reduced below the quiescent value I_C (as a result of a change in i_{BE}), the collapse of the flux in the transformer core causes the instantaneous collector voltage to rise above V_{CC}. The only restriction imposed upon the maximum collector voltage $V_{C(max)}$ is the breakdown voltage of the transistor. $V_{C(max)}$ is usually limited to a few volts below this value. In valves, the maximum anode voltage is restricted only by the quality of the insulation.

Given the operating conditions shown in Fig. 8.12, with the Q-point on the $P_{C(max)}$ hyperbola, the maximum collector voltage is $V_{C(max)} = 2V_{CC}$, which may be verified by drawing a tangent to the $P_{C(max)}$ curve at point Q. That is, the maximum possible collector voltage is twice the supply voltage. If saturation effects can be ignored, the minimum collector voltage is zero, giving a maximum possible collector current of

$$\frac{V_{C(max)}}{R'_L} = 2I_C$$

where R'_L is the reflected value of the load resistance. Owing to the non-linear nature of the characteristics, it is necessary to constrain the voltage and current swings, otherwise excessive harmonic distortion is introduced. Suppose that the maximum and minimum values of collector voltage are $v_{C(max)}$ and $v_{C(min)}$, respectively, and the respective values for collector current are $i_{C(max)}$ and $i_{C(min)}$, then the r.m.s. collector voltage is

$$V_{ce} = \frac{v_{C(max)} - v_{C(min)}}{2\sqrt{2}} \tag{8.6}$$

and the r.m.s. collector current is

$$I_{ce} = \frac{i_{C(max)} - i_{C(min)}}{2\sqrt{2}} \tag{8.7}$$

Since the load is resistive, the r.m.s. power P_{ac} delivered into the load is

$$P_{ac} = V_{ce}I_{ce} = \frac{(v_{C(max)} - v_{C(min)})(i_{C(max)} - i_{C(min)})}{8}$$

The average power P_{dc} drawn from the supply is

$$P_{dc} = V_{CC}I_C \tag{8.8}$$

Hence, the conversion efficiency for a class A power amplifier is

$$\eta = \frac{P_{ac}}{P_{dc}} = \frac{(v_{C(max)} - v_{C(min)})(i_{C(max)} - i_{C(min)})}{8V_{CC}I_C} \tag{8.9}$$

The maximum possible theoretical efficiency can be predicted by assuming that the whole of the voltage and current swing is available, when

$$v_{C(max)} = 2V_{CC} \quad v_{C(min)} = 0$$
$$i_{C(max)} = 2I_C \quad i_{C(min)} = 0$$

when

$$\eta_{max} = \frac{2V_{CC}2I_C}{8V_{CC}I_C} = 0 \cdot 5 \text{ per unit or 50 per cent}$$

This figure can be approached in transistor stages, since the minimum possible voltage is $V_{CE(sat)}$ for the transistor, which is of the order of $0 \cdot 2$ V to $0 \cdot 7$ V. Unfortunately, efficiencies of this magnitude can only be obtained at the expense of an increase in harmonic distortion.

In the case of valves, the anode voltage and current swing are limited to much smaller proportions of the theoretical values. Suppose that the maximum voltage swing is only $1\cdot4\,V_{CC}$ and the maximum current swing is $1\cdot4I_C$, then the conversion efficiency is $1\cdot4^2\,V_{CC}I_C/8V_{CC}I_C = 0\cdot245$ per unit or 24·5 per cent.

In a transistor amplifier, the power gain is useful in describing its operation, since a finite input power is required to develop an output power. In valve amplifiers, the input power is zero (unless grid current is allowed to flow), and the power gain is infinity. A convenient figure of merit for the valve amplifier is the *power sensitivity,* defined by the equation

$$\text{Power sensitivity} = \frac{\text{Signal power output}}{V_g{}^2}$$

where V_g is the r.m.s. value of the input signal voltage. The term in the denominator is derived from the fact that the input power is related to the square of the applied voltage. Thus, if $P_{ac} = 10\text{ W}$, and $V_g = 10\text{ V}$, then the power sensitivity is $10/10^2 = 0\cdot1\text{ W/V}^2$.

Example 8.4: A common-emitter class A transistor power amplifier uses a transistor with $h_{FE} = 60$. The load has a resistance of $81\cdot6\ \Omega$, which is transformer-coupled to collector circuit. If the peak values of the collector voltage and current are 30 V and 35 mA, respectively, and the corresponding minimum values are 5 V and 1 mA, determine:

(a) The approximate value of the quiescent current.
(b) The quiescent base current.
(c) P_{dc}.
(d) P_{ac}.
(e) The conversion efficiency.
(f) The turns ratio of the transformer.

Solution: (a) The quiescent collector current is approximately half-way between the maximum and minimum values of collector current.

$$I_C = \left(\frac{35 - 1}{2}\right) + 1 = 18\text{ mA}$$

Similarly,

$$V_C = \left(\frac{30 - 5}{2}\right) + 5 = 17\cdot5\text{ V}$$

Since the load is transformer coupled, then $V_{CC} \simeq 17\cdot5\text{ V}$

(b). $I_B = I_C/h_{FE} = 18/60 = 0\cdot3\text{ mA}$

(c). $P_{dc} = V_{CC}I_C = 17\cdot5 \times 18 = 315\text{ mW}$

(d). $V_{ce} = (30 - 5)/2\sqrt{2} = 8.84$ V

$$I_c = (35 - 1)/2\sqrt{2} = 12 \text{ mA}$$

$$P_{ac} = 8.84 \times 12 = 106 \text{ mW}$$

(e). $\eta = 106/315 = 0.337$ per unit or 33.7%

(f). The a.c. resistance in the collector circuit is determined from the slope of the load line. Thus,

$$-\frac{1}{R'_L} = \frac{35 - 1}{5 - 30} = -\frac{34}{25} \text{ mS}$$

Therefore,

$$R'_L = 25/34 \text{ k}\Omega = 0.735 \text{ k}\Omega \text{ or } 735 \text{ }\Omega$$

Now,

$$R'_L = N^2 R_L$$

hence,

$$N = \sqrt{(R'_L/R_L)} = \sqrt{(735/81.6)} = 3:1$$

8.9.1 Sources of distortion in the common-emitter amplifier

The main sources of distortion in the common-emitter amplifier are the non-linear regions of the input and output characteristics.

Consider the characteristics in Fig. 8.13, which are typical of a low frequency power transistor with a rating of about 3 A, 20 V. It is clear that, unless the input signal is small, distortion of the output waveform will result from the curvature of either or both of the characteristics. In the figure, two types of signal source are considered—a *voltage source* and a *current source*. It has already been shown that the two are fundamentally different in nature, since a pure voltage source has zero output impedance, and a pure current source has an infinitely large output impedance. In both instances, the quiescent point Q on the input characteristic is chosen to give a base current of 30 mA and a collector current of 1.65 A.

Considering firstly the voltage source input signal, the peak input voltage of 190 mV takes i_{CE} to 3 A, an upward swing of $(3 - 1.65) = 1.35$ A. The negative voltage swing of 190 mV takes i_{CE} to 0.4 A, or a negative-going swing of $(1.65 - 0.4) = 1.25$ A. It is seen that, in this case, there is a flattening of the negative peak of the collector current waveform.

Upon the application of a signal from a current source which provides a peak current of 24 mA, the peak positive collector current has a value of 2.5 A, and the minimum value is 0.4 A. This gives a positive-going collector current swing

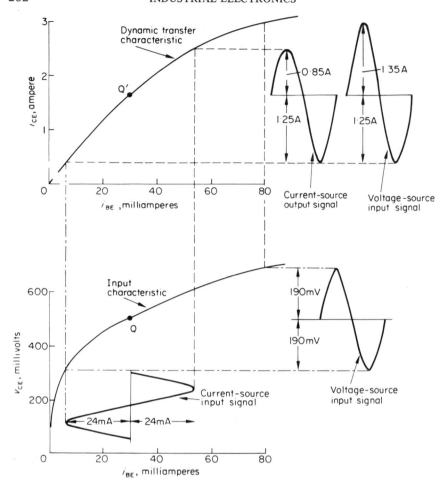

Fig. 8.13   Distortion can be introduced by using 'ideal' signal sources.

of 0·85 A, and a negative-going swing of 1·25 A, resulting in a flattening of the positive half-cycle of the collector current waveform.

It is clear that, for minimum distortion, the impedance of the signal source must be a compromise between the two types. Thus, the turns ratio of T1 in Fig. 8.11 is chosen to present the amplifier with a signal source of the correct impedance to give minimum distortion. The optimum source resistance can be computed by the graphical method outlined in Fig. 7.2 (page 143).

In Fig. 8.13, if the quiescent collector voltage of the transistor is 12 V, then

$$P_{dc} = 12 \times 1·65 = 19·8 \text{ W}$$

Assuming that optimum source resistance gives a peak-to-peak collector current swing of 2·4 A (e.g., from 0·45 A to 2·85 A) into an effective collector load of 10 Ω, then

$$P_{ac} = V_{ce}I_{ce} = I_{ce}^2 R'_{L} = \left(\frac{2·4}{2\sqrt{2}}\right)^2 \times 10 = 7·2 \text{ W}$$

giving a conversion efficiency of

$$\eta = 7·2/19·8 = 0·364 \text{ per unit or } 36·4 \text{ per cent}$$

8.10   Class A push-pull operation

Where more power is required by the load than can be supplied by one transistor, alternative circuits are adopted. If two transistors are connected directly in parallel, it is difficult to ensure load sharing between the two. Parallel operation is adopted in some instances, e.g., voltage regulators, where other techniques are not always convenient.

The most common method used in signal amplifiers is the *push-pull* connection; the basic circuit is shown in Fig. 8.14. This circuit has an advantage over parallel-connected circuits, in that even harmonics are cancelled out if the current gains of the two transistors are of the same order. In the circuit shown, the quiescent point is set by V_{BB}; V_{CC} is a common collector supply to both transistors. A feature of the push-pull amplifier is that the two primary windings of the output transformer carry current in opposite directions, giving zero net magnetization. This minimizes the possibility of distortion due to magnetic saturation of the iron circuit, resulting in a physically smaller transformer than is required for a *single-ended* class A stage. An amplifier is said to be single-ended when it employs only one transistor (or valve), as, for example, Fig. 8.11.

The transformers in Fig. 8.14 are marked with dots according to the *dot notation* for induced e.m.f.s. In this notation, if the instantaneous polarity of

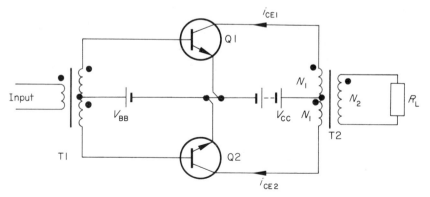

Fig. 8.14   A class A push-pull amplifier circuit.

any terminal marked with a dot is positive, then the induced e.m.f.s in all other windings on the same transformer are such as to make the ends of the windings marked with dots of like instantaneous potential, i.e., positive. Thus, if the upper terminal of the input transformer T1 in Fig. 8.14 becomes positive relative to the lower terminal, it induces an instantaneous positive potential at the upper ends of both secondary windings. Hence, a positive-going signal applied to the upper primary terminal of T1 causes the collector current of Q1 to increase, while that in Q2 decreases. A negative-going input signal reverses the effects on the collector currents.

The operation of the circuit in Fig. 8.14 with a sinusoidal input signal is illustrated in Fig. 8.15. The characteristics shown are the *dynamic mutual characteristics*, which give the relationship between the base-emitter voltage and the collector current of the transistor for given values of V_{CC} and R_L; the characteristics are deduced from a knowledge of the input characteristics, the source resistance, and the dynamic transfer characteristic. The bias point chosen on the characteristics results in the introduction of harmonic distortion (mainly second harmonic distortion) in the collector current waveforms. As will be seen from an inspection

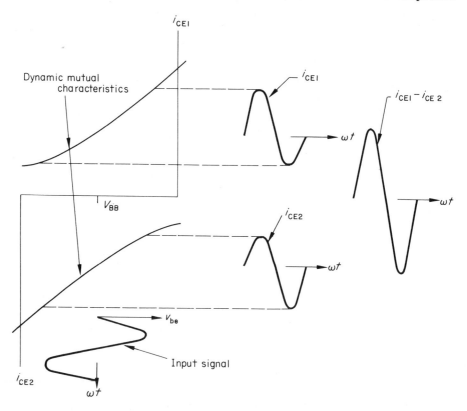

Fig. 8.15   Collector current waveforms in a class A push-pull amplifier.

of the dot notation applied to the output transformer T2, the current flowing in the load is proportional to $(i_{CE1} - i_{CE2})$. When the two current waveforms are subtracted from one another the resultant waveform is practically sinusoidal, which is due to the fact that the second harmonic terms in the separate waveforms are in opposition to one another. This is verified by the simplified analysis which follows. Suppose that the collector current of transistor Q1 is represented by the expression

$$i_{C1} = I_C + I_1 \sin \omega t + I_2 \sin 2\omega t + I_3 \sin 3\omega t + \cdots \qquad (8.10)$$

Where I_C is the quiescent current, I_1 is the peak value of the fundamental harmonic component, I_2 is the peak value of the second harmonic component, etc. The expression for i_{C2} is generally similar to the above, but having ωt replaced by $(\omega t + 180°)$ due to the phase-inversion of the base voltage signal by T1.

$$i_{C2} = I_C + I_1 \sin(\omega t + 180°) + I_2 \sin 2(\omega t + 180°)$$

$$+ I_3 \sin 3(\omega t + 180°) + \cdots$$

$$= I_C - I_1 \sin \omega t + I_2 \sin 2\omega t - I_3 \sin 3\omega t + \cdots \qquad (8.11)$$

Subtracting eq. (8.11) from eq. (8.10) gives

$$i_{C1} - i_{C2} = 2(I_1 \sin \omega t + I_3 \sin 3\omega t + \cdots) \qquad (8.12)$$

The d.c. components cancel as do the even harmonic terms, leaving the fundamental and odd harmonics, i.e., the 3rd, 5th, 7th harmonics, etc. When operating in class A the amplitude of odd harmonics is normally small enough to be ignored, and

$$i_{C1} - i_{C2} \simeq 2I_1 \sin \omega t$$

If the current gains of the two transistors are not equal, the even harmonic terms do not cancel out and some distortion is introduced.

8.11 Bias arrangements in push-pull circuits

A single-battery version of Fig. 8.14 can be evolved if the bias voltage is developed across a resistor. The battery would need to be connected between the centre-taps of the two transformers, and the resistor between the centre-tap of the input transformer and the common-emitter point. In this event, the resistor carries the sum of the two collector currents, which, from eqs. (8.10) and (8.11), is

$$i_{C1} + i_{C2} = 2(I_C + I_2 \sin 2\omega t + I_4 \sin 4\omega t + \cdots)$$

Under Class A operating conditions, the harmonic distortion is relatively small, and the emitter current is approximately $2I_C$. From this, it would appear that it is possible to use an unbypassed resistor in the emitter lead, but there is a snag.

The phase relationships of the even-order harmonics with the input signal are such that they generate *positive feedback conditions* (see chapter 9), which can lead to unstable operating conditions. For this reason, it is usual to bypass the emitter resistor with a capacitor. In many transistor amplifiers, an independent bias supply is provided.

8.12   Class B and class C operation

In *Class B* operation, the transistor or valve is biased so that collector (or anode) conduction occurs during one-half cycle of the input signal, as illustrated in Fig. 8.16. Owing to the curvature of the mutual characteristic, waveform distortion is introduced at the commencement and completion of the half-cycle.

There is, also, an additional class of operation known as *class AB,* in which collector (or anode) current flows for more than 180 degrees of the input cycle, but less than the whole of it. Class AB operation is generally a compromise between increased efficiency when compared with class A, and reduced harmonic distortion when compared with class B. In thermionic valve circuits, this class is further divided into two sub-classes AB1 and AB2. In class AB1, grid current does not flow, but in class AB2 it is permitted.

When operating in class C, the bias applied to the circuit is greater than that in class B, so that collector (or anode) current flows for less than one half-cycle,

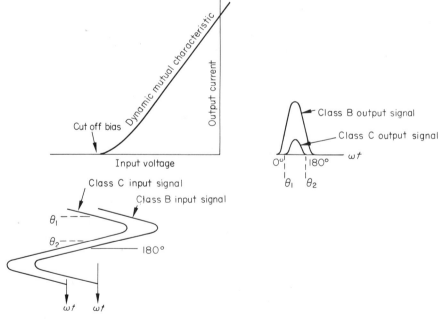

Fig. 8.16   Collector current waveforms in class B and class C circuits.

as shown in Fig. 8.16. In the figure, collector current flows over the range θ_1 to θ_2. Class C amplifiers are commonly used to drive tuned circuits which develop an oscillatory voltage between their terminals. It is only necessary in such circuits to provide a pulse of energy during each cycle, in order to supply the power losses of the circuit. The pulse of current supplied by the transistor (or valve) operating in class C fulfils this function.

As the amplifier classification changes from class A to class C a progressively greater conversion efficiency is achieved. Thus, a class AB stage is more efficient than a class A stage, a class B stage is more efficient than a class AB stage, and so on.

8.13 Class B push-pull operation

The principal disadvantage of class A working is the high quiescent power loss. The conversion efficiency is improved by class B working, but some form of push-pull circuit must be employed to overcome the problem of the harmonics which are generated in this mode of operation.

The circuit configuration of a class B push-pull stage is generally similar to that of the class A stage shown in Fig. 8.14. Ideally, the base bias of a class B stage is zero, but this has the unfortunate effect of introducing another form of distortion known as *crossover distortion*, which is illustrated in Fig. 8.17(a). In the region of zero collector current on the dynamic mutual characteristic of the transistor, the curve flattens and causes waveform distortion. The net effect is that odd harmonics of the signal frequency are introduced into the output waveform. This effect can be minimized by applying a small forward bias to the transistors, modifying the characteristics as shown in Fig. 8.17(b). The overall characteristic of the amplifier is then approximately linear.

It is advisable in push-pull circuits to choose two transistors which have closely matched characteristics, otherwise one half of the output waveform will have a larger peak value than the other due to the difference in amplification. This leads to an increase in distortion.

A typical class B transistor push-pull amplifier circuit is shown in Fig. 8.18. Resistors R_1, R_2, and R_E are chosen to provide the correct bias conditions and temperature stability. Resistor R_E normally has a very small value, often as low as a fraction of an ohm. Capacitor C is a harmonic bypass capacitor. A diode can, conveniently, be incorporated in series with R_2 to improve thermal stability.

The maximum conversion efficiency of a class B push-pull stage is estimated as follows. If the peak value of i_{C1} is I_{CM}, then the peak value of i_{C2} is $-I_{CM}$, and the r.m.s. value of the effective current in the primary of T2 (Fig. 8.18) is $I_{CM}/\sqrt{2}$. Since the peak excursion of the collector voltage is equal to V_{CC}, the r.m.s. value of the voltage change across T2 is $V_{CC}/\sqrt{2}$, so that the r.m.s. power developed is

$$P_{ac} = \frac{V_{CC}I_{CM}}{\sqrt{2}\,\sqrt{2}} = \frac{V_{CC}I_{CM}}{2} \qquad (8.13)$$

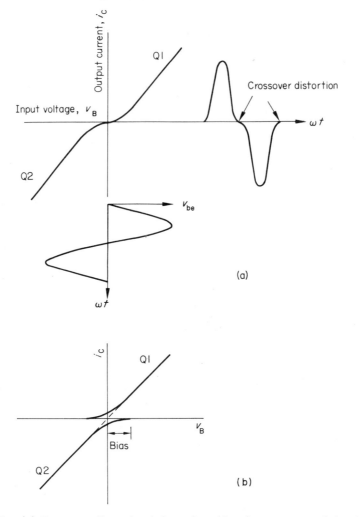

Fig. 8.17   (a) Crossover distortion is introduced by the curvature of the dynamic mutual characteristic. One method of counteracting this defect is shown in (b).

Since the power supply must provide a series of half-sine pulses of current to the two transistors in turn, the average value of the supply current is $2I_{CM}/\pi$; and since the average supply voltage is V_{CC}, then the average power supplied is

$$P_{dc} = 2V_{CC}I_{CM}/\pi \qquad (8.14)$$

giving a maximum theoretical conversion efficiency of

$$\eta = P_{ac}/P_{dc} = \pi/4 = 0\cdot785 \text{ per unit or } 78\cdot5 \text{ per cent}$$

Owing to the effect of saturation in transistors, the full voltage swing of V_{CC} cannot be realized, and the maximum efficiency in practice is about 75 per cent.

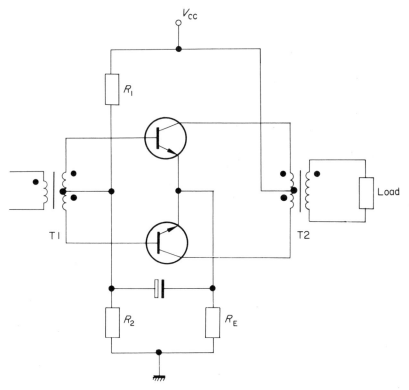

Fig. 8.18   A class B transistor push-pull amplifier.

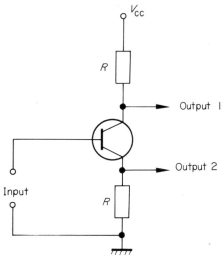

Fig. 8.19   A single-transistor phase-splitting amplifier.

8.14   Class B driver stages

Class B push-pull stages require complementary input signals to the two transistors for correct operation. The simplest method is to use a single-ended class A stage to energize the primary winding of the input transformer in Fig. 8.18. Alternatively, a long-tailed pair amplifier (Figs. 8.3. and 8.4) can be used to provide complementary signals.

The circuit in Fig. 8.19 is also used as a phase-splitter. The transistor is operated in Class A, so that the emitter potential is always within about 0·2 V to 0·5 V of the base potential. Thus, output 2 is in phase with the input signal, while output 1 is antiphase to it. If the resistors in the emitter and collector leads are of equal value, then the voltage gains between the input and each of the two outputs are equal. One snag with this circuit is that the output impedance at the emitter output is much lower than the output impedance at the collector output (see section 10.3), with the result that the application of similar loads to each output terminal gives different output voltages.

8.15   Transformerless class B stage

A transistor circuit which has no valve equivalent is shown in Fig. 8.20. It is known as a *complementary symmetric push-pull amplifier,* and does not require an output transformer.

The circuit uses a complementary pair of n-p-n and p-n-p transistors, both of which are driven from a common input signal. When the input signal is positive,

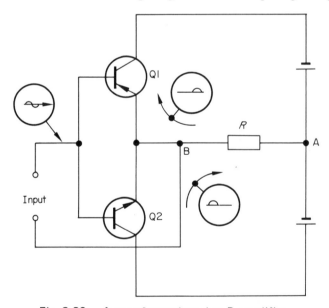

Fig. 8.20   A transformerless class B amplifier.

Q1 is cut off and Q2 is driven into conduction. Under this condition, the flow of current through Q2 causes point B to be driven positive with respect to point A. During the negative half-cycle of the input signal, Q1 conducts and Q2 is cut off; the direction of flow in this half-cycle makes point A positive with respect to point B. By this artifice, the waveshape of the output waveform developed across resistor R is a replica of the input waveform.

Problems

8.1 The input power P_1 applied to a circuit is 10 mW. If the output power P_2 is (a) 10 W, (b) 1 W, (c) 100 mW, (d) 10 mW, (e) 1 mW, (f) 1 μW, calculate the power gain in decibels in each case.

8.2 The voltage gains of a number of cascaded amplifiers and attenuators, expressed in decibels, are 22 dB, $-6\cdot2$ dB, $-10\cdot1$ dB, and 4·5 dB. Compute the overall gain in decibels.

8.3 If a voltage of 10 mV is applied to the network in question 8.2, calculate the output voltage.

8.4 If the voltage gain of the network in question 8.2 is reduced by a further 20 dB, calculate the overall gain in decibels. What then is the output voltage with an input of 50 mV?

8.5 Express in decibels a power of 10 W with reference to standard power levels of (a) 1 mW, (b) 100 mW, (c) 1 W, and (d) 100 W.

8.6 A constant input voltage of 25 mV r.m.s. of variable frequency is fed to an amplifier, and as the frequency is changed the output voltages recorded in the table are obtained.

Frequency (Hz)	50	100	250	500	1 k	2·5 k	5 k	10 k	20 k
Output (V r.m.s.)	0·07	0·22	0·59	0·75	0·78	0·71	0·30	0·27	0·11

Assuming the input and output impedances of the amplifier to be the same, plot on log-linear graph paper the output voltage in decibels against frequency. State the two frequencies between which the output is constant within ± 2 dB.

The output from the amplifier is then fed to a second amplifier having the same characteristic as the first. Plot the overall characteristic for the combination and again state the two frequencies between which the overall gain is constant to within ± 2 dB. What will be the output voltage from the combination at 6 kHz?

(C & G)

8.7 (a) Draw the constant voltage equivalent circuit of a resistance loaded triode amplifier which is R-C coupled to the following stage. The triode parameters are $\mu = 50$, $r_a = 10$ kΩ, and $R_L = 25$ kΩ. The capacitor in the coupling circuit has a capacitance of 0·02 μF, and the input impedance of the following stage is 100 kΩ. Bias is obtained by means of a resistor in the cathode lead which is shunted by a very large value of capacitance.

Estimate the output voltage at a frequency of 2 kHz when the input voltage is 1 V peak-to-peak. Justify any assumptions made.

Explain how the removal of the capacitor in the cathode bias circuit would affect the output.

(b) Sketch a graph showing how you would expect the gain to vary as the frequency of the constant voltage input is varied from a low value to a high value. Account, briefly, for the shape of the curve.

(C & G)

8.8 Explain the meaning and causes of 'drift' in d.c. amplifiers. Why are transistors inferior to vacuum valves with regard to drift?

Draw a circuit diagram of a transistor d.c. amplifier with one balanced stage employing negative feedback. Show how the arrangement minimizes drift.

(C & G)

8.9 Outline the special difficulties met in the amplification of small direct e.m.f.s and hence state why the simple direct-coupled amplifier is not widely used.

Draw a circuit diagram and explain the operation of a more suitable type having a balanced stage and a high input impedance.

Quote an industrial application.

(C & G)

8.10 Compare qualitatively common-base and common-emitter amplifiers in respect of (a) input and output resistances, (b) cut-off frequency, and (c) suitability for being cascaded.

Explain in general terms how the differences arise. Explain what is meant by the term 'alpha cut-off frequency'.

(C & G)

8.11 An ideal transformer has an input voltage of 200 V and an output voltage of 50 V. If the load is a pure resistance of 50 Ω, calculate the values of the primary current and the secondary current. Calculate also the input power to and the output power from the transformer. What is the turns ratio of the transformer?

8.12 Draw to scale the phasor diagram of the transformer in question 8.11. Show on the diagram the effect of the no-load current of a practical transformer.

8.13 A load of resistance 10 Ω is connected to the secondary of an ideal transformer with a voltage step-down ratio of 15. Compute the effective a.c. resistance presented to the primary winding.

8.14 What is meant by the terms *class A* and *class B* when applied to valve or transistor amplifiers?

Explain, with the aid of diagrams, the operation of a push-pull amplifier and list possible reasons for its use. (Give *either* a valve *or* a transistor amplifier push-pull circuit, indicating the input and output connections.)

Show how the bias is obtained for (a) class A operation and (b) Class B operation.

(C & G)

8.15 (a) Explain what is meant by phase (phase/frequency) distortion and frequency (amplitude/frequency) distortion in the output of an amplifier.

Illustrate your answer with either graphs or waveform diagrams.

What are the possible causes of these forms of distortion in a voltage amplifier and how may they be kept to a minimum?

(b) State why an a.f. power amplifier is usually transformer-coupled to its load and explain how the transformer may introduce distortion.

(C & G)

8.16 State the essential differences between voltage and power amplifiers with reference to:

(a) type and value of load,
(b) the class of operation (A, B, or C),
(c) input and output impedances,
(d) typical frequency response curves.

Give, complete with circuit connection diagram, a typical application of each type of amplifier.

8.17 A p-n-p power transistor has the following linear characteristics

i_B (mA)	i_C (A) for a collector voltage of	
	−2·5	−17·5 V
−10	−0·67	−0·70
−30	−1·45	−1·65
−50	−2·33	−2·64

For a collector voltage of −8 V, estimate the value of the common-emitter current gain for the transistor and hence calculate the value of the common-base current gain α.

The transistor is to be used as a common-emitter type amplifier having a collector load resistance of 6 Ω. If under no-signal conditions the base current is 30 mA and the collector voltage is −8 V, draw a suitable load line and estimate

(a) the collector supply voltage required,
(b) the current gain of the amplifier,
(c) the voltage gain of the amplifier assuming that the input resistance is 5 Ω.

(C & G)

8.18 The common-emitter characteristics of an n-p-n power transistor are linear between the following points.

v_{CE} (V)	4	30	
i_C (A)	0·05	0·1	at $i_B = 0$
i_C (A)	0·6	0·75	at $i_B = 10$ mA
i_C (A)	1·2	1·4	at $i_B = 20$ mA
i_C (A)	1·8	2·05	at $i_B = 30$ mA
i_C (A)	2·45	2·75	at $i_B = 40$ mA
i_C (A)	3·1	3·45	at $i_B = 50$ mA

The transistor is used in a common-emitter, single-ended class A amplifier with an ideal 5:1 step-down transformer and a 0·3 Ω load. The quiescent point is determined by the current in a resistor connected between the base of the transistor and the collector supply voltage.

If the collector dissipation is not to exceed 25 W, determine graphically a suitable point which will allow maximum undistorted output power to be developed. Determine also the collector supply voltage and the value of the bias resistor. State the maximum and minimum values of collector voltage swing.

8.19 Define the decibel.

Two amplifiers, having the gain/frequency responses given in the table below, are connected in tandem with a 10 dB resistive attenuator. Calculate and plot the overall gain of the combination, expressed in decibels, assuming that all input and output impedances are equal.

Frequency (kHz)	60	66	72	78	84	90	96	102	108
Voltage gain (Amp. 1)	29·8	34·5	38·0	38·9	38·5	38·0	38·0	34·7	29·0
Voltage gain (Amp. 2)	28·2	37·2	37·6	36·7	36·5	36·7	38·7	39·1	29·8

(C & G)

8.20 What is a constant current generator?

A pentode has a mutual conductance of 3·0 mA/V and an anode slope resistance of 600 kΩ. It is transformer coupled to a non-reactive load of 10 Ω. If the transformer has a turns ratio of 30, draw the equivalent circuit showing the valve as a constant current generator.

What will be the r.m.s. current in the load and the power supplied to it when a voltage 1·0 sin ωt is applied to the grid of the pentode?

What is the source of output power?

(C & G)

8.21 Draw a circuit diagram of a two stage transistor audio-frequency amplifier and explain its operation. Explain the precautions taken to stabilize against the effects of temperature variation.

8.22 Draw a circuit diagram of an audio-frequency transistor amplifier comprising a common-emitter stage in tandem with an emitter follower stage. Show clearly the biasing arrangements and briefly explain their operation.

State the order of magnitude of the input and output impedances of the amplifier.

Say what factors will affect the low-frequency response of the circuit you have drawn.

(C & G)

9. Feedback amplifier theory

9.1 The objects and methods of application of feedback

Up to this point in the book, basic amplifier circuits, sometimes known as *open-loop* amplifiers, have been considered. They are known as open-loop amplifiers because the output signal is not monitored and used to control the amplifier. A change in conditions, say a sudden demand for load current, results in the output voltage being reduced. In open-loop amplifiers, no attempt is made to compensate for changes in circuit conditions or in load conditions, e.g., changes in load resistance, supply voltage, component values and parameters, etc.

If the output is monitored, and the information is fed back to the input, correcting action can be taken. Amplifiers using some method of information feedback are known as *feedback amplifiers* or *closed-loop amplifiers,* since the feedback loop is closed. Generally speaking, feedback is applied only where it is beneficial and can improve some of the qualities of the circuit. In some cases, however, feedback has undesirable effects, adversely affecting the performance of the circuit.

In many amplifiers, either a part or all of the output signal is fed back and included as a part of the input signal. If the signal is fed back deliberately, then the results can be predicted by circuit theory. If the output signal is fed back as a result of parasitic components, e.g., by leakage capacitance, then the results are not easily predictable.

If the signal fed back is in phase with the input signal, it is referred to as *positive feedback* or *regenerative feedback*; if it is antiphase to the input signal, it is known as *negative feedback* or *degenerative feedback. Series feedback* is applied when the output signal, or a part of it, is connected in series with the input signal. In *shunt feedback,* the input and output signals are converted into electric currents by applying the voltage signals to resistances or impedances; the sum (or difference) of the two currents is used to develop a voltage across a further impedance (which may be the input impedance of the amplifier), which is then amplified.

Voltage feedback is said to be applied when the signal fed back is proportional to the output voltage of the amplifier. When the signal fed back is proportional to the output current, *current feedback* is applied. Generally speaking, the principal properties affected by feedback are the *input impedance,* the *output impedance,* and the *gain.* Other properties, such as *distortion, bandwidth,* etc. are also affected.

Hybrid feedback circuits, combining both current and voltage feedback are used in practice to give the correct combination of gain, input impedance, output impedance, bandwidth, etc. In general, positive feedback results in the opposite effects to those achieved by negative feedback.

The block diagrams of several basic forms of feedback circuit are shown in Fig. 9.1.

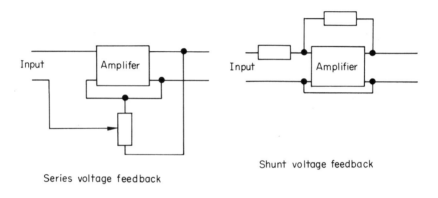

Series voltage feedback

Shunt voltage feedback

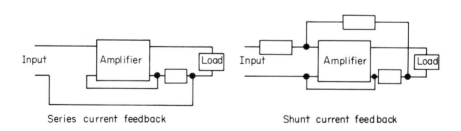

Series current feedback

Shunt current feedback

Fig. 9.1   Some examples of feedback amplifiers.

9.2   Series positive voltage feedback

The block diagram of a series positive voltage feedback amplifier is shown in
Fig. 9.2. The amplifier, of gain m, is non-phase-inverting, that is to say the output signal is in phase with the input signal. A proportion βV_2 of the output
signal is fed back and added in series with the input signal. The *feedback
network* or *β-network*, in this case a potentiometer, has a transfer ratio β which

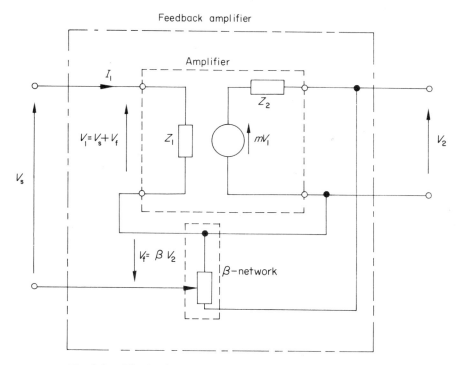

Fig. 9.2   The basic series positive voltage feedback amplifier.

has a value less than, or equal to, unity. With the connections shown, the signal
fed back is in phase with the input signal, and overall positive feedback is
applied.

In order to distinguish the complete circuit in Fig. 9.2 from the *amplifier*
itself, the complete network (including the amplifier) is described as a *feedback
amplifier* or *closed-loop amplifier,* since the 'loop' around the amplifier between
input and output is complete or closed.

With the configuration in Fig. 9.2, it is impossible to earth one of the input
terminals and one of the output terminals simultaneously, since this would result
in the output of the β-network becoming short-circuited, rendering the feedback
circuit inoperative. For this reason, this form of series positive voltage feedback
only finds application in a few practical cases. The negative feedback versions

(cathode and emitter follower circuits) find a wide range of applications. For this reason, the circuit in Fig. 9.2 is analysed in detail here, since it provides a convenient basis for the analysis of feedback amplifiers in later chapters.

In Fig. 9.2, the output voltage is

$$\left.\begin{array}{c} V_2 = mV_1 \\[2mm] V_1 = \dfrac{V_2}{m} \end{array}\right\} \tag{9.1}$$

or

where V_1 is the net voltage applied to the amplifier terminals. Also, from Fig. 9.2

$$V_1 = V_s + \beta V_2 \tag{9.2}$$

hence, from eqs. (9.1) and (9.2)

$$\frac{V_2}{m} = V_s + \beta V_2$$

or

$$V_2 = \frac{mV_s}{1 - m\beta}$$

The closed-loop no-load gain m' is, therefore

$$m' = \frac{V_2}{V_s} = \frac{m}{1 - m\beta} \tag{9.3}$$

The product $m\beta$ is known as the *loop gain*, and m is known as the *forward gain*. It will be seen from the following example that series positive voltage feedback effectively increases the gain of the feedback amplifier above that of the amplifier itself.

Example 9.1: Calculate the loop gain and the closed-loop gain of an amplifier similar to that in Fig. 9.1, if $m = 10$ and $\beta = 0.09$.

Solution:
    Loop gain = $m\beta$ = 10 x 0·09 = 0·9
    Closed-loop gain = m' = $m/(1 - m\beta)$ = 10/(1 − 0·9)
                    = $\underline{100}$

That is, the gain of the closed-loop amplifier is ten times greater than that of the amplifier itself.

The physical reason for the increase in the gain is as follows. When series positive voltage feedback is applied, the net input voltage V_1 applied to the amplifier is greater than the signal voltage V_s applied to the closed-loop amplifier, since a voltage βV_2 is added to V_s before being applied to the amplifier. Signal V_1 is then amplified m times, giving an output voltage which is significantly greater than would be the case if V_s was alone amplified. In example 9.1, if the feedback network was disconnected the gain would be 10, and, for an output of 1 V, the input signal applied to the amplifier would be 0·1 V. When positive series voltage feedback is applied, an output of 1 V from the closed-loop amplifier results in 0·09 V being fed back by the β-network. The signal voltage in this case is the difference between 0·1 V and 0·09 V, i.e., 0·01 V. That is, a signal voltage of 0·01 V gives an output of 1 V, corresponding to a gain of 1 V/0·01 V = 100.

The net input voltage V_1 is computed by eliminating V_2 between eqs. (9.1) and (9.2), giving

$$V_1 = \frac{V_s}{1 - m\beta} \tag{9.4}$$

There are clearly limitations to the theory developed here, since if the product $m\beta = 1$, then from eqs. (9.3) and (9.4),

$$m' = m/(1 - 1) = \infty$$

and

$$V_1 = V_s/(1 - 1) = \infty$$

This condition arises in the amplifier in problem 9.1 if the forward gain m is increased to $m = 1/\beta = 11·111$ (or if β is increased to $\beta = 1/m = 1/10 = 0·1$). This condition is said to be unstable, since sufficient output signal is fed back through the β-network to sustain the output voltage, with no input signal applied to the closed-loop amplifier. The problem of instability is discussed more fully in section 9.9.

9.3 Series negative voltage feedback

If the amplifier in Fig. 9.2 is replaced by a phase-inverting amplifier having gain $-m$, other connections remaining unchanged, the closed-loop gain is evaluated by replacing m by $-m$ in eq. (9.3), giving

$$m' = \frac{-m}{1 - (-m)\beta} = \frac{-m}{1 + m\beta} \tag{9.5}$$

Thus, if $-m = -100$, and $\beta = 0·09$, the amplifier gain is, from eq. (9.5),

$$m' = \frac{-100}{1 - (-100 \times 0·09)} = -10$$

The principal features of eq. (9.5) are

(a) The output voltage is antiphase to the input voltage V_s, indicated by the negative sign in the numerator.
(b) The gain of the closed-loop amplifier is less than the gain of the amplifier itself, since the denominator has a value greater than unity.

Feature (b) is explained as follows. An increase in the amplifier input voltage by an amount $+V_1$ results in an output voltage of $-V_2$. The voltage fed back by the

β-network is $\beta(-V_2) = -\beta V_2$. Since the connections to the input circuit are as shown in Fig. 9.2, the amplifier input voltage is

$$V_1 = V_s + (-\beta V_2) = V_s - \beta V_2$$

Comparing this equation with the equivalent expression for the positive feedback case, eq. (9.2), it is seen that V_1 is reduced when negative feedback is applied. Since the output voltage is $-mV_1$, the overall gain is reduced. This argument is supported by the following example.

If an amplifier has a gain of -1000, an input V_1 of 1 mV supports an output of -1 V. If the amplifier is used as a part of a negative series voltage feedback amplifier with a feedback coefficient of $\beta = 0 \cdot 1$, the voltage fed back to the input is $0 \cdot 1 \times (-1) = -0 \cdot 1$ V.

Now

$$V_1 = V_s + \beta V_2 = V_s - 0 \cdot 1$$

or

$$V_s = 0 \cdot 001 + 0 \cdot 1 = 0 \cdot 101 \text{ V}$$

This gives an overall gain for the feedback amplifier of

$$m' = V_2/V_s = -1/0 \cdot 101 = -9 \cdot 901$$

As a result of the application of negative feedback, the input voltage to the amplifier has to be increased in order to maintain the same output voltage, resulting in an overall reduction of gain. Using eq. (9.5)

$$m' = -1000/(1 - [-1000 \times 0 \cdot 1]) = -9 \cdot 901$$

If, in eq. (9.5), the loop gain $m\beta$ has a value which is much greater than unity, then

$$m' \simeq -m/-(-m)\beta = -1/\beta$$

That is, the closed-loop gain is set by the β-network, and is independent of the actual value of the amplifier gain. In the above illustrative example $-m\beta = 100$, which is much greater than unity, and the approximation given above can be used. This gives a gain of $m' = -1/0 \cdot 1 = -10$, which is in error by only 1 per cent.

An alternative method of applying series negative voltage feedback is to use a non-phase-inverting amplifier, and a phase-inverting β-network. This technique is readily applicable to systems in which the input and output are electrically isolated from one another, e.g., control systems, and electronic circuits using isolating transformers in the feedback path. The equation for the gain of these amplifiers is obtained by replacing β in eq. (9.3) by $-\beta$, giving

$$m' = m/(1 + m\beta) \tag{9.6}$$

The resulting closed-loop amplifier is non-phase-inverting with a gain less than m.

A study of the principal properties of series negative voltage feedback amplifiers follows.

9.3.1   Stabilization of gain

In the absence of feedback, a change in the amplifier gain, as a result of some internal change, results in a proportional change in the output voltage. In the feedback amplifier, this change of output is fed back to the input, causing V_1 to change. As a direct result of this the output voltage is brought back to a level which is nearly equal to its original value.

Example 9.2: Calculate the per-unit change in the closed-loop gain of a series negative voltage feedback amplifier, with $\beta = 0\cdot1$, when the amplifier gain falls from -1000 to -500.

Solution: When the gain is -1000

$$m' = -1000/(1 - [-1000 \times 0\cdot1]) = -9\cdot901$$

and when -500,

$$m' = -500/(1 - [-500 \times 0\cdot1]) = -9\cdot805$$

and the per-unit reduction in m', when m changes by $0\cdot5$ per-unit, is

$$(9\cdot901 - 9\cdot805)/9\cdot901 = 0\cdot0097 \text{ per unit or } 0\cdot97 \text{ per cent}$$

A physical explanation of this phenomenon is obtained by utilizing the results of eq. (9.4). Suppose that the input voltage to the feedback amplifier in example 9.2 is maintained at 1 V. The amplifier input voltage, for $m = -1000$, is

$$V_1 = 1/(1 - [-1000 \times 0\cdot1]) = 0\cdot009901 \text{ V or } 9\cdot901 \text{ mV}$$

and the amplifier output voltage is

$$V_2 = -1000 \times 9\cdot901 \times 10^{-3} = -9\cdot901 \text{ V}$$

When the gain is reduced to -500, the amplifier input voltage rises to

$$V_1 = 1/(1 - [-500 \times 0\cdot1]) = 0\cdot01961 \text{ V or } 19\cdot61 \text{ mV}$$

giving an output voltage of

$$V_2 = -500 \times 19\cdot61 \times 10^{-3} = -9\cdot805 \text{ V}$$

Thus, in order to compensate for the 50 per cent reduction in amplifier gain, V_1 practically doubles to give an overall gain for the feedback amplifier which is not significantly different from that which existed before the amplifier gain was reduced.

Alternatively, an expression for the change in closed-loop gain can be deduced by differentiating eq. (9.3) with respect to m, which gives the per-unit change in gain as $dm/m(1 - m\beta)$, where dm/m is the per-unit change in gain m. In example 9.2, this yields $-0\cdot5/(1 - [-1000 \times 0\cdot1]) = -0\cdot00495$, a reduction of $0\cdot495$ per cent. This should be compared with the figure of $0\cdot97$ per cent in example 9.2.

The reason for the difference in the two solutions is that the equation developed here is valid for small changes in m (up to, say, 5 per cent), and it does not give accurate results for large changes in gain.

9.3.2   Input impedance Z_1'

Practically all electronic devices have a finite input impedance, its value being high in the cases of thermionic and field effect devices, and relatively low in bipolar transistors. The input impedance of the circuit is modified by feedback, the actual change being dependent on the form of feedback used, and the way in which it is applied. In series negative voltage feedback amplifiers, the net input voltage V_1 applied to the amplifier is less than the signal voltage V_s, and the current drawn from the signal source is thereby reduced. This gives the effect of an increase in input impedance above that of the amplifier itself.

For example, if an amplifier has a gain of -100 and an input impedance of 1 kΩ, a 10 mV signal applied to the amplifier results in an output voltage of -1 V and an input current of 10 mV/1 kΩ = 10 μA. If a feedback network with $\beta = 0.09$ is used with the above amplifier, the voltage fed back is $-1 \times 0.09 = -0.09$ V, and the signal voltage V_s required to maintain the output of 1 V is

$$V_s = V_1 - \beta V_2 = 0.01 - (0.09 \times -1) = 0.1 \text{ V}$$

Under feedback conditions, the voltage applied to the amplifier terminals (V_1) remains at 10 mV, the current drawn from the signal source remains at 10 μA. The apparent input impedance, seen from the signal source, is

$$Z_1' = V_s/10 \times 10^{-6} = 0.1/10^{-5} = 10^4 \ \Omega \text{ or } 10 \text{ k}\Omega$$

that is, an apparent increase in input impedance by a factor of 10, from 1 kΩ to 10 kΩ.

The theoretical input impedance is

$$Z_1' = Z_1(1 - m\beta) \tag{9.7}$$

and in the above example this yields $Z_1' = 1(1 - [-100 \times 0.09])$ kΩ = 10 kΩ. In positive feedback amplifiers, when the product $m\beta$ has a positive value, the input impedance is reduced below that of Z_1.

9.3.3   Effect of loading the feedback amplifier

When current is drawn from the amplifier, the output voltage is reduced as a result of the p.d. in the output impedance Z_2 of the amplifier. The initial effect is an apparent reduction in voltage gain, and the general effects of load current on the feedback amplifier are broadly those resulting from a reduction of gain. However, negative feedback tends to compensate for reduction in gain, and the effects of loading on a negative feedback amplifier are generally less than in the case where no feedback is applied.

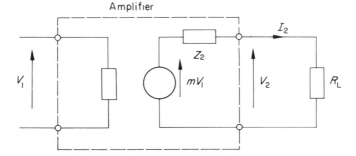

Fig. 9.3   A method of determining the effect of loading the amplifier.

The effects of loading may be analysed with the aid of Fig. 9.3.
By potential divider action, $V_2 = mV_1R_L/(Z_2 + R_L)$. The gain of the amplifier
itself, when loaded, is therefore

$$m_l = V_2/V_1 = mR_L/(Z_2 + R_L) \tag{9.8}$$

The effective constants of a voltage feedback amplifier when it is loaded are
obtained by substituting the value of m_l in place of m in the appropriate no-load
equations. Thus,

$$m' = \frac{m_l}{1 - m_l\beta} = \frac{m\dfrac{R_L}{Z_2 + R_L}}{1 - m\dfrac{R_L}{Z_2 + R_L}\beta} = \frac{mR_L}{Z_2 + R_L - m\beta R_L}$$

$$= \frac{mR_L}{Z_2 + R_L(1 - m\beta)} \tag{9.9}$$

In addition, the load applied to the amplifier has a significant effect upon the
input impedance of the amplifier. Any variation in the load current causes the
output voltage to change, and with it the voltage which is fed back. For reasons
which have already been given, this causes V_1 to alter, causing the current drawn
from the signal source to change. This results in a change in the input impedance
of the closed-loop amplifier. In the case of a negative feedback amplifier, a
reduction of output voltage causes V_1 to increase, resulting in an increased input
current, and a reduction in input impedance.

An expression for the input impedance of the feedback amplifier when it is
loaded is developed by substituting eq. (9.8) into eq. (9.7), when

$$Z_1' = Z_1(1 - m_l\beta) = \frac{Z_1(Z_2 + R_L[1 - m\beta])}{Z_1 + Z_2} \tag{9.10}$$

To illustrate the effects of eqs. (9.9) and (9.10), consider the case of an amplifier
with a gain of -100, $\beta = 0\cdot09$, $Z_2 = 1$ kΩ, $Z_1 = 1$ kΩ, and $R_L = 4$ kΩ. The results

of eqs. (9.9) and (9.10) and those for an unloaded amplifier (when $R_L = \infty$) are given below.

	Unloaded amplifier	Loaded amplifier
Voltage gain	-10	$-9\cdot1$
Input impedance (kΩ)	10	$5\cdot5$

So far, the effects of the β-network itself have been neglected. This network constitutes a load which is permanently connected to the amplifier, and it cannot always be ignored. The effects of the β-network are generally the same as those of an external load resistor.

9.3.4   Output impedance Z'_2

It has already been shown that the output voltage of a negative series voltage feedback amplifier remains substantially constant over a very wide range of loading conditions. That is, the output impedance of the feedback amplifier is small when compared with that of the amplifier itself.

The effective output impedance of the closed-loop amplifier is given by

$$Z'_2 = \frac{V'_2}{I'_2} = \frac{Z_2}{1 - m\beta \dfrac{Z_1}{Z_1 + R_s}} = \frac{Z_2(Z_1 + R_s)}{R_s + Z_1(1 - m\beta)} \tag{9.11}$$

In the case of an amplifier with $Z_1 = 1$ kΩ, $Z_2 = 1$ kΩ, $-m = -100$, $\beta = 0\cdot09$, and $R_s = 1$ kΩ, the closed-loop output impedance is 182 Ω, which is much lower than the 1 kΩ value for the amplifier itself.

The minimum value of output impedance is achieved when the source resistance R_s is zero, when

$$Z'_2 = \frac{Z_2}{1 - m\beta} \tag{9.12}$$

9.3.5   Noise in feedback amplifiers

Spurious signals, known as *noise* signals, appear at the output of amplifiers in addition to the wanted signals. Noise signals are either generated in circuit components by effects like microphony and 'shot' noise, or they may be induced from neighbouring signal sources such as the alternating mains supply.

If the noise originates at an early stage in the amplifier, e.g., in the first stage, negative feedback has no significant effect in reducing it, since the noise signal and the wanted signal are amplified equally. If the noise is generated at a late stage in the amplifier, the wanted signal receives more amplification than the noise, and the signal-to-noise ratio is improved. To keep the noise level at its lowest value, great care must be taken in designing the early stages of feedback amplifiers. A detailed analysis is beyond the scope of this book, but it is given elsewhere*.

9.3.6   Effect of feedback on distortion

It was shown in chapter 8 that non-linear characteristics can result in the introduction of harmonic distortion in the output waveform from an amplifier. The harmonics generated by this means fall within the definition of noise, given in

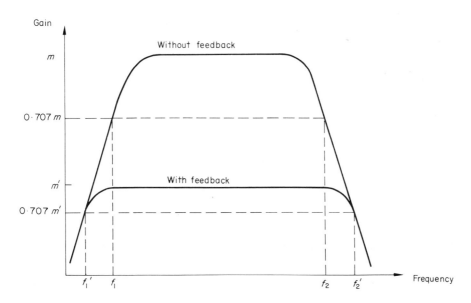

Fig. 9.4   Frequency response of an a.c. amplifier before and after feedback is applied.

section 9.3.5. Thus, if an amplifier of gain m has an input signal V_1, and the 'noise' signal or harmonic distortion introduced by the circuit is V_n, then the output voltage in the absence of feedback is given by

$$V_2 = mV_1 + V_n$$

giving a signal-to-noise ratio at the output of mV_1/V_n. When feedback is applied in accordance with Fig. 9.2, the expression for output voltage becomes

$$V_2 = \frac{m}{1 - m\beta} V_s + \frac{V_n}{1 - m\beta}$$

giving a signal-to-noise ratio at the output of mV_s/V_n.

To maintain the same value of V_2 both with feedback and without it, V_s is much larger than V_1 when negative feedback is applied. Thus, if the two values

* Morris, N. M., *Control Engineering*, McGraw-Hill (1968).

of signal-to-noise ratio are compared at the same value of output signal, it is seen that the signal-to-noise ratio is improved by negative feedback in the ratio of V_s/V_1. From eq. (9.4), this is seen to be a factor of $(1 - m\beta)$.

It can also be shown that negative voltage feedback also has the effect of reducing the phase distortion in amplifiers.

9.3.7   Effect of feedback on bandwidth

A typical gain-frequency curve for an a.c. amplifier is shown in Fig. 9.4 (upper curve). The reduction in gain at high frequency occurs, usually, as a result of the change in the reactance of stray capacitors which shunt the output of the amplifier (see chapter 8). At high frequency, the reactance of these capacitors falls to a low value, effectively short-circuiting the load. The reduction in gain at low frequencies is a direct consequence of the coupling methods used in a.c. amplifiers (see chapter 8).

The bandwidth of the amplifier in the absence of feedback is $f_2 - f_1$, and its gain is m, giving an open-loop gain-bandwidth product of $m(f_2 - f_1)$. When series negative voltage feedback is applied, the cut-off frequencies become f_1' and f_2', respectively, where $f_1' = f_1/(1 - m\beta)$ and $f_2' = f_2(1 - m\beta)$*. The gain also falls to m' $(= m/[1 - m\beta])$. The closed-loop gain-bandwidth product is $m'(f_2' - f_1')$. Provided that the bandwidth is reasonably large, then the two gain-bandwidth products are found to be approximately equal. That is

$$m(f_2 - f_1) \simeq m'(f_2' - f_1') \qquad (9.13)$$

For example, if the gain of the amplifier is reduced by a factor of ten by the application of negative voltage feedback, then the bandwidth is increased tenfold.

9.4   The effects of series positive voltage feedback

Since a detailed treatment of the effects of negative feedback has been given above, it is not necessary to treat the equivalent positive feedback cases in depth. In general, the effects of positive feedback are opposite to those of negative feedback, and a summary of the more important results of the two forms of feedback are given in Table 9.1.

Table 9.1
A summary of the more important effects of series voltage feedback

	Form of feedback used	
	Negative	Positive
Gain	Reduced	Increased
Input impedance	Increased	Reduced
Output impedance	Reduced	Increased
Bandwidth	Increased	Reduced

* See Morris, N. M., *Control Engineering*, McGraw-Hill (1968).

9.5 Voltage summation using a resistive network

Voltage summation can be achieved by converting voltages into currents, and then summating the currents. The voltage developed across a resistor carrying the sum of the currents is then proportional to the sum of the voltages. This technique is utilized in the circuit in Fig. 9.5, where V_s would normally be the signal voltage source, V_2 the output voltage of the amplifier, and V_1 the input voltage to the amplifier. R_1 and R_f are resistors, and Z_1 is the input impedance of the circuit to which V_1 is applied.

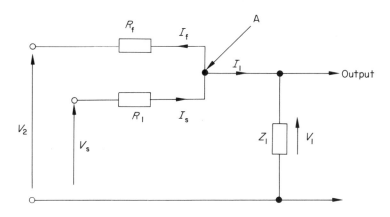

Fig. 9.5 Voltage summation using a resistive network.

Applying Kirchhoff's first law to the *summing junction* A in Fig. 9.5 yields

$$I_1 + I_f = I_s$$

or

$$\frac{V_1}{Z_1} + \frac{V_1 - V_2}{R_f} = \frac{V_s - V_1}{R_1}$$

or

$$V_1 \left(\frac{1}{Z_1} + \frac{1}{R_f} + \frac{1}{R_1} \right) = \frac{V_2}{R_f} + \frac{V_s}{R_1}$$

If we let

$$Y = \frac{1}{Z_1} + \frac{1}{R_f} + \frac{1}{R_1},$$

then

$$V_1 = \frac{V_2}{R_f Y} + \frac{V_s}{R_1 Y} \qquad (9.14)$$

9.6   Shunt voltage feedback amplifiers

In the opening paragraphs of this chapter, it was stated that shunt feedback is applied by converting the input and output signals into currents, which are summated to form the amplifier input signal. This is achieved using the resistive network in Fig. 9.5, and its method of application to amplifiers is shown in Fig. 9.6.

A simplified explanation of the application of this method of feedback is as follows. The voltage V_1 developed at the input to the amplifier is $I_1 Z_1$, and since

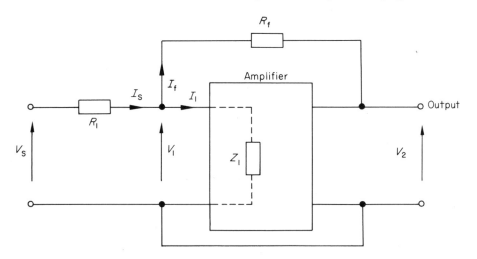

Fig. 9.6   Shunt voltage feedback.

$I_1 = (I_s - I_f)$ then $V_1 = Z_1 (I_s - I_f)$. That is to say, the amplifier input voltage is proportional to the difference between the input current I_s and the current I_f in the feedback path. In turn, these are related to the input and output voltages, respectively, so that overall voltage feedback is applied around the amplifier.

9.6.1   Voltage gain m'

A detailed analysis of this type of circuit is more complex than for the series voltage feedback case, and a simplified analysis is given here. The basic assumptions made here are (1) the gain of the amplifier is large and it has a negative value, and (2) the input impedance is very large. The first assumption may be valid in almost any type of amplifier, but the second is not always true in the case of bipolar transistor circuits.

Now,

$$V_1 = -V_2/m \tag{9.15}$$

and if m is very large then V_1 is very small, e.g., if $V_2 = -10$ V, and $m = -10^6$ (as it may be in analogue computing amplifiers), then $V_1 = 10\,\mu$V. In basic calculations, therefore, we may assume that $V_1 \simeq 0$. For this reason, the junction of the summing resistors is referred to as a *virtual earth*. Hence, in Fig. 9.6,

$$I_s = I_f \quad \text{or} \quad \frac{V_s}{R_1} = -\frac{V_2}{R_f} \tag{9.16}$$

$$m' = V_2/V_s = -R_f/R_1 \tag{9.17}$$

For example, if $R_f = 1$ MΩ and $R_s = 0{\cdot}1$ MΩ, then $m' = -10$.

A detailed analysis yields an expression for gain of

$$m' = \frac{m}{R_1 Y} \bigg/ \left(1 - \frac{m}{R_1 Y}\right) \tag{9.18}$$

$$= m_s/(1 - m_s \beta_s) \tag{9.19}$$

where Y is defined in section 9.5. In eq. (9.19) m_s is the forward gain $(= m/R_1 Y)$ of the shunt feedback amplifier, and β_s $(= R_1/R_f)$ is its effective feedback factor. The similarity between eqs. (9.19) and (9.3) should be noted. The effect of loading can be dealt with in much the same way as for the series feedback case.

Amplifiers of this type are sometimes known as *see-saw amplifiers*, since the input and output voltages appear to pivot about the virtual earth point.

9.6.2   Input impedance Z_1' of the shunt feedback amplifier

Assuming that the summing junction is a virtual earth, the input resistance is seen to be approximately equal to R_1. A detailed analysis yields the expression

$$Z_1' = R_1 + Z_1 R_f/(R_f + Z_1[1 - m]) \tag{9.20}$$

9.6.3   Output impedance Z_2' of the shunt feedback amplifier

To a first approximation, the output impedance is the parallel combination of Z_2 and R_f (since R_f is 'earthed' at one end by the virtual earth).

$$Z_2' \simeq Z_2 R_f/(Z_2 + R_f) \tag{9.21}$$

Example 9.3: Determine the voltage gain, the input impedance, and the output impedance of a shunt voltage feedback amplifier in which $R_1 = 10$ kΩ, $R_f = 200$ kΩ, gain $= -200$, $R_s = 1$ kΩ, $Z_1 = 1$ kΩ, and $Z_2 = 20$ kΩ.

Solution:

$$m' = -200/10 = -20$$

(the true gain is $-9{\cdot}5$, due largely to the low value of Z_1)

$$Z_1' \simeq R_1 = 10 \text{ k}\Omega$$

(to be compared with the more accurate value of $10 \cdot 5 \text{ k}\Omega$)

$$Z_2' = 20 \times 200/(20 + 200) = 18 \cdot 2 \text{ k}\Omega$$

(the true value is $17 \cdot 9 \text{ k}\Omega$).

9.7   Series current feedback

In a current feedback amplifier, the signal fed back is proportional to the output current. As with any feedback amplifier, it is the signal which is fed back that is maintained at a constant level, and in this case the output current is maintained

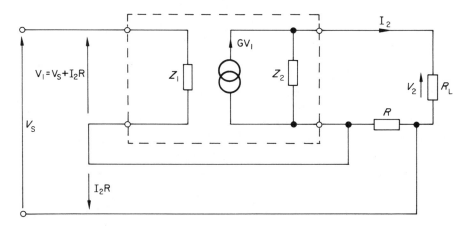

Fig. 9.7   The basic series positive current feedback amplifier.

at a value set by the input voltage signal. As a consequence of this, the series current feedback amplifier has a very high output impedance, a fact which is verified in section 9.7.4.

Figure 9.7 shows a block diagram of a series current positive feedback amplifier. The current flowing in the output circuit develops a voltage $I_2 R$ across resistor R, which is added in series with the input signal V_s. The net voltage applied to the amplifier input is

$$V_1 = V_s + I_2 R \tag{9.22}$$

Since the forward gain or *transconductance* of the amplifier is G A/V, then the current generated by the internal generator within the amplifier is $G V_1$ A.

9.7.1   No-load gain G'

No-load conditions correspond to the case where the power delivered to the load is zero. A current amplifier theoretically continues to deliver current to the load

so long as an input signal is applied. Thus, no-load conditions in a current amplifier correspond to the case where the voltage developed across the load is zero,
i.e., when $R_L = 0$. Since the resistance of resistor R is usually much less than the
value of Z_2, the following relationship holds good

$$I_2 = GV_1 \quad \text{or} \quad V_1 = I_2/G$$

Eliminating V_1 between the above expression and eq. (9.22) yields the no-load
closed-loop gain G'

$$G' = \frac{I_2}{V_s} = \frac{G}{1 - GR} \tag{9.23}$$

If this equation is compared with equivalent equations for the two voltage feedback cases, eqs. (9.3) and (9.19), it is seen that the 'β' value for Fig. 9.7 is equal
to R, and has the dimensions of resistance.

In the negative feedback case, using an amplifier of gain $-G$, if the product
GR has a value much greater than unity, then

$$G' = \frac{-G}{GR} = \frac{-1}{R}$$

That is,

$$\frac{I_2}{V_s} = \frac{-1}{R} \quad \text{or} \quad I_2 = -V_s/R \tag{9.24}$$

By this means, the output current is dependent on the input voltage and the
feedback resistor R, and is independent of the transconductance of the amplifier.

9.7.2   No-load input impedance Z'_1

The no-load input impedance is calculated with $R_L = 0$. The current drawn from
the signal source is

$$I_s = \frac{V_1}{Z_1} = \frac{V_s + I_2 R}{Z_1}$$

Eliminating I_2 between the above equation and eq. (9.23) yields the input
impedance

$$Z'_1 = \frac{V_s}{I_s} = Z_1(1 - GR) \tag{9.25}$$

If positive feedback is applied (i.e., the product GR has a positive value), then Z'_1
is less than Z_1; with negative feedback Z'_1 is greater than Z_1.

9.7.3 Effect of an external load on current feedback amplifiers

When the load resistor has a finite value, the load current falls below the theoretical value of GV_1, reducing the effective gain of the amplifier. This modifies the closed-loop constants as follows:

$$G' = G_l/(1 - G_l R) \quad \text{and} \quad Z_1' = Z_1(1 - G_l R)$$

where

$$G_l = GZ_2/(Z_2 + R_L + R)$$

9.7.4 Output impedance Z_2'

The output impedance is calculated by the same method used hitherto, that is the system is energized at its output terminals, and the signal source is replaced by its internal impedance. This gives an expression for the output impedance of the form

$$Z_2' = Z_2 + R - \frac{GRZ_1 Z_2}{R_s + Z_1} \tag{9.26}$$

which may be simplified if $Z_2 \gg R$ and $Z_1 \gg R_s$, when

$$Z_2' \simeq Z_2(1 - GR) \tag{9.27}$$

If positive current feedback is employed (i.e., the product GR has a positive value), then Z_2' has a value which is less than Z_2. By altering the value of the product GR, the output impedance can be made zero, or even negative. In the latter case, an increased value of load resistance results in an increased load current. If negative feedback is employed, Z_2' is greater than Z_2.

Example 9.4: Calculate the closed-loop transconductance or gain of an amplifier similar to that in Fig. 9.7 if $G = -50$ mA/V, $Z_1 = 10$ kΩ, $Z_2 = 50$ kΩ, $R_L = 2$ kΩ, $R = 1$ kΩ, and $R_s = 5$ Ω. What is the magnitude of V_s to ensure a load current of 10 mA?

Solution: The load modifies the amplifier gain to

$$G_l = -50 \times 50/(50 + 2 + 1) = -47 \cdot 2 \text{ mA/V}$$

Hence,

$$G' = -47 \cdot 2/(1 - [-47 \cdot 2 \times 1]) = -0 \cdot 98 \text{ mA/V}$$

(The value given by the no-load equation (eq. (9.23)) is 1 mA/V.)

$$Z_1' = 10(1 - [-47 \cdot 2 \times 1]) = 482 \text{ k}\Omega$$

(Equation (9.25) gives value of 510 kΩ.)

The output impedance, from eq. (9.26), is

$$Z'_2 = 50 + 1 - \frac{-50 \times 1 \times 10 \times 50}{0 \cdot 005 + 10} = 2551 \text{ k}\Omega$$

(the result from the approximate equation, eq. (9.27), is 2550 kΩ).

Since the closed-loop gain is $-0 \cdot 98$ mA/V, the source voltage to maintain an output current of 10 mA into a load of 2 kΩ is

$$V_s = \frac{10}{-0 \cdot 98} = -10 \cdot 2 \text{ V}$$

Since the output impedance of this circuit is greater than 2·5 MΩ, the load current will be substantially constant at 10 mA over a range of load resistance values from zero to about 200 kΩ.

9.7.5   Application of current feedback to a voltage amplifier

In some applications (see for example section 10.3), it is desirable to apply current feedback to a voltage amplifier. The block diagram remains as in Fig. 9.7, but the output from the circuit is V_2 (the voltage across R_L) and not I_2. Since series current feedback is applied, then $V_s \simeq -I_2 R$. Now $V_2 = I_2 R_L$, so that $I_2 = V_2/R_L$, hence $V_s = -V_2 R/R_L$. The voltage gain is, therefore,

$$m' = V_2/V_s = -R_L/R$$

9.8   Measurement of system constants

It is frequently necessary to measure such constants as gain, input impedance, output impedance, and bandwidth. These constants can be measured by relatively simple methods using only an oscillator, a high impedance voltmeter and a resistor, preferably in the form of a calibrated decade resistance box.

The voltage gain at any frequency is measured simply by determining the ratio of the output voltage to the source voltage at that frequency. The bandwidth is the band of frequencies between the points at which the gain falls to 0·707 of the mid-band gain (see Fig. 9.4).

To determine the input impedance, it is first necessary to measure the voltage gain of the stage in the manner described above. Let its value be m_1. A resistance R_1 is then connected in series with the input of the amplifier, and the gain is again measured. Let its value be m_2. In the latter case, the voltage applied to the amplifier terminals is

$$V'_s = \frac{Z_1}{Z_1 + R_1} V_s$$

where Z_1 is the input impedance of the amplifier. The output voltage from the amplifier is

$$V_2 = m_1 V_s' = m_1 \frac{Z_1}{Z_1 + R_1} V_s$$

hence,

$$m_2 = m_1 \cdot \frac{Z_1}{Z_1 + R_1}$$

therefore,

$$Z_1 = \frac{R_1}{\dfrac{m_1}{m_2} - 1}$$

If, for example $m_1/m_2 = 1\cdot1$, $R_1 = 1$ kΩ, then $Z_1 = 10$ kΩ.

The output impedance can be measured by determining the gain with a load resistor R_2 connected between its output terminals. Let its value be m_3. If the gain in the absence of R_2 is m_1, then the additional load reduces the output voltage to

$$V_2 = m_1 V_s \frac{R_2}{R_2 + Z_2}$$

where V_s is the input signal voltage applied to the amplifier, and Z_2 is the output impedance of the amplifier.

Hence,

$$m_3 = \frac{V_2}{V_s} = m_1 \frac{R_2}{R_2 + Z_2}$$

therefore,

$$Z_2 = R_2 \left(\frac{m_1}{m_3} - 1 \right)$$

In an actual circuit, if $m_1/m_3 = 1\cdot1$ and $R_2 = 50$ kΩ, then $Z_2 = 5$ kΩ. These techniques can be applied to amplifiers both with, and without, feedback applied.

9.9 Stability

It was stated in section 9.2 that unstable operation could occur in feedback amplifiers under certain conditions. To deduce the general conditions for instability, the equations for the no-load gain of the positive feedback amplifiers discussed in this chapter are rewritten here

$$m' = m/(1 - m\beta) \tag{9.3}$$

$$m' = m_s/(1 - m_s \beta_s) \tag{9.19}$$

$$G' = G/(1 - GR) \tag{9.23}$$

An inspection shows, in each case, that the closed-loop gain becomes infinity when the loop gain ($m\beta$ or GR) is unity. This constitutes an unstable operating state, since an output can exist when $V_s = 0$.

The equivalent equations for negative feedback circuits, using phase-inverting amplifiers, are

$$m' = -m/(1 + m\beta)$$

$$m' = -m_s/(1 + m_s\beta_s)$$

$$G' = -G/(1 + GR)$$

Instability occurs in these amplifiers when the loop gain is -1. The conditions for instability for both positive and negative feedback amplifiers are seen to be the same, if the sign of m (or G) is borne in mind.

Instability thus occurs when the loop gain is unity and the loop phase-shift is zero, simultaneously.

The critical condition is the loop phase-shift. If the loop gain is less than unity when the loop phase-shift is zero, then the closed-loop amplifier is stable, but if the loop gain is equal to, or greater than, unity then the closed-loop amplifier is unstable, and the output is oscillatory.

The general procedure to determine the stability of a feedback amplifier is as follows. Firstly, the feedback loop is broken or opened at one point, a convenient point for this operation being at the input to the amplifier. The normal signal source V_s (see Fig. 9.2) is removed and replaced by its internal resistance. A signal V_x is then injected into the amplifier, and the magnitude and phase-shift of the signal after it has been amplified around the loop are measured. This test is then repeated over a range of frequencies. If the magnitude of the signal after it has passed round the loop is V_y, and its phase-shift with respect to V_x is ϕ, then the loop gain is

$$m\beta = \frac{V_y}{V_x} \angle \phi$$

The locus of $m\beta$, as it changes with frequency, is then plotted on polar graph paper, and the resulting curve is known as a *Nyquist plot.* Alternatively, if the gain and phase-shift are plotted separately to a base of frequency, the combined graphs are known as *Bode diagrams.* If, at some frequency ω_p, the loop gain is greater than unity when the loop phase-shift is zero, then the amplifier will be unstable when the loop is closed and the output voltage will oscillate at that frequency. Frequency ω_p is known as the *phase crossover frequency.* If the loop gain is less than unity at the phase crossover frequency, then the loop may be closed and the amplifier will operate in a stable manner.

Circuits which depend for their operation on unstable conditions existing are known as *oscillators,* which are dealt with in chapter 10. The detailed theory of frequency response tests, together with the application of Nyquist and Bode diagrams, is beyond the scope of this book, and is available elsewhere*.

9.10   Simplified valve and transistor equations

In this section, we deduce simplified equations relating valve and transistor parameters to the terms m, G, Z_1, and Z_2 used in this chapter.

For the valve, the input impedance is very high and is assumed to be infinity. Its output impedance is equal to r_a, and its voltage gain is equal to $-\mu$ (the negative sign implies phase-inversion). Using the voltage-to-current generator conversion for the valve

$$G = \frac{\text{Voltage gain}}{\text{Output impedance}} = \frac{-\mu}{r_a} = -g_m$$

In the case of the transistor, it is convenient to deduce the constants in terms of the general h-parameters h_i, h_r, h_f, and h_o. The resulting equations can then be used in any of the transistor configurations. Also, to simplify the computations, the reverse parameter h_r is neglected. The input current is given by $I_1 = V_1/h_i$, hence $Z_1 = h_i$. Similarly the output impedance is $1/h_o$; since the internal current generator develops a current of $-h_f I_1 = -h_f V_1/h_i$, the effective transconductance of the transistor is $-h_f/h_i$. When converted into its constant voltage equivalent, the value of m for the transistor is found to be $-h_f/h_i h_o$. The results are summarized in Table 9.2.

Table 9.2

	Valve	Transistor
Z_1	∞	h_i
Z_2	r_a	$\dfrac{1}{h_o}$
m	$-\mu$	$\dfrac{-h_f}{h_i h_o}$
G	$-g_m$	$\dfrac{-h_f}{h_i}$

Problems

9.1 (a) Voltage negative feedback is applied to an amplifier. Explain how it affects

(i) the frequency response, and
(ii) variation in stage gain due to fluctuating supply voltage.

* See Morris, N. M., *Control Engineering*, McGraw-Hill (1968).

(b) Sketch separate typical circuit diagrams to show how current and voltage feedback may be obtained for valve amplifiers. For each connection state how the feedback fraction may be determined from the component values.

(c) When voltage gain is applied to an amplifier of gain 100 the overall stage gain falls to 50. Calculate the fraction of the output voltage fed back. If this fraction is maintained, calculate the value of the amplifier gain required if the overall stage gain is to be 75.

(C & G)

9.2 Prove from first principles:
  (a) Series negative voltage feedback reduces the output impedance of an amplifier.
  (b) Series negative current feedback increases the output impedance of an amplifier.

9.3 If the gain of an amplifier without feedback is represented by A, derive an expression for the gain when a fraction β of the output voltage is fed back in opposition to the input.

In an amplifier with a constant input of 1 volt the output falls from 50 to 25 volts when feedback is applied. Calculate the fraction of the output which is fed back. If due to ageing, the amplifier gain fell to 40, find the percentage reduction in stage gain:
  (i) without feedback
  (ii) with the feedback connection.

(C & G)

9.4 An amplifier has a voltage amplification μ and a fraction, β, of its output is fed back in opposition to its input. If $\beta = 0\cdot1 \underline{/0°}$ and $\mu = 100\underline{/0°}$, calculate the change in the gain of the system if μ falls 6 dB due to ageing. Thence, or otherwise, give one advantage arising from the introduction of negative feedback. What other advantages result?

(C & G)

9.5 If the gain of an amplifier stage without feedback is represented by A, derive an expression for the gain when a fraction β of the output voltage is fed back in opposition to the input.

An amplifier has a gain of 1000 without feedback. Calculate the gain when 0·9 per cent of negative feedback is applied. If due to ageing, the gain without feedback falls to 800, calculate the percentage reduction in gain
  (a) Without feedback.
  (b) With feedback.
Comment on the significance of the results of (a) and (b), and state two other advantages of negative feedback.

(C & G)

9.6 Prove that series negative voltage feedback (a) stabilizes the no-load gain of an amplifier against change in the amplifier gain, (b) reduces the output impedance, and (c) increases the input impedance of the amplifier.

If the gain of the amplifier is 45 and it has input and output impedances of 200 and 0·5 Ω, respectively, in the absence of feedback calculate the input and output impedance if 20 per cent of the output voltage is fed back. What is the percentage change in the no-load closed-loop gain if the forward gain of the amplifier is changed by 5 per cent?

9.7 An amplifier has a constant signal input of 2 V and an output of 40 V when series negative voltage feedback is applied. The output increases to 60 V when the feedback signal is removed. Calculate the fraction of the output signal that is fed back.

If the amplifier gain fell to 20, calculate the percentage reduction in gain, (i) without feedback, and (ii) with the feedback connection.

(C & G)

9.8 Discuss the effects of negative and positive series current feedback on the steady-state output impedance of amplifiers.

A d.c. amplifier with an output impedance of 1 kΩ supplies a current of 3 mA at a terminal voltage of 7 V when the input to the amplifier is 0·2 V. Calculate the effective output impedance if this amplifier is used in a closed-loop amplifier of the type shown in Fig. 9.7 when using (a) negative current feedback, and (b) positive current feedback if $Z_2 = 1$ kΩ and $R = 10$ Ω. The input impedance of the amplifier is infinite.

If overall negative voltage feedback is then applied in conjunction with positive current feedback, determine the value of β (the amount of voltage feedback used) to reduce the net output impedance to 50 Ω.

9.9 (a) List the reasons why negative feedback is frequently employed in amplifiers which form part of instrumentation systems.

(b) An amplifier has a voltage gain A without feedback. Derive an expression for the overall gain when a fraction β of the output voltage V_0 is fed back in series with the external signal voltage v_i.

(c) What are the special properties of an emitter follower? Give a circuit diagram and quote a typical example of its use in instrumentation.

(C & G)

Note: see chapter 10 for part (c).

9.10 Explain briefly why negative feedback is used on carrier system line amplifiers.

An amplifier has a gain of 60 dB without feedback and 30 dB when feedback is applied. If the gain without feedback changes to 55 dB calculate the new gain with feedback.

(C & G)

9.11 Show that the gain of an amplifier with negative feedback may be given by the expression:

$$\mu_F = \frac{\mu}{1 + \mu\beta}$$

where

μ_F = gain with feedback,
μ = gain without feedback,
β = fraction of output voltage fed back to input.

An amplifier has a gain without feedback of 54·8 dB. What will be the gain when a feedback path, having a β of 0·0082, is connected? If the gain without feedback rises by 6 dB, what will be the new gain with feedback, expressed in decibels?

(C & G)

10. Feedback amplifier circuits and oscillators

10.1 Emitter follower and cathode follower circuits

These are perhaps the simplest to understand of all of the feedback amplifier circuits. Examples of practical circuits for a.c. applications are shown in Fig. 10.1. The bias current for the emitter follower, Fig. 10.1(a), is bled from the power supply via R_B; this ensures that a steady potential appears at the emitter under no-signal conditions.

If the potential of the base region with respect to the common line is v_B, then

$$v_B = V_2 + v_{BE}$$

where v_{BE} is the total instantaneous base-emitter voltage, which normally has a value between 0·2 V and 0·8 V. Since v_{BE} is generally small compared with V_2, then

$$V_2 \simeq v_B$$

That is to say, the output voltage 'follows' variations in the input signal, hence the name emitter follower. In the cathode follower, v_{GK} is small compared with V_2, and the cathode potential 'follows' the grid potential. This argument also shows that output voltage is in phase with the input signal, since an increase in the input voltage V_s causes v_B (in the transistor circuit) and V_2 to increase by approximately the same amount.

Since the net input voltage v_{BE} to the amplifying device is the difference between the input and output signal voltages, the emitter follower is a series negative voltage feedback amplifier with 100 per cent feedback ($\beta = 1·0$). Emitter followers and cathode followers are also known as *common-collector amplifiers* and *common-anode amplifiers* respectively.

In the cathode follower, Fig. 10.1(b), the bias voltage is obtained by dividing the cathode resistor R_K into two parts, the grid being returned to the common connection by R_g, which usually has a value of several megohms. The function of this form of connection is to provide a very high input resistance, and to ensure that the grid remains at a negative potential with respect to the cathode at all times.

Two other features of importance in these circuits are their high input impedance and low output impedance. The former results from the fact that the output voltage is practically equal in magnitude to the input voltage;

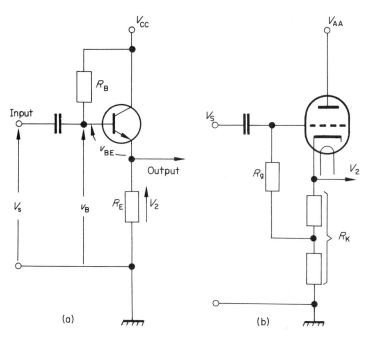

Fig. 10.1   (a) An emitter follower circuit, and (b) a cathode follower circuit.

negative feedback reduces the net input voltage to the active device to a small value, so that only a small current flows from the signal source. Suppose that the small-signal gain of an emitter follower is 0·95 and that the input impedance of the transistor is 1 kΩ. An input signal of 1 V gives rise to an output of 0·95 V, hence the signal voltage applied to the transistor is 0·05 V. The signal current flowing into the transistor base is, therefore, 0·05 V/1 kΩ = 50 μA, which is drawn from the input signal source. Since the actual input voltage is 1 V, the apparent input impedance of the circuit is 1 V/50 μA = 20 kΩ, which is much greater than the input impedance of the transistor.

The low output impedance results from the use of 100 per cent negative feedback, as follows. If the loading conditions are suddenly changed, say the load resistance is reduced, the output voltage tends to change—a reduction in this case. This causes the net input voltage ($V_s - V_2$) applied to the active device (transistor or valve) to change (an increase in the case of a bipolar transistor and a reduction in the case of a valve). This in turn causes the current flowing through the load to increase, so that the output voltage rises to a value which is nearly equal to the original value. Since the output voltage remains substantially constant over a wide range of loading conditions, the output impedance appears to have a low value.

To summarize, the principal features of interest are:

(a) The gain is approximately unity.
(b) The amplifier is non-phase inverting.
(c) The input impedance is high.
(d) The output impedance is low.

A simplified analysis of the emitter follower is now given in which it is assumed that the input impedance of the transistor is equal to h_{ie}, that $h_{fe} \gg 1$, that $R_E h_{fe}/h_{ie} \gg 1$, and that the source resistance can be neglected.

Voltage gain

$$I_e = (h_{fe} + 1)I_b \simeq h_{fe} I_b$$

$$V_2 = R_E I_e \simeq R_E h_{fe} I_b \simeq R_E h_{fe}(V_s - V_2)/h_{ie}$$

Collecting terms

$$m' = \frac{V_2}{V_s} = \frac{R_E h_{fe}/h_{ie}}{1 + R_E h_{fe}/h_{ie}} \simeq 1$$

Input impedance
From the work on voltage gain, $(V_s - V_2) \simeq V_2 h_{ie}/R_E h_{fe}$

$$Z_1' = \frac{V_s}{I_b} \simeq \frac{V_s}{(V_s - V_2)/h_{ie}} \simeq \frac{V_s}{V_2 h_{ie}/h_{ie} R_E h_{fe}} = R_E h_{fe} \frac{V_s}{V_2}$$

$$\simeq R_E h_{fe}$$

Output impedance
The open-circuit output voltage from the amplifier is

$$V_{oc} \simeq V_s$$

If the output terminals are short-circuited, then $V_2 = 0$ and the signal fed back is zero; in this mode of operation V_s is applied directly to the amplifier, and the base current is

$$I_{b(sc)} \simeq V_s/h_{ie}$$

and the current flowing in the emitter short-circuit is

$$I_{sc} = h_{fe} I_{b(sc)} = h_{fe} V_s/h_{ie}$$

hence,

$$Z_2' = V_{oc}/I_{sc} = V_s/(h_{fe}V_s/h_{ie}) = h_{ie}/h_{fe}$$

Thus, if $h_{ie} = 1\ \text{k}\Omega$, $h_{fe} = 50$, $R_E = 1\ \text{k}\Omega$, then $m \simeq 0.98$, $Z_1' \simeq 50\ \text{k}\Omega$, and $Z_2' \simeq 20\ \Omega$.

The current gain is given by

$$I_e/I_b = (1 + h_{fe})I_b/I_b = 1 + h_{fe}$$

One variation of the circuits in Fig. 10.1 is known as the *bootstrap amplifier*, in which the input signal is applied between the base and emitter (the grid and cathode in a valve) via an isolating circuit. The name of the circuit indicates that a change in base (grid) potential 'pulls' the input source voltage up by an amount equal to the output signal, i.e., it is 'pulled up' by its own bootstraps.

A detailed analysis of the circuits in Fig. 10.1, based on the work in chapter 9, gives the following results for bipolar circuits

$$m' \simeq \frac{h_{fe}}{h_{ie}h_{oe}}\bigg/\left(1 + \frac{h_{fe}}{h_{ie}h_{oe}}\right) \simeq 1$$

$$Z_1' \simeq h_{ie} + R_E h_{fe} \simeq R_E h_{fe}$$

$$Z_2' \simeq (h_{ie} + R_s)/h_{fe}$$

The equivalent expressions for the valve are

$$m' \simeq \mu R_K/(r_a + R_K[1 + \mu])$$

$$Z_1' \to \infty$$

$$Z_2' \simeq \frac{r_a}{1 + \mu} \simeq \frac{1}{g_m}$$

Owing to the type of feedback used, cathode follower and emitter follower stages have a high input impedance and low output impedance when compared with common-cathode and common-emitter amplifiers, respectively. The gain of the feedback amplifiers is also much lower. These features make the circuits particularly useful as buffer stages between a high impedance driver source and a low impedance load. An application of the circuits described here is as input stages of electronic measuring equipment. A more common application of both emitter follower and cathode follower circuits is in series voltage regulators (see chapter 13).

The performance of bipolar transistor circuits is further improved if a Darlington-connected pair of transistors is used to replace the single transistor. Field effect devices can be used in a circuit generally similar to Fig. 10.1(b), giving an input impedance of the same order as a valve stage. These circuits are known as *source followers* or *common-drain amplifiers.*

Example 10.1: A transistor, whose h_e parameters are given below, is used in an emitter follower stage; the signal source resistance is 600 Ω and the load resistance is 5 kΩ. Evaluate the voltage gain, the current gain, the input resistance, and the output resistance of the stage.

$$h_{ie} = 1500 \ \Omega$$
$$h_{fe} = 60$$
$$h_{oe} = 25 \times 10^{-6} \ \text{S}$$

Estimate the appropriate values for a triode using a valve with $\mu = 30$, $r_a = 12 \ \text{k}\Omega$.

Solution:

$$h_{fe}/h_{ie}h_{oe} = 60/1500 \times 25 \times 10^{-6} = 1600$$

Hence,

$$m' = 1600/(1 + 1600) \simeq 1$$
$$\text{Current gain} = 1 + 60 = 61$$
$$Z_1' = 1 \cdot 5 + (5 \times 60) \ \text{k}\Omega = 302 \ \text{k}\Omega$$

(a more accurate figure, not making any simplifying assumptions, is 289 kΩ).

$$Z_2' = (1500 + 600)/60 = 35 \ \Omega$$

In the case of the triode, the respective solutions are

$$m = 0 \cdot 9$$
$$\text{Current gain} \rightarrow \infty$$
$$Z_1' = \text{Many megohms (depending upon the input circuit)}$$
$$Z_2' = 387 \ \Omega$$

10.2   Series negative voltage feedback amplifier

A circuit diagram of a common-emitter form is shown in Fig. 10.2. This corresponds to the block diagram in Fig. 9.2, which is one of the fundamental building blocks of electronic feedback theory. In Fig. 10.3, the signal fed back is antiphase to the input signal, and is derived from potentiometer RV1.

An unfortunate feature of this type of circuit is that it is not possible to earth *both* the input line A *and* the transistor common line B, otherwise the signal which is fed back will be short-circuited. As a result of this, other feedback circuits are adopted where a gain greater than unity is required (see for example shunt voltage feedback—section 10.4). The type of feedback described here is generally adopted where the output can be electrically isolated from the input, as in the case where the load is transformer-coupled to the amplifier.

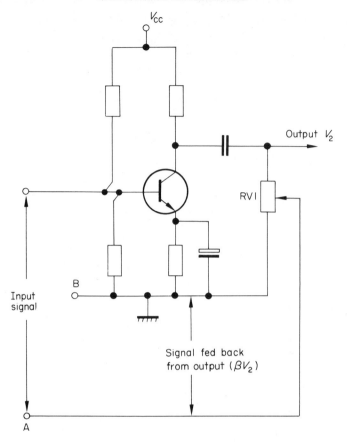

Fig. 10.2   A common-emitter negative feedback amplifier.

10.3   Current feedback amplifiers

One form of current feedback amplifier, the *paraphase amplifier* or *single-stage phase splitter,* is shown in Fig. 10.3. This circuit provides two anti-phase outputs V_2 and V_2' which can be used to drive a push-pull output stage. In this application, $R_L = R_E$, and since the collector and emitter currents are of the same order of magnitude, the voltage gains at each of the outputs relative to the input are equal.

Output V_2' is derived from an emitter resistor, and this part of the circuit is similar to an emitter follower. Consequently, V_2' is of the same order of magnitude as the input signal, and the output impedance is low by virtue of the emitter follower effect.

This amplifier is almost identical to the one dealt with in section 9.7.5, in which current feedback was applied to a voltage amplifier. In Fig. 10.3, we may

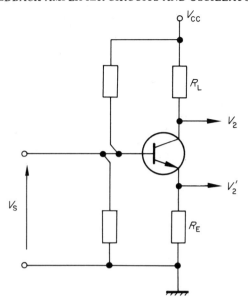

Fig. 10.3   A paraphase amplifier using only one transistor.

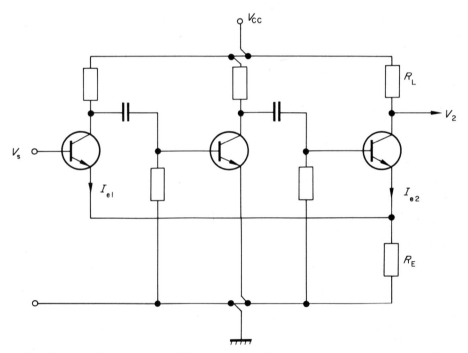

Fig. 10.4   A three-stage amplifier with overall series negative current feedback.

assume that $I_c = I_e = I$, say. Now, by emitter follower action, $V_s \simeq V_2' = IR_E$, hence, $I \simeq V_s/R_E$. Also, $V_2 = -IR_L = -V_sR_L/R_E$. Therefore,

$$\text{Voltage gain at the collector} = V_2/V_s = -R_L/R_E$$

and

$$V_2/V_2' = -R_L/R_E$$

In the case of the valve amplifier, the gain is approximately $-R_L/R_K$.

Current feedback can be applied over a number of stages, as shown in Fig. 10.4 in which current feedback is applied between the first and third stages of the amplifier. For clarity, the bias and temperature stabilization circuits have been omitted. The voltage applied to the first transistor is proportional to the difference between the input signal and a voltage proportional to the current in R_L. Provided that $I_{e1} \ll I_{e2}$ and the gain of the three transistor stages without feedback is large, then the gain of the circuit in Fig. 10.4 is approximately

$$\frac{V_2}{V_s} = -\frac{R_L}{R_E}$$

10.4 Shunt voltage feedback amplifiers

Shunt voltage feedback as applied to a pentode amplifier is illustrated in Fig. 10.5. In this circuit, R_K and C_K provide a steady grid bias voltage, while C_1 and C_2 are blocking capacitors which prevent d.c. potentials from being applied to the control grid.

Assuming for the moment that the gain of the pentode amplifier without feedback is very high, then the signal voltage at the grid need only be very small in order to sustain the normal value of output voltage. That is to say, the grid is a 'virtual earth' to signal voltages. This is reasonable in practice, since if $m = -100$, then a 10 mV grid signal gives a 1 V output. In this event, the signal current flowing in R_1 has approximately the same magnitude as the signal current in R_f, and

$$\frac{V_1}{R_1} \simeq -\frac{V_2}{R_f}$$

and the closed-loop gain is

$$m' = \frac{V_2}{V_1} = -\frac{R_f}{R_1}$$

Since the signal potential at the grid is virtually zero, then the input impedance of the amplifier is approximately equal to R_1.

The equations given here also apply to transistor circuits, provided that the gain is high. In the case of bipolar transistor circuits, the gain will be less than

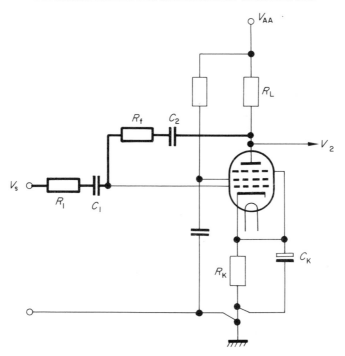

Fig. 10.5   A shunt voltage feedback amplifier.

that predicted here, due to the finite input impedance of the amplifier.

In Fig. 10.5, resistor R_1 includes the output resistance of the signal source, and in some circuits the source resistance may have a sufficiently high value to remove the need for a separate input resistor R_1. For example, if three stages are cascaded, then a feedback resistor R_f of 10 kΩ together with a signal generator of source impedance 1 kΩ ensure an overall voltage gain of -10.

10.5   Hybrid feedback circuits

In order to obtain a desired combination of features, it may be necessary to use the combined effects of more than one method of feedback. For example, to achieve a very high gain it may be necessary to incorporate some measure of positive feedback within the amplifier. However, positive feedback can lead to instability (see section 9.9), and it is usual in such cases to apply some overall negative feedback to give stability.

The results of feedback are modified to some extent by the method of applying the feedback. Two methods of applying feedback are illustrated in Fig. 10.6; the bias circuits and coupling capacitors have been omitted for clarity. In Fig. 10.6(a), a signal proportional to the output current (which flows in R_{E1}) is fed back to the input by the shunt feedback resistor R_{f1}. In

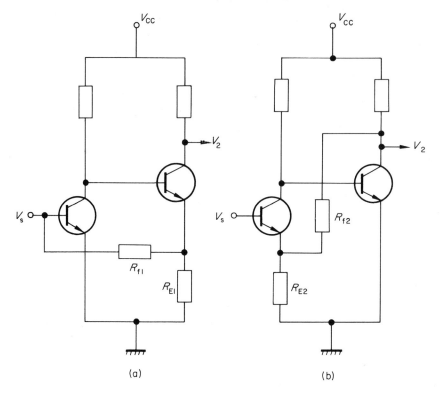

Fig. 10.6   Circuits employing hybrid feedback.

Fig. 10.6(b), the output signal is fed back by resistor R_{f2}, and is added to the input signal in emitter resistor R_{E2}. In both cases, the phase relationships are such as to apply overall negative feedback.

10.6   Phase-shift oscillators

Oscillators are generally feedback amplifiers which are deliberately designed to be unstable; they can work either as linear (class A) or non-linear (class C) networks. In the former case, the amplitude of the sinusoidal oscillations is so small that the excursion of the output voltage lies within the linear part of the characteristics; in the latter case, the amplitude of the output oscillations is large. As with amplifiers, the efficiency of an oscillator working in class A is lower than one working in class C.

High frequency oscillators generally use resonant L-C circuits, since the physical size of these elements is small at these frequencies. However, at low frequency (e.g., in the range 1 Hz to 100 kHz) the physical size of the elements in L-C resonant circuits becomes much larger and manufacturing difficulties present problems; *R-C phase-shift oscillators* are commonly used to generate this band of frequencies.

Broadly speaking, there are two main types of *R-C* oscillators

(a) Those with ladder feedback networks.
(b) Those with bridge feedback networks.

The type of ladder network used depends to some extent on the device used in the amplifier. Where the input impedance is high, as in a FET circuit or a valve circuit, a 'voltage' transfer stage of the type shown in Fig. 10.7(a) is used in the feedback network. Where the input impedance of the active device is

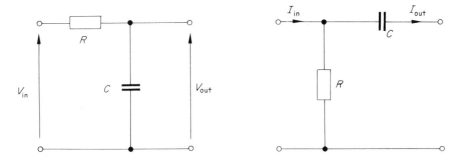

Fig. 10.7   (a) A voltage-transfer ladder network section, and (b) a current-transfer section.

low, as in a bipolar transistor circuit, the 'current' transfer network of Fig. 10.7(b) is used. An essential difference between the two networks is that while the output voltage of Fig. 10.7(a) lags behind the input voltage, the output current of Fig. 10.7(b) leads the input current.

A typical circuit diagram for an *R-C* oscillator using a bipolar transistor and a 'current' transfer ladder feedback network is shown in Fig. 10.8. The feedback network consists of three identical sections $(R_1 C_1)$, each contributing at the oscillatory frequency a phase-shift of approximately 60 degrees, giving 180 degrees of phase shift through the feedback network. Since the transistor provides a further 180 degrees of phase-shift, the loop phase-shift at the oscillatory frequency is 360 degrees (or 0 degrees) which is a condition for unstable operation. The theoretical oscillatory frequency of the circuit in Fig. 10.8 is

$$f_o = 1/2\pi R_1 C_1 \sqrt{6}$$

and the theoretical current gain of the transistor amplifier should be at least 29, since the ladder network provides this degree of attenuation at the oscillatory frequency. The equation above ignores the effect of the input impedance and output impedance of the transistor amplifier, with the result that the theoretical frequency is generally lower in value than the actual oscillatory frequency. Resistors R_2, R_3, and R_E, and capacitor C_E provide base bias and temperature stabilization. To ensure that an adequate current gain is available to cause oscillation, the transistor used should have an h_{fe} of about twice the theoretical value, i.e., $h_{fe} \geqslant 60$.

The value of R_1 chosen is a compromise between one which is less than h_{oe}, and greater than h_{ie}. The frequency of operation and the output

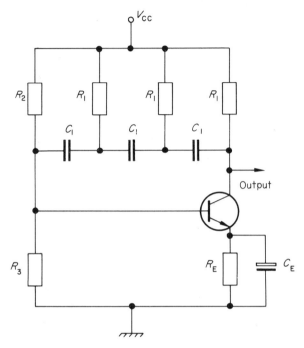

Fig. 10.8   A simple R-C phase-shift oscillator.

amplitude of this type of circuit are closely related to one another, and an alteration in the circuit loading significantly affects both. Amplitude control of the output signal can be obtained either by replacing R_E by a potentiometer, C_E being connected to the wiper, or alternatively by inserting an emitter follower stage between the oscillator proper and the load. In the former case, the effect of introducing a small unbypassed section of resistor in the emitter circuit increases the input impedance of the transistor amplifier and reduces the base current. In the latter case, the emitter follower presents an approximately constant load of high impedance to the oscillator, the emitter resistor being a potentiometer to allow the output voltage to be controlled.

Another form of R-C oscillator using two transistors and a ladder network is shown in Fig. 10.9. Since both transistors are connected in the common-emitter mode, the phase-shift between the base of Q1 and the collector of Q2 is theoretically zero. Clearly, instability occurs when the phase-shift introduced by the feedback network R_1C_1, R_2C_2 is zero. Resistor R_3 in the emitter circuit of Q2 ensures that the second stage has a high output impedance, resulting in Q2 appearing as an ideal constant-current source. The phase lead introduced at the oscillatory frequency by R_1 and C_1 is equal to the phase lag introduced by R_2 and C_2, ensuring the correct phase relationship for oscillatory

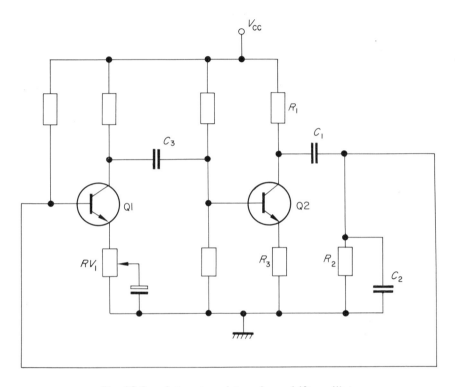

Fig. 10.9   A two-transistor phase-shift oscillator.

operation. Potentiometer RV_1 is used to control the amplitude of the oscillations; other resistors are required for bias purposes, and C_3 is an interstage coupling capacitor.

A block diagram of a *Wein-bridge oscillator* is shown in Fig. 10.10. This circuit follows the general principles of the oscillator in Fig. 10.8, in that two stages of amplification are employed which provide a phase-shift of 360 degrees (or 0 degrees), so that the feedback network must give 0 degrees of phase shift at the oscillatory frequency. In Fig. 10.10, components R_1C_1 introduce a phase lead at the oscillatory frequency which is equal to the phase lag introduced by R_2C_2. The circuit is known as a Wein-bridge oscillator because the configuration of the feedback network $R_1C_1R_2C_2$ resembles the frequency-sensitive Wein bridge. An analysis of the oscillator shows that the frequency at which the phase-shift between points A and B is zero is

$$f_0 = 1/2\pi\sqrt{(R_1 R_2 C_1 C_2)}$$

This equation also applies to Fig. 10.9. If $R_1 = R_2 = R$ say, and $C_1 = C_2 = C$, say, then

$$f_0 = 1/2\pi RC$$

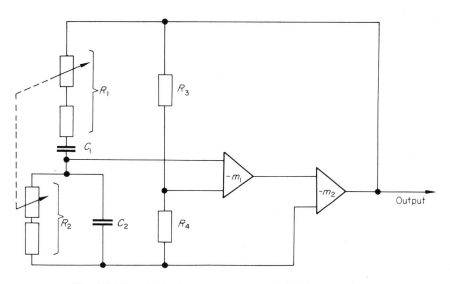

Fig. 10.10   A block diagram of a Wein-bridge oscillator.

It can also be shown that, for instability, the forward gain (= product $m_1 m_2$) should have a numerical value of 3, and that $R_3 = 2R_4$. In all unstable systems, the magnitude of the output oscillations tends to build up until it is limited by the non-linear characteristic of a circuit element. To bring about this limitation in valve circuits, R_4 is generally a tungsten-filament lamp; the resistance of the lamp increases with flow of current, so that as the r.m.s. amplitude of the output signal tends to rise it causes the ohmic value of R_4 to rise, which in turn reduces the input signal to the first amplifier. This causes the output voltage to be maintained at a constant level. Amplitude stability is generally achieved in semiconductor circuits by using a thermistor for R_3.

In a variable frequency oscillator, R_1 and R_2 each consist of a fixed resistor in series with a variable resistor, the variable sections being ganged together. If the fixed resistors have a value of 1·8 kΩ, say, and the variable sections are 20 kΩ each, then with a given value of capacitance in the bridge the frequency range covered by one complete sweep of the resistors is (20 + 1·8) : 1·8 = 12·1 : 1. Thus, if $C_1 = C_2 = 0·5\ \mu F$, the minimum frequency is 14·6 Hz and the maximum frequency is $12 \cdot 1 \times 14 \cdot 6 = 177$ Hz. By changing the values of C_1 and C_2, the frequency range can readily be altered.

10.7   Another look at unstable operation

So far, we have regarded an oscillator as a regenerative amplifier in which the input signal is derived from its own output. It is, however, possible to regard oscillators in other ways.

When a closed-loop amplifier becomes unstable, an oscillatory current flows in the output circuit even if no external signal is applied. For this condition to exist, it is clear that the apparent impedance of the circuit must be zero to allow current flow with zero externally applied voltage. That is to say, the active device in the circuit appears as an impedance with a negative value, which is equal to the impedance of the external circuit.

10.8   Radio-frequency oscillators

At radio frequencies L-C circuits have advantages over R-C circuits; a number of typical r.f. oscillator circuits are shown in Fig. 10.11. The circuit in Fig. 10.11(a) is a common-emitter version of *Colpitt's oscillator,* named after its valve equivalent. Capacitor C_f is a decoupling capacitor, and base bias is supplied by a resistive network. Both common-base and common-collector versions of the circuit exist. Using the ideas developed in section 10.6, it can be shown that the oscillatory frequency of the Colpitt's oscillator is

$$\omega_0 = \sqrt{\left[\frac{1}{L}\left(\frac{1}{C_1} + \frac{1}{C_2}\right)\right]} \text{ rad/s}$$

and

$$f_0 = \omega_0/2\pi$$

The circuit in Fig. 10.11(b) is the common-emitter version of the *Hartley oscillator,* in which the feedback is derived from a tapped coil. The frequency of oscillation of the Hartley circuit output is

$$\omega_0 \simeq 1/\sqrt{[(L_1 + L_2 + 2M)C]} \text{ rad/s}$$

The choice between Colpitt's and Hartley circuits in any given application is often a question of the advantages or disadvantages of the use of a tapped coil when compared with a two-terminal coil. In both circuits, amplitude stability is maintained by the bias arrangement, and the frequency stability is dependent on the parameters of both the resonant circuit and the amplifier. The secondary effects of temperature variation, and other variable factors result in a variation of output frequency of a few percent. If the frequency is to be accurately

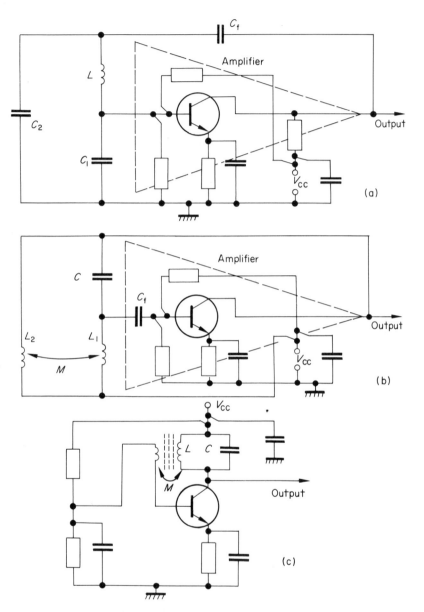

Fig. 10.11   (a) A Colpitt's oscillator, and (b) a Hartley oscillator.
A tuned-collector oscillator is shown in (c).

stabilized, then a crystal-controlled oscillator (see section 10.11) should be used.

Another form of circuit, which uses a tuned collector load, is shown in Fig. 10.11(c). Here, the input signal is derived by the mutual coupling M which exists between the input and output circuits, the interconnections being such as to provide positive feedback. Alternatively, the base circuit can be tuned, the drive being obtained. by mutual coupling with the output circuit.

10.9   Class C bias circuits

In the oscillators so far considered, a fixed bias was employed and the circuits may be regarded as class A oscillators. The current flowing in the load circuit of a class C amplifier is zero for more than one-half of the time, and for this reason the main bias circuit is generally not included in the emitter of a transistor, or in the cathode of a valve. One method of developing a large enough bias voltage to ensure class C operation in valve circuits is by means of *grid leak bias*. A tuned anode oscillator employing grid leak bias is shown in Fig. 10.12.

When the circuit in Fig. 10.12 is first switched on, capacitor C_G is uncharged and the grid bias is zero. Oscillations of current commence in the anode circuit, inducing a voltage in the grid circuit. During positive half-cycles of the grid circuit waveform, the grid-cathode region of the valve is forward biased and grid current flows. This charges capacitor C_G with the polarity shown. Since a discharge path is provided through R, the voltage developed across C_G falls slightly when the valve is cut off, and rises slightly when it is conducting. As a

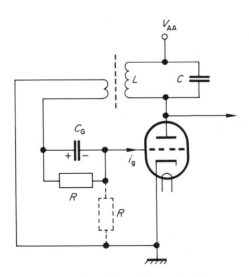

Fig. 10.12   Class C bias circuits.

result of this process, a unidirectional bias voltage is developed across capacitor C_G.

The value of the grid time constant RC_G is very critical to the operation of class C oscillators, since too small a value results in a small bias voltage, and too large a value allows the bias voltage to continue to build up in successive cycles. The reason for the latter is as follows. In the charging period, the current is restricted only by the input impedance of the device used in its conducting mode (which is about 1 kΩ in the case of valves), resulting in a fairly rapid build-up of bias voltage. In the non-conducting period, the capacitor discharge is controlled by the value of R. If this has a large value, then the voltage across the capacitor falls at a very low rate. The bias then continues to build up in succeeding cycles, with a consequent reduction in amplifier gain as the working point on the characteristics changes. Ultimately, a condition will be reached when, in oscillators, the loop gain falls below unity and oscillations cease. When this occurs the input signal falls to zero, and the bias voltage falls relatively slowly, its rate depending upon the time constant of the bias circuit. This results in periodic bursts of oscillations, and is known as *squegging*. In this mode of operation, the charging time constant is much smaller than the discharge time constant, and the amplitude of each burst builds up more rapidly than it decays.

Upon initial switch-on, the bias voltage in the circuit is zero and the load current can have a value which may exceed the rating of the device. In such cases, a circuit breaker may be connected in the supply line to protect it. It is also an advantage in these cases to include a cathode bias circuit, as in class A amplifiers, as a protection against excessive current.

A feature of the parallel R-C bias circuit in Fig. 10.12 is that only a small amount of the signal voltage appears across resistor R, since it is shunted by C_G. As an alternative arrangement, the resistor can be connected between the grid and cathode of the valve (shown by the dotted connection in Fig. 10.12), but this circuit requires more power to drive it since practically all the input signal voltage is developed across the grid leak resistor.

10.10   Negative resistance oscillators

It was shown in section 10.7 that oscillations could occur if the device used in the amplifier appeared as a negative resistance. A few devices have an inherent negative resistance region in their characteristics, and do not require feedback techniques to make the output oscillatory. One semiconductor device with this type of characteristic is the tunnel diode; a thermionic device with a negative slope resistance at one point on its characteristics is the tetrode.

The *dynatron oscillator* employs a tetrode with a parallel L-C circuit in its anode, the resistance of the parallel circuit at resonance being L/CR, where R is the resistance of the coil. By adjusting the operating conditions of the tetrode until it works at a point on the output characteristics at which $r_a = - L/CR$, the circuit is forced into unstable operation. The circuit has the merit of being extremely simple to construct, but suffers from the disadvantage that the secondary emission effects vary over the life of the valve, and also vary between valves of the same type. As a result, dynatron oscillators are rarely used in commercial oscillators.

The suppressor grid of a pentode can be used as a control electrode, since its effects are generally similar to those of any control grid. Since the cathode current of the pentode is substantially constant, an increase in the (negative) voltage applied to the suppressor grid causes the screen grid current to increase, thereby causing the screen grid potential to fall (assuming a resistive connection to the power supply). By introducing capacitive coupling between the suppressor grid and the screen grid, conditions for positive feedback are achieved. This results in a negative resistance characteristic between the suppressor grid and the cathode over a range of anode voltages. Thus, by using the suppressor grid as the output electrode, and by coupling a parallel L-C circuit between the suppressor grid and a negative bias rail, the output voltage at the suppressor is oscillatory. This type of oscillator is known as a *transitron oscillator.*

Another type of oscillator which uses the inherent Miller feedback effect of the triode is known as a *tuned-anode tuned-grid oscillator.* Miller feedback (see section 1.12.1) between the anode and grid of a triode is due to the grid-anode capacitance of the valve. If the anode load of the valve is resistive, then the variation in anode voltage is antiphase to the variation in input voltage, and the signal fed back to the grid circuit via the Miller capacitor is at 90 degrees to the input voltage. This modifies the input impedance of the valve, but it does not make it negative. However, with an inductive anode load, the phase relations in the circuit are such that the signal fed back to the grid circuit has a component which is in phase with the input signal. Conditions for positive feedback now exist, resulting in a negative resistance characteristic between the grid and the cathode. In this type of oscillator, the tuned circuit which controls the oscillatory frequency is connected between the grid and the cathode, so that the negative input impedance of the valve compensates for the dynamic resistance of the circuit at resonance. To provide an inductive anode load, a parallel L-C circuit which has a resonant frequency slightly greater than that in the grid circuit is used (see also Fig. 10.13). In order to prevent spurious operation at any frequency other than that of the resonant frequency of the grid circuit, the two parallel circuits are isolated from one another to prevent any form of mutual coupling.

It has already been shown that the *tunnel diode* has a negative resistance over a restricted range of forward bias voltages (between about 0·1 V and 0·3 V). By connecting the diode in series with a d.c. power supply and a parallel L-C circuit, an oscillatory condition is sustained when the circuit reactance is zero and the net circuit resistance is negative.

10.11 Crystal-controlled oscillators

Certain crystalline substances exhibit what is known as the *piezo-electric effect,* which results in either a mechanical strain in the crystal in the form of elongation or contraction when an electrical signal is applied to opposite faces of the crystal, or a mechanical stress produces a p.d. between opposite faces of the crystal.

The crystal acts as an electrical resonant circuit, the frequency of oscillations being dependent on the substance used and on the way in which the crystal slice is cut and mounted. The most popular substance for oscillator crystals is

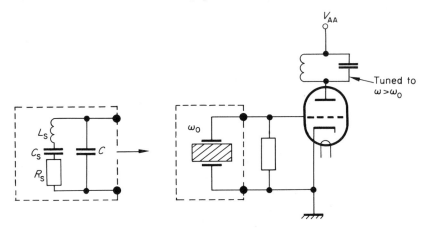

Fig. 10.13   One form of crystal controlled oscillator.

quartz, since it is less affected by moisture and temperature change than other substances, but it has a rather lower piezo-electric effect than some other crystalline materials such as Rochelle salt. When the crystal is set into mechanical oscillation, it has the equivalent electrical circuit shown in the inset in Fig. 10.13, where L_S, R_S, and C_S represent the inertia, friction losses, and stiffness of the crystal and its mounting, while C is the self-capacitance of the crystal. The resistance R_S may have a value up to about 100 Ω, but the other component values are such that the Q-factor of the crystal may be 10^5. To achieve a very high frequency stability, the crystal and its mount may be enclosed in a temperature-controlled oven, enabling a frequency stability of the order of 1 part in 10^{10} to be obtained.

Since the equivalent circuit of the crystal contains three reactive components, it has two resonant frequencies, the lower resonant frequency ω_S being series-type, the upper resonant frequency ω_P being parallel-type. The two frequencies are given by the following equations:

$$\omega_S = \frac{1}{\sqrt{L_S C_S}} \qquad \omega_P = \sqrt{\left[\frac{1}{L_S}\left(\frac{1}{C_S} + \frac{1}{C}\right)\right]}$$

Generally speaking, C has a much greater value than C_S, and both resonant frequencies are very close to one another.

Oscillators can be designed to use either the series resonant frequency ω_S, or the parallel resonant frequency ω_P, or the narrow band of frequencies ($\omega_P - \omega_S$) in which the crystal appears to behave as an inductance. One popular circuit using the parallel resonant frequency is the crystal-controlled valve oscillator shown in Fig. 10.13 which is a version of the tuned-grid tuned-anode oscillator. Many forms of crystal-controlled transistor oscillator are also used; an example of one type is a Colpitt's oscillator with the inductor replaced by a crystal.

10.12 Oscillator frequency stability

In order that the frequency of oscillation is maintained at a stable value, it is important to ensure that the effects of temperature variation, load variation, etc., are minimized. A change in frequency can result from many factors including (1) a change in the parameters of the tuned circuit, (2) a change in valve or transistor parameters, (3) variation in loading conditions, and (4) variation in supply voltage.

For good frequency stability, the components of the tuning circuit should be mechanically stable; where a very stable frequency is required a crystal-controlled oscillator is used. To isolate the effects of loading in a high stability oscillator, a 'buffer' amplifier is used between the oscillator and the load. It is also necessary in class A oscillators to stabilize the operating point on the characteristics, since a change in the quiescent point gives rise to frequency variation, and may affect the output amplitude and waveshape.

10.13 Multivibrators

Multivibrators are a class of electronic *switching circuits* which depend for their operation on regenerative feedback. They are also a class of *relaxation oscillators,* in which the active devices in the circuit are driven well beyond cut off for a period of time. The output waveform of all forms of relaxation oscillators is basically non-sinusoidal, and is generally in the form of either rectangular pulses or square pulses. The name 'multivibrator' is derived from the fact that a harmonic analysis of the output waveform shows that it is rich in harmonics, or multiple vibrations, of the fundamental frequency. There are three main types of multivibrator:

1. The *monostable multivibrator* or *one-shot circuit,* which has one stable operating state and one quasi-stable state. The application of an input signal triggers the circuit into its quasi-stable condition, in which it remains for a period of time which is dependent upon the value of the circuit constants. After this period of time, the circuit conditions return to their stable state; the process is repeated upon the application of each trigger pulse.

2. The *astable multivibrator* or *free-running multivibrator*, which has two quasi-stable operating states, one following the other in succession.

3. The *flip-flop* or *bistable multivibrator,* which has two stable operating states. The application of a control signal (or signals) causes the output to change from one state to the other. The design of flip-flops is discussed in detail in chapter 12.

All three types include two stages, and at any one time one of the active devices is saturated and the other is cut off.

10.13.1   The monostable multivibrator

The most popular form of monostable multivibrator is illustrated in Fig. 10.14(a), with typical waveforms in Fig. 10.14(b). In the stable operating mode of the circuit, transistor Q2 is saturated, with v_{BE2} and v_{CE2} both having very small values; Q1 is cut off, having its emitter junction reverse biased and $i_{CE1} \simeq 0$. The circuit is triggered into its quasi-stable state either by turning Q1 on or by turning Q2 off. This is usually accomplished either by applying a positive-going pulse to input A, or by applying a negative-going pulse to input B or at the collector of Q2.

Prior to the application of a trigger pulse, capacitor C is charged to V_{CC} with the polarity shown, since the base of Q2 is approximately at earth potential and the collector of Q1 is at $+V_{CC}$ (since $i_{CE1} = 0$). When a positive-

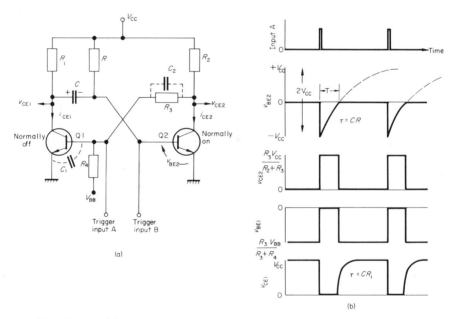

Fig. 10.14   (a) A popular form of monostable multivibrator circuit, and (b) typical waveforms.

going pulse of sufficient amplitude is applied to input A, it drives Q1 into saturation and effectively connects the left-hand plate of capacitor C to earth. As a result, the base of Q2 is driven to $-V_{CC}$ by the charge held on C. This reverse biases the emitter junction of Q2 and its base current and collector current fall to zero. Capacitor C now has to discharge through R; since one end of R is at $+V_{CC}$ and the other end is taken to $-V_{CC}$ when the trigger pulse is applied, the voltage across R at this instant is $2V_{CC}$. The voltage at the base of Q2 thus rises in an exponential fashion from $-V_{CC}$ towards $+V_{CC}$ with a time constant of CR. As v_{BE2} approaches zero, Q2 begins to conduct and its collector potential falls rapidly to zero. This drop in potential is communicated to the base of Q1 by the feedback network R_3R_4, causing the emitter junction of Q1 to be reverse biased. This forces Q1 to be cut off, and the circuit resumes its stable operating state. The period of time that the circuit remains in its quasi-stable state is given by T (Fig. 10.14(b)), which can be determined from the time taken for the voltage across the capacitor to change from $-V_{CC}$ to zero, when

$$V_{CC} = (2V_{CC})e^{-T/RC}$$

or

$$T = RC\ln 2 \simeq 0\cdot693RC \tag{10.1}$$

Thus, if $C = 0\cdot01~\mu F$ and $R = 15~k\Omega$ (see example which follows), the pulse width is

$$T = 0\cdot693 \times 0\cdot01 \times 10^{-6} \times 15 \times 10^3 = 104 \times 10^{-6}~s$$

$$= 104~\mu s$$

The rate of rise of the collector voltage is constrained by the charging time constant R_1C, and the waveform at the collector of Q1 does not have a very good rise time. For this reason, the output from Q1 is generally not used. It can be improved by modifications to the circuit, one example of which is described later in this section.

In practice, the waveforms at the base and collector of Q2 differ very slightly from those shown in Fig. 10.14(b) since, when Q2 turns on, the base potential of Q2 rises rapidly to a positive potential equal to V_{BE2sat}. At the same instant, the collector potential of Q2 falls to V_{CE2sat}, which has a value of about $0\cdot2$ V.

A factor which has been overlooked so far is the effect of the input capacitance C_1 of the transistor on the switching performance. When Q2 is suddenly turned off, the rate of rise of the voltage at the base of Q1 is restricted by the charging time constant of the circuit containing C_1. To compensate for this delay, resistor R_3 is shunted by capacitor C_2, known as a *speed-up capacitor*, which has a value of a few hundred picofarads.

To estimate the values of the components in the circuit it is necessary to know the supply and bias voltages, the transistor parameter h_{FEsat}, the output pulse duration, and the maximum collector current. Let us now consider a design in which $V_{CC} = 10$ V, $V_{BB} = -6$ V, $h_{FEsat} = 50$, and the maximum value of i_{CE} is 10 mA. Under steady operating conditions, Q2 is saturated, and we may assume that $v_{CE2} \simeq 0$ and $v_{BE2} \simeq 0$. This gives a maximum collector current of

$$i_{CE2(max)} \simeq V_{CC}/R_2$$

or

$$R_2 = V_{CC}/i_{CE2(max)} = 10/10 = 1 \text{ k}\Omega \tag{10.2}$$

In this condition,

$$i_{BE2(max)} = i_{CE2(max)}/h_{FEsat} = 10/50 = 0.2 \text{ mA}$$

This is the minimum value of current which must flow through R to hold Q2 in the saturated state, hence

$$R \leqslant V_{CC}/i_{BE2(max)} \tag{10.3}$$
$$= 10/0.2 = 50 \text{ k}\Omega$$

In modern transistors, the base current can approach that in the collector circuit, so that R can have any value between about 5 kΩ and 50 kΩ. Suppose we let $R = 15$ kΩ.

When Q2 is saturated, the resistor chain $R_3 R_4$ must apply a reverse bias to the base of Q1 to hold it in the non-conducting condition. With a bias supply voltage of -6 V, this could mean a reverse bias of about -1.5 V at the base of Q1. Since the collector of Q2 is approximately at earth potential,

$$v_{BE1} \simeq \frac{R_3}{R_3 + R_4} V_{BB} \tag{10.4}$$

therefore,

$$-6 \frac{R_3}{R_3 + R_4} = -1.5$$

or

$$R_4 = 3R_3$$

After the application of a trigger pulse, the operating conditions of the two transistors change so that Q1 becomes saturated and Q2 is turned off, forcing i_{CE2} to fall to zero. Resistor chain $R_2 R_3 R_4$ must now provide sufficient current to saturate Q2. The conditions for Q1 are now similar to those for Q2 under which eq. (10.2) was computed, hence

$$R_1 = V_{CC}/i_{CE1(max)} \tag{10.5}$$

and

$$i_{BE1(max)} = \frac{i_{CE1(max)}}{h_{FEsat}} = \frac{V_{CC}}{R_1 h_{FEsat}} \tag{10.6}$$

Assuming that both transistors are generally similar, then the maximum base current in Q1 is 0·2 mA. In order to satisfy the reverse bias conditions in eq. (10.4) as well as being capable of supplying the maximum base drive given by eq. (10.6), the resistor chain $R_2 R_3 R_4$ must carry a current which is greater than $i_{BE1(max)}$. If this current is twice $i_{BE1(max)}$, then

$$\frac{V_{CC} - V_{BB}}{R_2 + R_3 + R_4} \geq 2 i_{BE1(max)} \tag{10.7}$$

or

$$\frac{10 - (-6)}{1 + R_3 + R_4} \geq 0·4 \text{ mA}$$

hence,

$$R_3 + R_4 \leq 39 \text{ k}\Omega$$

but, from eq. (10.4)

$$R_4 = 3R_3$$

hence,

$$R_3 \leq 9·75 \text{ k}\Omega \qquad\qquad\qquad \text{say } 8·2 \text{ k}\Omega$$

and

$$R_4 = 3R_3 = 24·6 \text{ k}\Omega \qquad\qquad \text{say } 22 \text{ k}\Omega$$

In the quasi-stable state, when Q2 is cut off, the output voltage is given, approximately, by the relationship

$$v_{CE2(max)} \simeq \frac{R_3}{R_2 + R_3} V_{CC} \tag{10.8}$$

$$= \frac{8·2}{1 + 8·2} 10 = 8·9 \text{ V}$$

The application of an external load will, of course, reduce the output voltage below this level.

A simple variation which is frequently used to improve the rise-time at the collector of Q1 is shown in Fig. 10.15(a), and is known as the Rozner modification. The circuit includes a diode and resistor R_1' in addition to the normal circuit components. The diode effectively isolates the recharging exponential from the collector waveform of Q1, as follows. At the end of the quasi-stable condition, the emitter junction of Q1 is reverse biased and its collector potential begins to rise; this causes the diode to become reverse biased, so isolating capacitor C from transistor Q1, and the capacitor then recharges through R_1'. If an external load resistor R_L with a high value is connected to the collector of Q1, the output waveform will be a clean square wave with a small rise-time, as shown in Fig.

10.15(b). With a low value of load resistance, the collector potential of Q1 rises rapidly to $V_{CC}R_L/(R_L + R_1)$ until the capacitor has charged to this value. The diode then begins to conduct and the final waveform and voltage level is as shown in Fig. 10.15(b).

Valve circuits are similar in most respects to their transistor equivalents, with some minor variations which are now described. In the valve circuit described here, Q1 is replaced by V1, and Q2 by V2, otherwise the circuit is generally similar to Fig. 10.14. If the cut-off bias of V1 is $-V_K$, then under stable operating conditions the grid bias applied to V1 exceeds this value. Assuming that both valves are similar, and that capacitor C is fully charged,

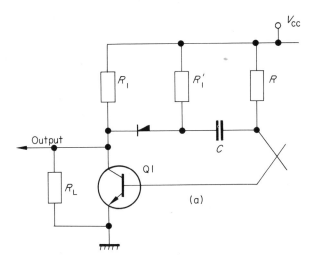

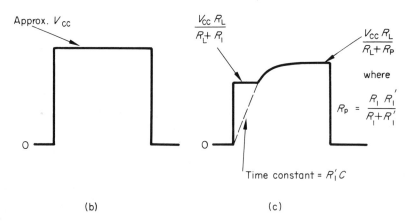

Fig. 10.15   The Rozner modification to give an improved output waveform.

then the lowest steady value of output voltage is equal to V_L (see Fig. 10.16(a)) when $v_{GK} = 0$. This is the steady output voltage at the anode of V2. Upon the application of a trigger pulse to the circuit, the grid voltage of V2 is depressed by an amount equal to V_S, shown in Fig. 10.16(b), when the voltage across R rises to $(V_S + V_{AA})$. The grid voltage then rises exponentially with a time constant of $(R + R_A)C$, where R_A is a source resistance comprising R_1 in parallel

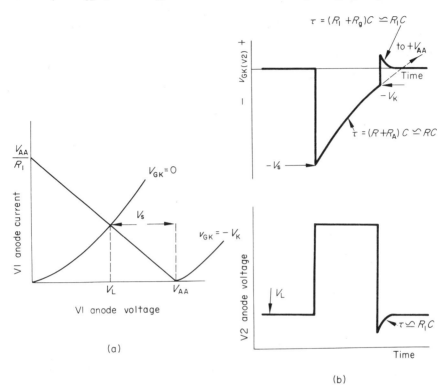

Fig. 10.16   Waveforms in a valve monostable multivibrator circuit.

with the d.c. resistance of V1. In most circuits, $R \gg R_A$, and the time constant associated with v_{GK2} is approximately equal to RC. The grid voltage continues to rise until v_{GK2} reaches the cut-off voltage $-V_K$ of V2, after which time regenerative feedback takes place, and the circuit returns to its stable state. When this occurs, the grid potential of V2 rises rapidly to a small positive potential, which is limited by the conducting resistance ($\simeq 1\ k\Omega$) of the forward biased grid-cathode region of V2. The net result is that C recharges through a circuit with a time constant of $(R_1 + R_g)C$, where R_g is the slope resistance of the forward biased grid-cathode region of V2. Since $R_1 \gg R_g$, the time constant of this part of the waveform is R_1C, as shown in Fig. 10.16(b). Consequently, there is a small overshoot on the grid voltage waveform, and a corresponding undershoot on the anode voltage waveform. The periodic time of the waveform is computed in a similar manner to that for the transistor as follows.

$$V_{AA} + V_K = (V_{AA} + V_S)e^{-T/\tau}$$

where

$$\tau = RC$$

whence,

$$T = \tau \ln(V_{AA} + V_S)/(V_{AA} + V_K)$$

Thus, if

$$V_{AA} = 300 \text{ V}, \quad V_S = 200 \text{ V}, \quad V_K = 8 \text{ V}, \quad \text{and} \quad \tau = 500 \text{ } \mu\text{s},$$

then

$$T = 79 \text{ } \mu\text{s}.$$

10.13.2   The astable multivibrator

There are many forms of astable multivibrator, and we restrict ourselves to the most popular form which is shown in Fig. 10.17(a). It consists of two stages which are generally similar to the first stage of the monostable multivibrator in section 10.13.1. The d.c. design follows the same general pattern as that of the monostable circuit, allowing the values of R_1, R_2, R_3, and R_4 to be calculated.

The operation of the circuit is as follows. When Q2 is saturated, the charge on C_2 causes Q1 to be cut off until v_{BE1} rises to zero, when Q1 saturates. The drop in potential at the collector of Q1 then causes Q2 to be cut off. The circuit remains in the second quasi-stable state until v_{BE2} rises to zero, when Q2 saturates and the circuit is returned to the first quasi-stable state. The process is then repeated indefinitely. The period of time for which Q1 is cut off is given by T_1 in Fig. 10.17(b); T_2 is the period of time for which Q2 is cut off. Their approximate values are

$$T_1 \simeq 0{\cdot}693R_4C_2 \qquad T_2 \simeq 0{\cdot}693R_3C_1$$

The period of time for one complete cycle is $T = T_1 + T_2$, and the frequency is

$$f = 1/T \text{ Hz}$$

For example, if $R_3 = R_4 = 15$ kΩ, and $C_1 = C_2 = 0{\cdot}01$ μF, then the periodic time for one cycle is 208 μs, and the frequency is 4$\cdot$81 kHz. The ratio T_1/T_2 is known as the *mark-to-space ratio* or the on-to-off ratio, and it has unity value when $R_3 = R_4$ and $C_1 = C_2$. Accurate control over the mark-to-space ratio is generally difficult to realize with 5 or 10 per cent tolerance components. One method of achieving a given mark-to-space ratio is to connect the top ends of R_3

and R_4 to opposite ends of a potentiometer, with the wiper taken to V_{CC}. Adjustment of the wiper position alters the net values of resistance in the two collector circuits, giving a degree of control over the mark-to-space ratio. However, the simple circuit in Fig. 10.17(a) is unsuitable for mark-to-space ratios greater than about 10:1, and other circuits must be used.

It is clear from Fig. 10.17(b) that the output waveforms at both collectors are not good square waves, and are unsuitable for use in circuits where a short

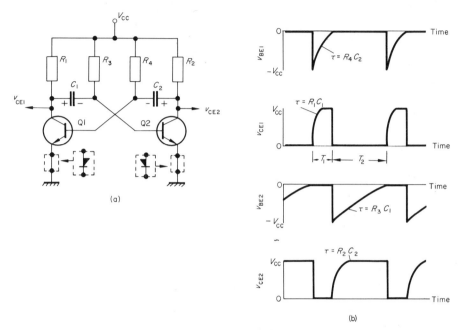

Fig. 10.17   (a) One form of astable multivibrator circuit, and (b) typical waveforms.

rise-time is required. A simple method of improving the rise-time is by using the Rozner modification (Fig. 10.15(a)) in both collector circuits.

Since it almost impossible to guarantee two identical halves of the circuit in Fig. 10.17(a), the circuit will almost certainly be self-starting, that is to say the output will be oscillatory. However, the circuit is designed to work so that the transistors are driven heavily into saturation. In this state of operation, the current gain of transistors is low, and upon switching it is just possible that the transistors become so heavily saturated that the loop gain is less than unity, and unstable conditions are not achieved. In cases of this kind, it is necessary to take steps to ensure self-starting*.

* See the paper by G. H. Stearman 'Transistor astable multivibrators for general purpose logic elements' in *Electronic Engineering,* Dec., 1965.

Where the peak negative base-emitter voltage at the instant of switching is large enough to damage the emitter junction of the transistor, a diode can be connected in series with the emitter (shown in the insets in Fig. 10.17(a)) as a means of inverse voltage protection.

10.13.3   Synchronizing the astable multivibrator

The simple astable multivibrator is a *free-running oscillator,* and its output frequency is dependent upon the circuit constants. If accurate control over the frequency is required, it is necessary to provide a means of synchronizing the multivibrator output with an accurate frequency source. A common method of achieving this end is shown in Fig. 10.18, in which a series of positive-going pulses are applied to the base of one of the transistors. Whenever the externally applied synchronizing pulse forces the base voltage to become positive (in n-p-n transistors), the transistor is switched to its saturated state. As a result, the output duration is T_1', compared with T_1 in the free-running circuit. By this means, the multivibrator may be used as a frequency divider.

For accurate frequency stability, the frequency of the synchronizing signal should not be greater than about ten times the natural frequency of the multi-

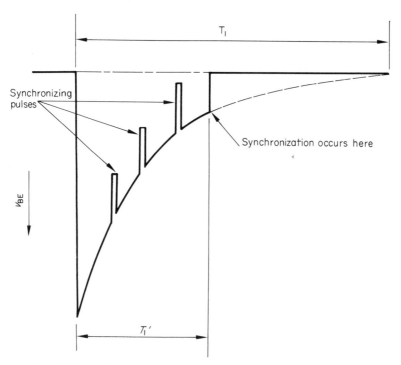

Fig. 10.18   A simple method of synchronizing multivibrators.

vibrator, and the peak of each pulse which is superimposed on the bias voltage should rise above the commencing point of the following pulse by about 20 per cent of the amplitude of the pulse. A smaller trigger pulse (or a larger one for that matter) may cause loss of synchronization.

10.14   The blocking oscillator

An *astable blocking oscillator* or free-running blocking oscillator is shown in Fig. 10.19(a), together with its output waveform in Fig. 10.19(b). It is a single-transistor oscillator with a pulse transformer as the feedback element.

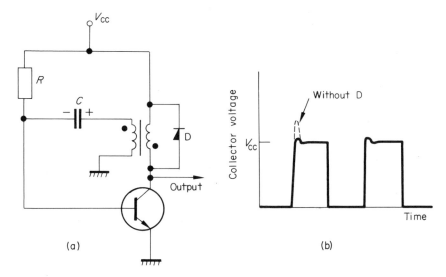

Fig. 10.19   One form of astable blocking oscillator.

The interesting characteristics of the pulse transformer are that the inductances of both the primary winding and the secondary winding are large, and that the mutual coupling coefficient between the windings is also large. In addition, the self-capacitance of the windings is such that a rapid change of current in the winding causes the voltage across it to 'ring' or rapidly oscillate.

Initially, the base current supplied through R causes a collector current to flow, with a consequent reduction of collector voltage and increase in the voltage across the primary winding of the pulse transformer. The transformer secondary winding is so connected that the induced voltage due to the change in primary voltage causes the base current to increase. This in turn causes the collector current to increase, with consequent further increase in base current. The regenerative feedback rapidly forces the transistor into saturation, when the collector voltage becomes approximately zero. In this event, the primary winding of the transformer supports the whole of V_{CC} across it, and the current through it increases at a constant rate. During the regenerative period, the direction of the current through the capacitor causes it to charge with the polarity indicated. The collector current continues to increase until it is limited by some circuit parameter (e.g., current gain). This causes the voltage fed back to fall to zero. The charge on the capacitor forces the transistor to turn off, the rate of which it turns off being accelerated by transformer feedback from the collector.

When the collector current falls to zero, the voltage across the primary winding 'rings', giving the voltage overshoot which is shown by the broken line in Fig. 10.19(b). The amplitude of this overshoot is limited by diode D, which becomes forward biased when the primary winding voltage exceeds V_{CC} in the transient period. The diode may, alternatively, be replaced by a resistor.

Once the transistor has been cut off, the output voltage remains at V_{CC} until the charge on C has decayed through R to a point where conduction and regeneration again occur.

Synchronization of the astable blocking oscillator can be achieved in much the same way as in the astable multivibrator, described in section 10.13.3. *Monostable blocking oscillators* are less common than astable circuits, but are often encountered in thyristor and triac control units.

In the design of any transistor blocking oscillator, the transistor must (1) have a cut-off frequency capable of dealing with the desired rise-time and fall-time of the output pulse, (2) be capable of handling the load power, and (3) have an adequate collector voltage rating.

Valve versions of both the astable circuit and the monostable circuit are used, the advantage of valve circuits being that the limitations on the anode voltage swing are minimal.

Problems

10.1   Describe the operation of (a) an emitter follower circuit, and (b) a cathode follower circuit, and enumerate features of importance. Suggest practical applications of both types of circuit.

10.2   Show that the output resistance of a cathode follower which employs a valve with $g_m = 5$ mA/V is approximately 200 Ω, and that the input impedance is high.

10.3   Sketch a connection diagram for *either* a triode amplifier *or* a transistor amplifier with voltage negative feedback. Label (a) input and output connections and (b) the source of the voltage which is fed back. Explain why the phase of this feedback voltage will be in opposition to the input signal.

Indicate how the fraction of the voltage fed back may be estimated from the component values.

A voltage amplifier with a gain of 100 has 1/50th of the output fed back in opposition to the input. Find the new overall gain.

(C & G)

10.4   If the voltage gain of a resistance loaded pentode amplifier without feedback is $-m$, show that the voltage gain of the amplifier in Fig. 10.5 is

$$-m/(1 + R_1(1 + m)R_f)$$

10.5 Draw circuit diagrams of *two* different types of resistance-capacitance oscillators and describe the basic action in each case. Compare and contrast the two circuits, particularly from the point of view of convenience of control. Explain why *R-C* oscillators have generally displaced *L-C* oscillators for the generation of audio-frequency test signals.

(C & G)

10.6 Draw the circuit diagram of a Colpitts oscillator and explain its operation. By what is the approximate frequency of oscillation determined?

10.7 Draw a circuit diagram for a tuned-grid oscillator, having a self-biasing circuit, and explain:

(a) How oscillations start and are maintained.
(b) The operation and purpose of the self-biasing circuit.

In such an oscillator the tuned circuit comprises a 100 mH coil in parallel with a 500 pF capacitor. Calculate the approximate frequency of oscillation.

(C & G)

10.8 Draw a circuit diagram and explain the operation of a single-valve oscillator producing a sinusoidal output of medium frequency. List the factors which affect the operating frequency.

What is meant by *grid-current bias* and explain why it is widely used in oscillator circuits.

(C & G)

10.9 Draw the circuit diagram and explain the operation of a tuned-circuit oscillator employing a transistor. Explain how the output can be obtained from the oscillator.

10.10 Show how a valve or transistor may be used to form a circuit element having a negative resistance characteristic.

10.11 Explain, with the aid of a circuit diagram and waveforms, the action of a free-running multivibrator.

State an industrial application.

(C & G)

10.12 Design an astable multivibrator using the principles outlined in this chapter using transistors with $h_{FE} = 100$, $V_{CC} = 6$ V, $I_{CM} = 5$ mA. The mark-to-space ratio is to be unity, and the periodic time is 100 μs.

10.13 Discuss the special problems in astable multivibrators of generating (a) square waves with very long periodic times (i.e. several seconds), and (b) a large mark-to-space ratio (say 100:1).

10.14 Explain, with the aid of waveform sketches, the basic action occurring in a blocking-oscillator type of pulse generator. What factors determine the duration and repetition rate (or recurrence frequency) of output current pulses? Give two possible applications of blocking oscillator, stating in each one reason why it is appropriate for the purpose given.

(C & G)

10.15 Explain in detail, with the aid of a diagram, the operation of a transistor one-shot blocking oscillator.

Draw the waveforms to be expected at the input, output and base and collector of the transistor, showing the relationship between them.

(C & G)

10.16 Draw a circuit diagram of *either* a Hartley *or* a Colpitts type oscillator using a transistor and briefly explain its operation.

List the factors affecting the frequency stability.

(C & G)

10.17 Sketch the anode current/anode voltage characteristic of a tetrode valve and from this derive curves showing the variation with anode voltage of the static and dynamic anode resistances.

Thence, or otherwise, explain how such a valve could be used as an oscillator and give a circuit diagram.

What factors affect the output power from such an oscillator?

(C & G)

10.18 By means of a circuit diagram, describe in detail how a transistor may be used in conjunction with a tuned circuit to generate oscillations. Explain the operation of the circuit chosen and the function of each component.

Explain the factors which determine the stability of the frequency of the oscillations.

(C & G)

10.19 Explain the principle of operation of a transistor blocking oscillator, using a circuit diagram to illustrate your explanation.

Discuss the various factors affecting the design of such an oscillator.

(C & G)

10.20 With the aid of graphs, explain briefly the effect of negative feedback on the frequency and phase response of a single-stage resistance-capacitance coupled amplifier.

A pentode amplifier has $\mu = 500$, $r_a = 54$ kΩ, and the anode load is 6 kΩ. The input impedance at a certain frequency is 10 kΩ. If 0·1 of the output voltage is fed back as negative feedback in series with the input, calculate (i) the gain with feedback, (ii) the output impedance, (iii) the input impedance, and (iv) the gain of the feedback amplifier if the gain *without* feedback falls by 6 dB.

(C & G)

11. Electronic measuring instruments

11.1 The cathode-ray oscilloscope

The cathode-ray oscilloscope (C.R.O.) is perhaps the most versatile measuring instrument available for use with electronic circuits. With it, it is possible to measure voltage, current, phase-angle, and a whole host of other quantities, and it gives electronic engineers another eye to 'see' inside the circuit itself. By utilizing a range of probes and plug-in units, measurements from microvolts to megavolts can be made at frequencies from zero (d.c.) to several gigahertz (10^9 Hz).

The rudiments of the C.R.O. are shown in Fig. 11.1; the heart of the C.R.O. is the *cathode-ray tube* (C.R.T.), in which a controlled beam of electrons causes an illuminated spot to appear on the *fluorescent screen* of the tube. By controlling the movement of the illuminated spot in both the X-direction and the Y-direction, waveforms and other curves can be traced out on the screen; in some oscilloscopes it is possible to modulate the brilliance of the spot, giving what is known as Z *modulation*. In most modern oscilloscopes, there is a great deal of sophistication, particularly in the amplifiers and circuitry to ensure that the trace or *oscillograph* is a true representation of the input signal or Y *input*.

The input signal is first either amplified or attenuated to give the correct voltage levels to the rest of the circuit; from this section of the circuitry, a *synchronizing signal* or *trigger signal* is derived which initiates the movement of the spot on the face of the tube. Alternatively, the oscillograph may be synchronized with some other external signal by switching S1 to the upper position. The synchronizing signal (*sync. signal*) is applied to the *timebase generator* which causes the spot on the face of the tube to be deflected in the X direction. It is usual in most oscilloscopes to provide switch S2, which allows the operator to choose the sign of the slope at the point at which the trace commences. In some applications, it is necessary to apply signals from an external source directly to the X amplifier; this is achieved by means of switch S3. When the X amplifier derives its input from the timebase generator, it traces out a *Y-time graph*, which is a graph of the Y input signal to a base of time. When the input to the X amplifier is derived from an external signal source, it operates as an *X-Y oscilloscope*, since the Y input signal is graphed to a base of the X input. In the X-Y mode, the oscilloscope can be used as an accurate means of measuring phase-shift and frequency.

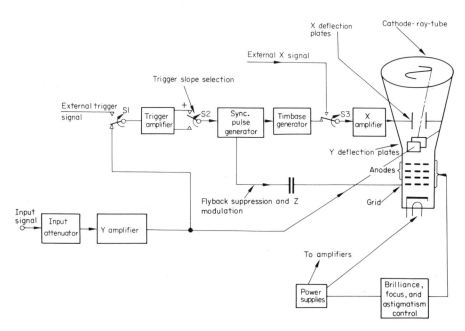

Fig. 11.1	A block diagram of a cathode-ray oscilloscope.

The power supplies to the grid and anode systems are derived from one or more power sources, and *brilliance, focus,* and *astigmatism* control of the illuminated spot are obtained by potentiometric adjustments. It is also necessary to provide *X shift* and *Y shift* controls to give the operator overall ability to position the trace at any part of the screen.

## 11.2	The cathode-ray tube (C.R.T.)

The C.R.T. is a funnel-shaped high vacuum valve with an *electron gun* at the end of the long neck of the valve, and a screen treated with a *phosphor* at the flared end. A schematic diagram of a C.R.T. for a typical instrument application is shown in Fig. 11.2. Between the cathode and the display is an array of electrodes which is used both to control the electron beam and to overcome any distortion which is introduced in the valve.

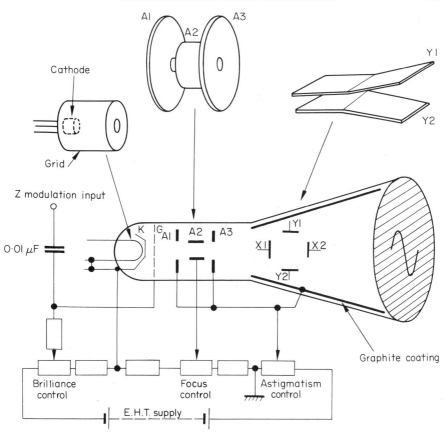

Fig. 11.2 The cathode-ray tube.

The cathode, which is in the form of a flat disc to give high electron emission from one face, is indirectly heated and is completely surrounded by a metallic control grid. The grid has a small hole in one end to allow electrons to escape in the longitudinal direction towards the anodes of the electron gun. As a general rule there are three anodes A1, A2, and A3, each at a positive potential with respect to the cathode. In simple terms, A1 and A3 are discs having a small centre hole, and A2 is cylindrical (possibly with internal diaphragms). The function of the anode system is twofold. Firstly, it accelerates the electron beam, and secondly it takes the convergent beam of electrons which emerge through the hole in the grid and forms it into a divergent beam by the time it has passed through the anodes. As a result of the latter function, this section is known as the *electron-lens system.* Focusing can be achieved by electromagnetic means, and is discussed in Section 11.3. The electron beam then enters the deflection system, where it is deflected first in the Y direction and then in the X direction by signals applied between pairs of specially shaped plates. Each pair of deflection plates is flared at the end to prevent the introduction of waveform distortion resulting from electrostatic fringing at the ends of the plates.

In order to ensure that the electron beam continues to travel at a constant longitudinal velocity when it leaves the final anode, the inside of the flared part of the tube is coated with a graphite conducting material known as *Aquadag*, which is maintained at the same potential as the final anode. The Aquadag layer also provides the return path for the current from the screen of the tube.

The general term used to describe the process of converting energy into light is known as *luminescence* (the special case of converting thermal energy into light is known as *incandescence*). When rays or beams of energy fall on certain substances they produce a form of luminescence known as *fluorescence*, which occurs when energy of one wavelength is absorbed by the substance and is emitted as light. In the case of the C.R.T., the cathode rays or beam of high speed electrons produce fluorescence in the film of material laid down on the face of the tube. Luminescence which continues for some time after the excitation energy has fallen to zero is known as *phosphorescence* or *after-glow*. After-glow is due to the release of stored energy in the form of light over a period of time. The *persistence* of the after-glow of a C.R.T. phosphor is generally defined as the time taken for the light output from the phosphor to decline to 10 per cent of the original intensity. Fluorescence and after-glow are always produced together, one generally predominating over the other (depending on the phosphor). The choice of phosphor depends upon the principal application of the C.R.O., a list of the more common phosphors being given in Table 11.1. In the table, short persistence is taken to be in the range 0·01 ms to 1 ms, medium

Table 11.1

Cathode-ray tube phosphor characteristics

Phosphor	Colour		Persistence	Application
	Fluorescence	After-glow		
P2	Yellow-green	Yellow-green	Long	General low speed applications
P4	White	White-yellow	Medium	High contrast displays
P7	Blue-white	Yellow-green	Long	Low speed applications, e.g., control systems
P11	Blue	Blue	Short	Fast writing rates, photographic applications
P31	Green	Green	Short	General purpose applications

persistence 1 ms to 100 ms, and long persistence 100 ms to several seconds; for example the P31 phosphor has a persistence of about 0·5 ms while the P7 phosphor has a persistence of about 3 s.

When the primary electron beam arrives at the screen, electrons are released from the phosphor by secondary emission. For a correct potential balance to exist on the screen, the number of primary electrons arriving at the screen must be equal to the net number leaving it. Thus, if m electrons arrive in a given time in Fig. 11.3(a), then m secondary electrons must be attracted to the Aquadag

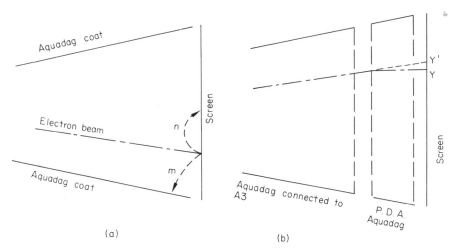

Fig. 11.3   (a) Return path for the screen current; (b) post-deflection-acceleration anode.

coat. It is possible that additional electrons may be released from the screen by secondary emission, leaving the screen with a net positive charge on it. To maintain the potential balance on the screen, the surplus electrons are attracted back to the screen by random routes, path n being an example in Fig. 11.3(a). The beam current is controlled by the brilliance control in Fig. 11.2, which controls the grid-cathode potential; if this control is set too high (particularly with a long persistence phosphor), it is possible to *burn* the face of the tube at the point of arrival of the beam. At high beam intensities, the spot on the screen is surrounded by a ring or halo of light, which is due to an excessive number of secondary electrons falling back on the screen to produce secondary illumination.

Despite the fact that the Aquadag coating is provided as a return path for the secondary electrons, there is a net accumulation of negative charge on the screen which leads to a reduction of brightness of the trace. One method adopted to overcome this problem is by backing the screen with a thin layer of aluminium which the electron beam can penetrate. This conducting layer is connected to the final anode, so preventing the formation of a negative charge on the screen. Such a tube is said to be *aluminized.*

One method which is sometimes adopted to increase *either* the brilliance *or* the deflection sensitivity is to use a *post-deflection-acceleration (p.d.a.) anode.* This is simply an Aquadag coat around the screen end of the tube which is isolated from the main Aquadag coat, and is maintained at a potential of 2 kV to 4 kV above the final anode. The net result, illustrated in Fig. 11.3(b), is that the electron beam is deflected. In the absence of the p.d.a. anode, the beam would reach the screen at point Y', but when p.d.a. is applied the additional longitudinal acceleration causes the beam to arrive at point Y. Clearly, if p.d.a. is to be used to increase the brilliance of the trace (due to the increased final velocity), then the deflection sensitivity is reduced since a smaller deflection results. On the other hand, for a given brilliance, p.d.a. allows a greater deflection sensitivity to be achieved since the final anode is run at a much lower voltage, allowing the beam to pass more slowly through the deflection system. In the latter case, the individual electrons stay under the influence of the deflecting voltage for a longer period of time, and suffer a greater deflection.

There are some interesting variations on the C.R.T. which are worthy of mention here. In some applications, it is necessary to study a series of simultaneous events, in which case multiple traces must be drawn out on the screen, and there are various ways of doing this. Firstly, the main beam can be divided into two by a *splitter plate,* each half of the beam then being processed by an independent electron-lens and Y deflection system (the X deflection system being common to the two beams). Secondly, two completely separate electron guns and Y deflection systems can be used. Thirdly, a control circuit can be used to cause a single beam to trace out a number of curves by switching each input signal to the Y amplifier in succession. By the latter means, it is not uncommon to have up to four independent traces on the face of the tube. While the C.R.O. is a versatile instrument, its measuring accuracy is usually not much better than 3 per cent, which is further reduced if the rectangular grid (called a *graticule*), which is placed in front of the tube for measuring purposes, is some distance from the screen. This introduces parallax errors. Where a high degree of accuracy is required, a tube with the graticule scribed on the inside face of the screen should be used.

The trace drawn out on the screen is subject to several forms of distortion peculiar to C.R.T.s, which include *trapezoidal distortion, pincushion distortion,* and *barrel distortion.* Trapezoidal distortion or trapezium distortion can be recognized as a change in deflection sensitivity as the spot sweeps across the screen, resulting in a signal of constant r.m.s. amplitude tracing out a trapezoidal area on the face of the screen. This defect is generally overcome by feeding opposite deflection plates (e.g., Y1 and Y2) from the outputs of a paraphase amplifier such as a long-tailed pair. Pincushion distortion and barrel distortion cause an otherwise rectangular display to be pincushion-shaped and barrel-shaped, respectively. One method of overcoming these defects is to place a shield between the X plates and the Y plates which is at the average potential of the two sets of plates.

11.3   Deflection systems and focusing systems

In the tube in Fig. 11.2, the beam is focused and deflected by electrostatic forces, and is of the type commonly used in electronic instruments. The velocity of the electrons leaving the electron gun is proportional to the final anode potential, and the greater the value of this voltage the smaller is the time that they dwell in the deflection system. Consequently, the deflection sensitivity is

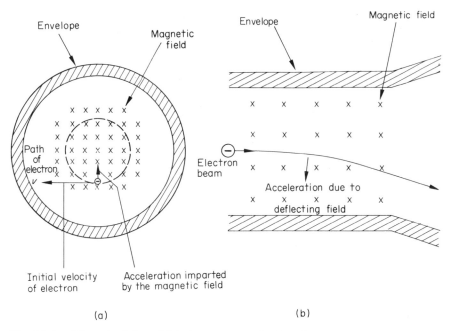

(a)   (b)

Fig. 11.4   The principles of (a) electromagnetic focusing, and (b) electromagnetic deflection.

inversely proportional to the final anode voltage, since the deflection is proportional to the length of time the electrons spend in the deflection system.

The principle of *electromagnetic focusing* can be understood by reference to Fig. 11.4(a). If a longitudinal magnetic field is set up along the axis of the tube by an electromagnet or a permanent magnet which is external to the tube, then electrons which travel in a direction parallel to the field (i.e., into or out of the page) do not experience any force to deflect them in one direction or the other, and will continue to travel along the tube in a straight line. However, in the general case, electrons leave the electron gun at an infinite variety of angles, with the result that they enter the longitudinal magnetic field with a transverse velocity v, illustrated in Fig. 11.4(a). Since this motion is perpendicular to the magnetic field, the electron experiences a force which is mutually perpendicular to the transverse motion of the electron and the direction of action of the magnetic field. In the case illustrated, the magnetic flux enters the page, and the electron experiences an upward force at the instant considered. A study of its motion at subsequent intervals of time reveals that the electron travels in a circular orbit so long as it remains in the magnetic field. The magnets commonly used for this purpose provide a stronger field near to the envelope than they do at the centre. As a result, electrons with a wide angle of divergence enter a stronger magnetic field than those with a small divergence, causing the former to be turned toward the axis more rapidly than the latter. This enables all electrons to be focused accurately, even with a wide range of angles of divergence.

The principle of *electromagnetic deflection* is much the same as that for electromagnetic focusing, except that the magnetic field is applied in a transverse direction *across* the neck of the tube, as shown in Fig. 11.4(b). After the electron leaves the focusing system, it enters the transverse magnetic field and is deflected, the deflection being dependent on the initial velocity of the electron and the current flowing in the deflecting coils.

Electromagnetic focusing is fundamentally more satisfactory than electrostatic focusing, since the electrons approach the screen from a variety of angles. In electrostatic focusing, the electrons approach the screen in the form of a beam, and the mutual repulsion between the electrons can cause defocusing. Electromagnetic deflection allows a much greater angle of deflection to be obtained without defocusing than is the general case with electrostatic deflection. However, electromagnetic deflection systems cannot handle high frequency signals (higher than about 20 kHz) due to the inductance of the coils. In some applications, notably domestic television tubes, a hybrid system with electrostatic focus and electromagnetic deflection is used.

11.4   Power supplies and controls

Owing to the wide variety of equipment used in oscilloscope circuits, a number of separate power supplies is required. The main transformer usually has several 6·3 V windings on it to supply energy to the heater circuits of the rectifier valves and amplifier valves. The C.R.T. heater is supplied from an isolated winding, so that it can be directly coupled to the cathode of the tube, thus avoiding applying an excessive voltage to the insulation which separates them.

The final anode voltage may be anything from +1 kV to +20 kV or more with respect to the cathode. In a typical tube with a p.d.a. voltage of +3 kV, the A3 anode voltage may be about +250 V, with an A2 anode voltage of −500 V, and a cathode potential of −1 kV. The main controls of the circuit are as shown in Fig. 11.2, the brilliance control effecting control of beam current and the brilliance of the spot. The focus control alters the potential of the A2 anode and modifies the electric field within the electron lens. The astigmatism control enables the operator to adjust the shape of the spot on the screen so that it remains circular over its full scan. An unfortunate feature of many simple oscilloscopes is that the brilliance and focus controls interact upon one another, and increasing the brilliance of the trace causes defocusing.

In addition, a supply must be maintained to the amplifiers. In sophisticated units, the power supply is obtained from a stabilized voltage source, the design principles of which are dealt with in chapter 13. In less complex oscilloscopes, the power supply to the amplifiers is smoothed but not regulated, and to account for mains voltage variation the gain of the amplifiers used can be made inversely proportional to the supply voltage over a limited range of supply voltages. It is common to provide a calibration signal output on the front panel of the oscilloscope, in order to permit easy calibration of the gain and time controls. This signal is usually of the order of 1 V peak-to-peak at mains frequency, and is more often than not generated by a shunt voltage regulator (see chapter 13) comprising two cascaded stages of either Zener diodes or gas-filled diodes.

11.5   Oscilloscope amplifiers

Almost without exception in oscilloscopes used as measuring instruments, the Y amplifiers consist of long-tailed pair stages. In a simple circuit, one stage of amplification would be used, and in more advanced equipment a number of

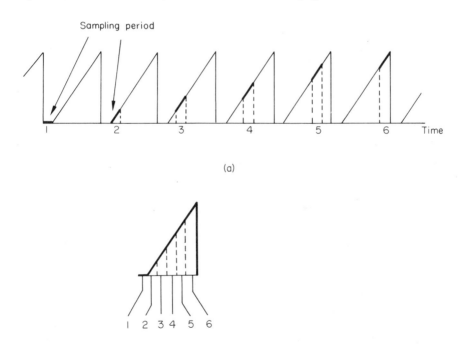

(a)

(b)

Fig. 11.5   Reconstitution of a waveform by a sampling oscilloscope.

cascaded stages are employed. The anode load used in these amplifiers is usually inductive to give optimum transient response, the value of the inductance being adjustable by means of a movable core.

In an X-Y oscilloscope the X amplifier and the Y amplifier should be identical in all respects. To increase the usefulness of the C.R.O., Y shift and X shift controls are added to allow the trace to be located at any point on the screen. This is generally achieved by adjusting the bias voltage applied to the amplifier, causing a shift in output voltage.

Conventional oscilloscopes use *real time* scanning, that is to say that only waveforms that exist within the bandwidth of the amplifier system can be accurately observed. At the present state of development, the maximum real time frequencies that may be observed are in the megahertz region. Developments in *sampling oscilloscopes* have, however, allowed repetitive events in the gigahertz region to be viewed. Amplifiers used in sampling oscilloscopes 'sample' the input waveform periodically, and display the magnitude of the sample on the screen. Thus, if the waveform takes the form in Fig. 11.5(a), and is sampled at the instants 1 to 6 shown, the waveform can be reconstructed in the form in Fig. 11.5(b). By this means, a waveform which has a frequency above the normal range of real time systems can be displayed.

11.6   Horizontal deflection circuits

The function of the horizontal deflection circuit is to provide a voltage which causes the electron beam to be deflected in the X direction. In this section, we shall consider several circuits which generate *sawtooth waveforms,* so providing suitable constant-velocity waveforms in order that periodic waveforms may be displayed on the screen. The circuits described here are mainly valve circuits since most present-day C.R.O.s are valve operated. The principal advantage of the valve over transistors in C.R.O. circuits is that they are capable of working with large voltage swings at their anodes, a limitation that is being rapidly overcome by advances in semiconductor technology.

The prime function of the timebase circuit is to generate a signal which causes the spot to scan across the face of the tube so that its displacement from the point of its commencement is proportional to the time which has elapsed from the start of the sweep. The *timebase* is the time taken for the spot to make one sweep in the X direction, and the circuit which develops the signal is known as the *timebase generator* or *sweep generator.*

In a tube with electrostatic deflection, the ideal timebase waveform is shown in Fig. 11.6(a). The signal applied to the X_1 plate rises linearly from zero to a maximum value, after which it falls to zero voltage in zero time. A signal of the opposite polarity is applied to the X_2 plate. An imperfect practical version of the waveform is shown in Fig. 11.6(b), in which the voltage rise is non-linear during the sweep time S, there is a finite *flyback time F* during which the spot

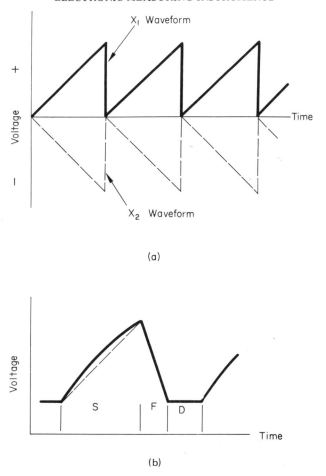

Fig. 11.6   (a) Idealized timebase waveform, and (b) a practical waveform.

returns to the commencement of the sweep, and there is a *delay time D* before
the commencement of the next sweep. In addition, the voltage does not fall to
zero due to the p.d. developed across some component in the circuit. Normally,
the flyback time is made as short as possible, but where the sweep time is short
(as is the case when high frequency phenomena are being investigated), the
flyback time takes up an increasing proportion of the timebase and may
represent as much as 20 per cent of the total time. The flyback causes a line
to be drawn on the face of the tube, in much the same way as the trace is drawn
during the sweep. To eliminate this line, since it has only nuisance value, a
flyback suppression pulse is generated and is applied to the grid of the C.R.T.
during the flyback period. This has the effect of reducing the beam current
during the flyback period, and eliminating the flyback trace.

Broadly speaking, timebase circuits generate the sweep signal by charging a capacitor, and the flyback signal by discharging it. The circuits may be divided into three categories:

(a) Those employing a voltage-sensitive switch.
(b) Miller-effect circuits.
(c) Blocking oscillator and similar circuits.

11.7 Simple timebase circuits

A relaxation oscillator circuit employing voltage-sensitive switching elements is shown in Fig. 11.7. Upon switch-on, the voltage across the capacitor rises exponentially until it reaches the ionization potential of the cold cathode tube, the time constant of the output waveform being that of the charging circuit, i.e., RC. After ionization the tube begins to conduct and the capacitor rapidly discharges through the tube which has slope resistance r_a. The periodic time of the

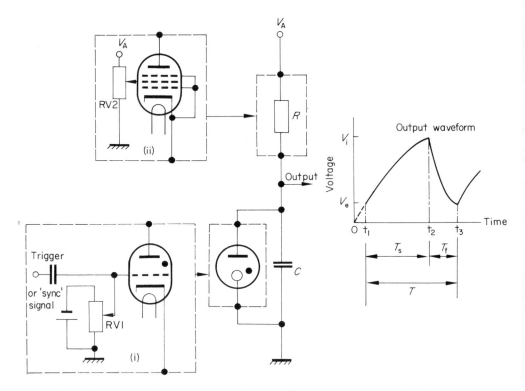

Fig. 11.7   A simple timebase circuit; inset (i) shows how the cold-cathode tube may be replaced by a thyratron, and inset (ii) shows a simple constant-current generator.

oscillation is computed as follows. The voltage across the terminals of the capacitor during the charging period is

$$v = V_A(1 - e^{-t/RC})$$

where v is the voltage at time t. At some time t_1 this voltage is equal to the extinction voltage V_e, and at t_2 it has risen to the ionization potential V_i, hence

$$V_e = V_A(1 - e^{-t_1/RC})$$

$$V_i = V_A(1 - e^{-t_2/RC})$$

hence,

$$V_A - V_e = V_A e^{-t_1/RC} \qquad (11.1)$$

$$V_A - V_i = V_A e^{-t_2/RC} \qquad (11.2)$$

Dividing eq. (11.1) by eq. (11.2), and solving for t_1 and t_2 gives

$$T_s = t_2 - t_1 = RC \ln \frac{V_A - V_e}{V_A - V_i}$$

and the flyback time is given by

$$T_f = t_3 - t_2 = r_a C \ln \frac{V_i}{V_e}$$

The periodic time for the complete cycle is $T = T_s + T_f$, but the circuit constants are generally chosen so that $T_f \ll T_s$, and the periodic time is approximately T_s.

One disadvantage of the cold cathode tube is the relatively small potential difference between V_i and V_e, resulting in a relatively small peak-to-peak voltage. In addition, there is a limit to the maximum current that may be drawn by the tube; to limit the current to a safe value, it may be necessary to include an additional resistance in series with the tube. Thus, if $V_i = 80$ V, and the normal tube drop is 65 V, the difference of 15 V must not cause an excessive current to flow through the tube. Suppose that the maximum tube current is 5 mA; the total resistance in the tube circuit must be 15 V/5 mA = 3 kΩ. Since the slope resistance of the tube may only be about 100 Ω, it is necessary to include a resistance of about 3 kΩ in the discharge circuit. This has the effect of increasing the flyback time. We have also neglected the ionization time and deionization time of the tube in our calculations. Although these are small compared with the timebase itself, they are finite, and add to the total time for each cycle. As a consequence of the factors involved, the maximum useful frequency of cold cathode tube relaxation oscillators is about 10 kHz to 50 kHz. Alternatively, the cold cathode tube can be replaced by a thyratron which shows an improvement in both ionization and deionization times, and is not subject to the same anode current limitations.

To improve the control of the circuit, the cold cathode tube can be replaced by the thyratron as shown in inset (i) in the figure. The thyratron is triggered into its conducting state by the application of a positive pulse at its control grid via a decoupling capacitor, allowing the timebase to be synchronized with the input waveform to the C.R.O. Since the anode voltage at which conduction commences in a thyratron is controlled by the grid voltage, then potentiometer RV1 can be used as a form of *amplitude control* of the sawtooth waveform.

The curvature of the sweep can be corrected by replacing resistor R in Fig. 11.7 by a device with a constant current characteristic, such as a pentode (see inset (ii)). The anode current of the pentode is controlled by its grid voltage, which is controlled in the figure by potentiometer RV2. The relationship between the current I flowing through the valve and the rate of change of voltage across C is given by

$$I = C\frac{dv_c}{dt}$$

or

$$\frac{dv_c}{dt} = \frac{I}{C}\ \text{V/s}$$

Thus, if I is maintained at a constant value, then the voltage across the capacitor rises linearly at the rate of I/C V/s. For a given ionization potential of the discharge tube, the scan time is dependent upon the rate of rise of the capacitor voltage, since the faster the rise time the shorter the scan time. Consequently, potentiometer RV2 can be used as a *velocity control.*

The circuit in Fig. 11.7 produces a positive-going sawtooth or *run-up sawtooth*, since the waveform increases from a low value to a high value during the scan time. A negative-going sawtooth or *run-down sawtooth* is generated across resistor R. Both types of waveform are used in timebases.

One form of transistor sawtooth generator is shown in Fig. 11.8, which uses a complementary pair of p-n-p and n-p-n transistors. The operation of the circuit is as follows. When capacitor C is discharged, the emitter junction of Q1 is reverse biased and its collector current is zero. Since the collector current of Q1 is the base current of Q2, it is also cut off at this instant of time. In effect, the circuit between the emitter of Q1 and the emitter of Q2 is open. This allows the voltage across C to rise exponentially until the emitter junction of Q1 becomes forward biased, allowing collector current to flow. In turn, this forces Q2 into conduction, and the two transistors rapidly switch to a saturated state, and capacitor C is rapidly discharged through R_E. The voltage across the capacitor falls rapidly until Q1 turns off, when the charging process recommences. Typical component values are as shown in the figure, and almost any complementary pair of transistors will work in the circuit. Typical transistors include ASY26 and ASY28 or OC200 and BSY10. It is possible in this circuit that the leakage current in Q1 under high ambient temperature conditions may just trigger Q2 into conduction. This may be avoided by connecting a 1 kΩ resistor between the collector of Q1 and earth, to allow an alternative conduction path.

In Fig. 11.8, RV1 is the velocity control and RV2 is the amplitude control. When the capacitor discharges, it produces a voltage pulse across R_E which can be used to generate a flyback suppression signal, since the pulse exists only during the flyback period. This circuit may, alternatively, be used to trigger thyristor* and triac power devices (see also chapter 12). Improved linearity of the scan signal is obtained by replacing resistor R with the constant current

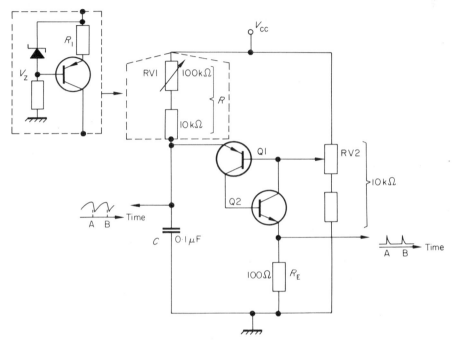

Fig. 11.8   A hook-connected transistor pulse generator circuit.

generator in the inset to Fig. 11.8 (see also chapter 13). The charging current resulting from the circuit in the inset is

$$I = V_Z/R_1$$

hence,

$$\frac{dv_c}{dt} = \frac{V_Z}{CR_1} \text{ V/s}$$

An alternative semiconductor circuit using a unijunction transistor, which also generates a sawtooth and a pulse, is described in section 12.9.2 (page 332). As with Fig. 11.8, the unijunction circuit generates a non-linear sweep signal, which can be linearized by the use of the constant current circuit in the inset in Fig. 11.8.

* See *Control Engineering* by N. M. Morris, McGraw-Hill (1968).

11.8   A Miller-effect timebase

It was shown in section 1.12.1 (page 26) that the capacitance existing between
the anode and control grid of a valve is effectively multiplied by a factor of
$(1 - m)$ so far as the input circuit is concerned, when the valve is used in an
amplifier stage. The so-called Miller circuits utilize this effect to increase the
effective capacitance of a capacitor connected between the grid and anode of the
valve.

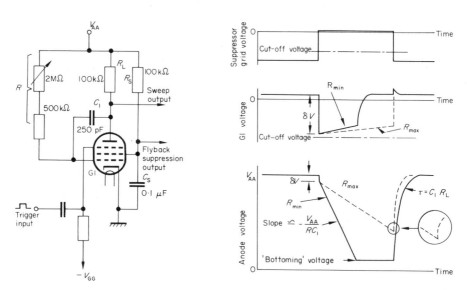

Fig. 11.9   A Miller-effect timebase circuit.

A circuit which gives a run-down sawtooth is shown in Fig. 11.9. A pentode
valve is used so that the capacitance which exists between the control grid and
the anode is effectively C_1 (since C_{g1a} is small). The *Miller capacitor* in this
circuit is therefore C_1; since the gain of the pentode amplifier is normally
greater than about -100, C_1 is multiplied by $(1 - m)$ while the circuit operates
as an amplifier.

Under quiescent operating conditions, the negative voltage V_{GG} applied to
the suppressor grid is sufficient to prevent any of the space current from
reaching the anode. As a result, all the cathode current flows to the screen grid,
giving a low screen voltage. Capacitor C_s acts as a decoupling capacitor to earth.
During this period of operation, the control grid is slightly positive with respect
to the cathode due to the ohmic connection to V_{AA} through resistor R. This
resistor is divided into two parts, the 500 kΩ fixed resistor preventing the
control grid from being connected directly to V_{AA} in the event that the variable
part is reduced to zero. In this circuit, the variable resistor acts as the velocity
control.

When a positive-going pulse of sufficient amplitude is applied to the trigger input connection, the suppressor grid voltage rises to zero and allows anode current to flow. At the instant this occurs, the anode voltage begins to fall. Since the control grid and the anode are coupled by C_1, the instantaneous reduction in anode voltage is transferred to the control grid, forcing the valve to approach cut off. Since the charge held by the capacitor cannot change instantaneously, the initial reduction of anode voltage is δV, which is approximately equal to the grid cut-off voltage of the valve. For this application, sharp cut-off valves are used in which δV is less than about 5 V. Since the grid-cathode region of the valve is now reverse biased, capacitor C_1 begins to discharge through resistor R and the anode-cathode path of the valve. Owing to the change in operating conditions, the majority of the cathode current flows to the anode, and the screen grid voltage rises rapidly to a high level.

A detailed analysis* shows that the anode run-down voltage takes place at a linear rate of approximately V_{AA}/RC_1 V/s, and at the same time the grid voltage rises by a few volts in a linear fashion. With the minimum value of resistance R_{min} in circuit, the rundown is fairly rapid and the valve bottoms or saturates before the trigger pulse is removed. When this happens, the gain of the pentode falls to zero since its amplifying action fails, and the input capacitance falls from $C_1(1-m)$ to C_1. This results in the control grid potential rising rapidly along an exponential path with a time constant of RC_1 to a small positive voltage. Consequent upon the valve bottoming, the screen current increases and the screen potential falls. Thus, the screen voltage is high while the anode rundown continues, otherwise it is low; this means that the screen voltage can be used as a means of flyback suppression. When the trigger input signal is finally reduced to zero, the space current flowing to the anode falls to zero, and C_1 charges through R_L and the forward biased grid-cathode path of the valve.

Increasing the value of the resistor R causes C_1 to discharge more slowly than before, giving a slower rundown and a longer timebase. However, if R is made too large (see curves for R_{max}) the anode rundown may not be completed before the trigger signal is reduced to zero, resulting in a reduction of sweep amplitude.

In pentode circuits, the supply voltage is of the order of 200 V to 300 V, and the saturated voltage is of the order of 10 V to 20 V, so that a run-down voltage of about 200 V is easily obtained. This voltage change is adequate for most purposes.

A variation of the Miller timebase is the *phantastron†*, which uses the combined effects of the Miller circuit and the transitron circuit. The circuit generates its own gating pulse, so that it can be triggered into operation by a very short pulse.

* See *Engineering Electronics* by J. D. Ryder, McGraw-Hill 1967.

† The name 'phantastron' is believed to be derived from the fact that early designers thought its operation to be *fantastic*.

11.9 Input circuits

The accuracy with which a signal is presented on the C.R.T. is only as good as the circuits which process it allow it to be; input circuits are no exception to this rule. The input impedance and capacitance on all input ranges must be standardized on all settings of the input attenuator, otherwise a change of input impedance alters the load that the C.R.O. presents to the circuit being monitored. If the input impedance is not standardized, a change in the attenuator setting leads to an apparent difference in several factors including the magnitude, the phase, and the waveshape of the input signal. As a general rule, the input impedance of an oscilloscope is standardized at about 1 MΩ shunted by a capacitance of 20 pF to 50 pF. Since the input amplifier generally has an input impedance greater than 1 MΩ, and an input capacitance which is less than the standard value, a resistor and a capacitor are connected between the amplifier input terminals to give the standard values.

A normal range of input attenuator settings is 0·1 V/cm to 50 V/cm, corresponding to the voltage required to give a deflection of 1 cm on the face of the C.R.T. Suppose that we are to design input attenuators to give calibrations of 0·1 V/cm, 0·2 V/cm, 0·5 V/cm, and then in decade multiples up to 50 V/cm, what procedure must we adopt? Firstly we must derive the basic equations of the simple attenuator in Fig. 11.10(a), and then design four attenuators based on these equations. The final network will then be as shown in Fig. 11.10(b). A prime requirement of each network is that its input impedance must be 1 MΩ when it is terminated in a load of 1 MΩ. In the following all values are in megohms. Since the input impedance of Fig. 11.10(a) is $R_1 + [R_2 \times 1/(R_2 + 1)]$ MΩ, and this must have a value of 1 MΩ, then

$$1 = R_1 + \frac{1 \times R_2}{1 + R_2} \text{ M}\Omega \qquad (11.3)$$

Now, if the ratio V_2/V_1 of the circuit is N, then

$$N = \frac{V_2}{V_1} = \frac{(1 \times R_2)/(1 + R_2)}{R_1 + (1 \times R_2)/(1 + R_2)} \qquad (11.4)$$

But, from eq. (11.3), the denominator of eq. (11.4) has unity value, hence

$$N = R_2/(1 + R_2)$$

or

$$R_2 = \frac{N}{1 - N} \qquad (11.5)$$

R_1 can then be evaluated from eq. (11.3)

Since the oscilloscope calibration is 0·1 V/cm when the input is directly coupled to the Y amplifier, then an attenuator with $N = 0·5$ calls for an input

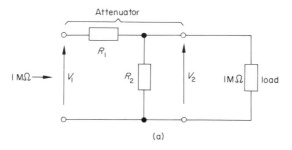

(a)

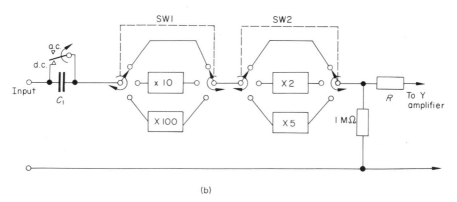

(b)

Fig. 11.10   (a) The basic attenuator circuit, and (b) the layout of a typical oscilloscope input circuit.

voltage of 0·2 V to give a deflection of 1 cm on the screen. Substituting $N = 0·5$ into eq. (11.5) gives

$$R_2 = 1 \text{ M}\Omega$$

and R_1 is evaluated by substituting this value into eq. (11.3).

$$R_1 = 0·5 \text{ M}\Omega$$

For our purposes, we will describe this as a 'times two' attenuator, since it doubles the input voltage required to produce a given deflection on the screen. The combinations of resistor values for the four switched attenuators in Fig. 11.10(a) are listed in Table 11.2. If accuracy and repeatability are required,

Table 11.2

Attenuator	N	R_1 (MΩ)	R_2 (MΩ)
x 2	0·5	0·5	1·0
x 5	0·2	0·8	0·25
x 10	0·1	0·9	0·111
x 100	0·01	0·99	0·0101

high stability resistors with a tolerance of 1 per cent or better must be used in the attenuators.

The circuit in Fig. 11.10(b) highlights two other aspects of oscillography. Firstly, it may be desirable to monitor either the total instantaneous input signal (i.e., the d.c. component + the a.c. component) or only its alternating component. In the former case, the input signal is connected directly to the attenuator networks, and in the latter case it is coupled through capacitor C_1 which blocks the d.c. component of the input signal. C_1 typically has a value of 0·1 μF. Secondly, it is possible to apply an excessive voltage to the input circuit; to prevent an excessive grid current in the first valve of the Y amplifier, a resistor R is included in series with the input line, its value being about 100 kΩ.

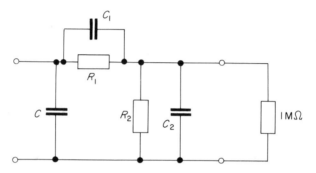

Fig. 11.11   A frequency-compensated attenuator.

In a practical attenuator, the input capacitance of the amplifier causes a reduction in input impedance at high frequency. In Fig. 11.11, this is represented by capacitor C_2. This must be compensated for in some way in order to reproduce high frequency phenomena faithfully. One method commonly used is to shunt the series resistor R_1 by capacitor C_1, which is adjusted in value until a square wave at the input is correctly traced out on the screen of the C.R.T. If C_1 is too small, the corners of the leading edge of the wave are rounded, and if too large the leading corners have overshoots or 'spikes'. In fact, much as R_1 and R_2 form a d.c. potential divider chain, C_1 and C_2 form an a.c. potential-divider chain. Capacitor C which is connected across the input of the attenuator is adjusted to give a total input capacitance which is equal to the standard value of input capacitance for that oscilloscope. When these adjustments have been made, the attenuator is described as a *frequency compensated attenuator*.

11.10   Oscilloscope probes

A *probe* is simply a test lead which contains either a passive network or an active network at its end or at some point along the lead.

A *simple voltage-divider probe* looks like the basic attenuator in Fig. 11.10(a), and may be used for voltages up to about 15 kV. In this type of probe, the input resistance is generally much greater than 1 MΩ, and a fairly complex frequency compensating network is employed to achieve a good high frequency response. A disadvantage of this arrangement is that the signal attenuation may be severe (up to about 1000:1 in some probes) which may be unacceptable in some instances due to the small size of the trace on the C.R.T.

If alternating voltages of the order of 50 kV are to be measured, a *capacitive potential-divider probe* is employed. In its simplest form, it consists of two series-connected capacitors as shown in Fig. 11.12. The input voltage divides between the two capacitors in inverse proportion to their capacitances.

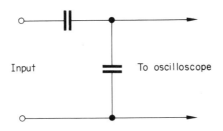

Fig. 11.12   A simple capacitive voltage-divider probe.

In other circuits, the 1 MΩ input impedance of the C.R.O. may not be large when compared with the output impedance of the circuit. One example where this may occur is the Miller timebase circuit in Fig. 11.9. In this case, a high input impedance probe is required. A *cathode follower probe* is commonly used in this application, the cathode follower being mounted in the probe itself. This type of probe presents an input impedance of the order of 30 MΩ, and has a voltage gain of the order of 0·8 (this should be compared with typical values for a resistive potential-divider probe, which may be 10 MΩ and 0·1, respectively).

The simplest method of measuring current is to connect a low value of resistance in series with the circuit, and monitor the voltage across the resistance. Alternatively, a *current probe* can be used, which consists of a clip-on transformer with a secondary winding which is connected to the oscilloscope input.

11.11   Electronic voltmeters

Electronic voltmeters fall into two main groups. The first group contains *electronic voltmeters* (E.V.M.) proper, which are analogue (i.e., continuous reading) instruments which usually use a moving coil voltmeter as the read-out device. E.V.M.s are, more often than not, built around a linear amplifier with a high input impedance and high gain. Alternative names used for the E.V.M. proper are the valve voltmeter (V.V.M.), vacuum-tube voltmeter (V.T.V.M.), transistor (bipolar) voltmeter (T.V.M.), and FET input voltmeter (F.E.T.V.M.).

The second group contains a range of instruments which give a digital readout, and are collectively known as *digital voltmeters* (D.V.M.)

11.11.1   E.V.M.s

Electronic voltmeters can be used to measure unidirectional (d.c.) quantities as well as alternating quantities. The vast majority of E.V.M.s use a moving-coil meter in which the torque acting on the movement is proportional to the average

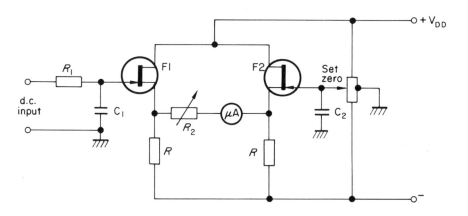

Fig. 11.13   One form of d.c. electronic voltmeter.

current flowing through it; consequently, E.V.M.s are basically instruments which read unidirectional quantities, but they can be calibrated to read r.m.s. alternating quantities if the input signal is a pure sinusoid. The readings may be considerably in error if the input signal is non-sinusoidal.

A popular form of FET circuit used in d.c. instruments is shown in Fig. ·11.13. It is in the form of a bridge circuit, with the two FETs forming one half of the bridge, and the source resistors R forming the other half. The meter (which usually has a full-scale deflection of between 200 μA and 1 mA) is coupled between the two sources which are the 'corners' of the bridge circuit. The network $R_1 C_1$ serves to attenuate any a.c. signals which may be super-imposed upon the steady input voltage. The value of R_1 lies between 100 kΩ and 5 MΩ, while C_1 has a value between about 0·025 μF and 0·1 μF. With zero input signal, the bridge is balanced by the set zero control which alters the bias applied to FET F2; capacitor C_2 decouples the gate of F2 to earth.

When the gate of F1 is raised to a positive potential, the potential of the source is raised by a similar amount and current flows through the meter from the source of F1 to the source of F2. In order to provide a means of calibrating the instrument, a variable resistor R_2 is connected in series with the meter; this resistance is adjusted at the time of the final calibration. The sensitivity of this type of instrument (i.e., the current flowing in the meter for a given input voltage) is very approximately $g_m/2$, where g_m is the mutual conductance of the FET. Thus, if $g_m = 1$ mA/V then the sensitivity of the instrument is approximately 0·5 mA/V.

Another circuit which is in common use in d.c. instruments is the long-tailed pair; in valve circuits, the meter is connected between the two anodes. Transistor versions of both the circuit in Fig. 11.13 and the long-tailed pair are very popular, FET-input circuits having a great advantage in terms of input impedance.

The circuits described above are prone to drift effects, however small, and require the set zero control to be adjusted periodically. This defect can be overcome by the use of a chopper amplifier (see also section 7.1, page 140); this type of amplifier converts the d.c. signal into an alternating signal, which is then amplified by conventional a.c. methods which are not prone to the effects of drift. After amplification, the signal is rectified and applied to a moving-coil meter. Full-scale deflection down to 1 mV can be achieved by this technique.

Alternating quantities can be measured by rectifying the signal at some stage. Broadly speaking, a.c. instruments are either of the rectifier-amplifier type or of the amplifier-rectifier type. In the former, a d.c. amplifier is used and the instrument can be used to measure either alternating or direct quantities. In the latter type, a straightforward a.c. amplifier is used, and rectification takes place at the output. In this type of instrument, which only measures alternating quantities, overall negative feedback is applied to overcome the effects of temperature variation, ageing, etc. The output rectifier is usually included in the feedback loop to reduce the effects of the non-linear diode characteristics on the reading of the output meter. These instruments give full-scale deflections between 1 mV and several hundred volts in the frequency range 10 Hz to 10 MHz.

Alternative forms of rectifier circuits used with rectifier-amplifier instruments are shown in Fig. 11.14. Provided that the product RC is greater than about 100 times the periodic time of the lowest frequency to be measured, the output voltage of either circuit approaches the peak a.c. voltage. However, circuit (a) is at a disadvantage when compared with circuit (b) for a number of reasons, including:

(a) There must be a conducting path for direct current through the supply source.

(b) Where thermionic diodes are used there are two additional problems which affect the operation. (i) The a.c. supply to the heater may induce voltages in the circuit. (ii) The anode 'splash' current at zero anode voltage causes the capacitor to charge, giving errors in readings; this effect also varies with source resistance, and it is difficult to cancel it out.

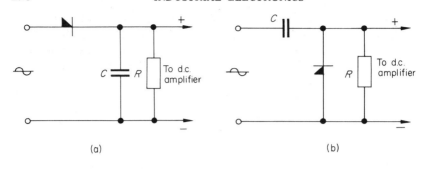

(a)   (b)

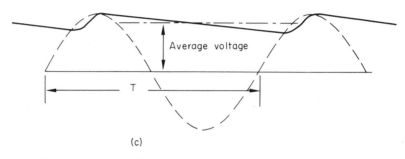

(c)

Fig. 11.14   Two forms of rectifier input circuit for use with E.V.M.'s.

For these reasons, the circuit in Fig. 11.14(b) is preferred. The high frequency limit of this type of circuit is set by the parameters of the diode, in particular the inductance of the leads and the anode-cathode capacitance. At very high frequencies, these cause resonance effects. Measurements up to several hundred megahertz can be made with a diode probe using a diode with a low interelectrode capacitance and low lead inductance.

One circuit, known as a *slide-back voltmeter*, has maintained its popularity over many decades. The basic circuit is shown in Fig. 11.15. The input terminals are short-circuited during the initial period of calibration, and the bias voltage V_1 is adjusted until the anode current can just be read on the milliameter. When the short-circuit is removed and the input signal is applied, the anode current increases. This current is then backed off to its initial value by sliding the wiper of RV1 towards the negative pole of the bias supply. The peak value of the input voltage is then equal to the difference between the two readings of V_1. It is not advisable to reduce i_A to zero since the mutual characteristic of the valve approaches the zero current axis very slowly, and the cut-off voltage is difficult to determine. The two capacitors C_1 and C_2 are for decoupling purposes.

The slide-back voltmeter is a very accurate instrument, its accuracy being practically independent of the valve used. Its calibration stability is very good, and it has an extremely high input impedance since it operates under cut-off conditions. A disadvantage of the slide-back voltmeter is that it is unable to follow variations in the peak input voltage since it must be manually adjusted on each occasion; to overcome this disadvantage, several forms of automatic slide-back voltmeter have been developed.

At the very high frequency end of the spectrum, several forms of h.f. a.c. E.V.M. are available which are basically calibrated tuned amplifiers—rather like a radio receiver with a meter as the output device.

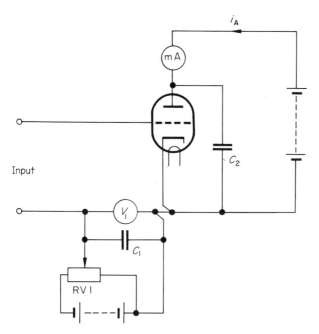

Fig. 11.15   The basic circuit of the slide-back voltmeter.

11.11.2   Digital voltmeters (D.V.M.)

An almost bewildering array of digital voltmeters exists today, and the problem of selecting the best model for any given application is difficult. In this section of the book, the more popular types are described. D.V.M.s are direct potential reading instruments, and require additional circuitry to enable them to read other quantities.

The terminology used in specifying the voltmeters can also be confusing. For instance, the maximum reading on a four digit voltmeter is theoretically 9999, but the design of the circuit may limit the useful range to a maximum reading of, say, only 1999; in this case, the most significant digit can only assume either of the values zero or unity. Ideally, the limit of the accuracy of the instrument corresponds to the smallest digit displayed. However, factors such as the design of the circuit, electrical noise, input impedance, etc., mean that the value of the least significant digit is suspect. Most manufacturers quote an accuracy of ± 1 digit; some prudent operators suggest that the least significant digit should be ignored.

A block diagram of one type of D.V.M. known as a *continuous balance voltmeter* is shown in Fig. 11.16. The counter is initially set to zero at the beginning of the measuring process. At a later point in the measuring process, the counter holds a succession of numbers, and these are converted into analogue (i.e.,

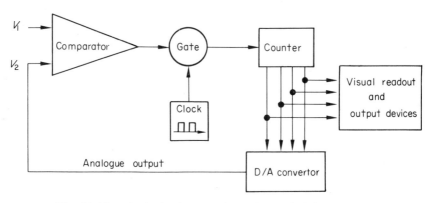

Fig. 11.16   A block diagram of one form of digital voltmeter.

continuous as distinct from digital) signals V_2 by the digital-to-analogue (D/A) convertor. The output from the D/A convertor is compared with the input signal V_1 in an electronic comparator. When $V_1 > V_2$, the output voltage from the comparator is at its highest level, causing the electronic 'gate' to be opened to allow the 'clock' pulses to be fed into the counter. So long as the gate remains open, the number stored in the counter continues to increase as it counts the clock pulses. This number is converted into an analogue voltage by the D/A convertor, giving a steadily increasing value of V_2. When $V_2 = V_1$, the output voltage from the comparator suddenly falls to zero, so closing the electronic gate and freezing the number held in the counter. The value of the input voltage is then displayed on the visual read-out device and other output devices which are connected to the D.V.M. The counter is usually a reversible counter so that it can change its value 'up' or 'down' as the input voltage changes.

Successive-approximation types of D.V.M. use a variant of the above principle. Let us first consider the operation of a D.V.M. of the type in Fig. 11.16. Suppose that we are using a two-digit D.V.M. with a maximum reading of 15 V, and the input voltage is 13 V. Since the counter commences at zero, 13 steps of 1 V are necessary to reach final balance. The essential difference between the continuous balance D.V.M. and the successive-approximation D.V.M. lies in the sequence of test voltages used to balance the voltmeter. Suppose that we were to try the sequence 8, 4, 2, 1. The first clock pulse in the successive-approximation type sets the counter to 8. This is compared with V_1, and since $V_1 > V_2$ the comparator allows the count to continue in the 'up' direction. The next clock pulse adds another 4 to the count, which now becomes 12; this is compared with V_1, and the count continues in the 'up' direction. The next clock pulse adds 2 to the count stored in the counter, making it 14. The comparator must then instruct the counter to count 'down' by the previous count, reducing the number in the counter to 12 again. The comparator now recognizes that $V_2 < V_1$, and the counter then counts 'up' by the next sample number which is unity, making the stored number equal to 13. At this point, after five steps, the D.V.M. balances. Clearly, the successive-approximation D.V.M. is potentially far more rapid in operation than the more simple continuous balance D.V.M., but only at an increase in complexity and capital cost.

In both types of D.V.M. described above, there is a danger of error due to induced voltages in the leads to the instrument. For this reason, a filter is usually included in the input circuit of the D.V.M; in some applications this may have the undesirable effect of reducing the response of the instrument.

In another type of D.V.M., known as an *integrating D.V.M.,* a capacitor is charged to the input voltage. The capacitor is then disconnected from the input, and connected to a constant-current discharge circuit. To determine the input voltage, clock pulses are gated into a counter during the discharge period. By a suitable choice of capacitance value, clock frequency, and sampling rate, the input voltage can be displayed directly.

Yet another type uses a *voltage-to-frequency convertor,* in which the input voltage controls the frequency of an oscillator. In this oscillator, the number of pulses per second is linearly proportional to the input voltage. By a suitable choice of the counting period and pulse frequency, the visual display is made equal to the input voltage.

Digital voltmeters often form the basis of *data processing systems* or *data logging systems.* In these systems, a number of analogue input signals are scanned sequentially by an electronic system; each signal is then converted into an equivalent digital value by the D.V.M. The digital value is then transmitted to an electro-mechanical printer, together with information about the input line from which the signal is derived, and the information is printed out. In this way, a large number of input signals can be automatically scanned or processed, and their values printed or logged.

Problems

11.1 Draw a block diagram of a cathode-ray oscilloscope and briefly explain the purpose of each portion. Detailed circuit diagrams are not required.

An oscilloscope is to be used to examine waveforms occurring in a high quality audio amplifier. Briefly discuss the requirements for the bandwidth of the deflection amplifier and the velocity of the timebase.

(C & G)

11.2 With the aid of a diagram, and starting with electron-emission from the cathode, describe the complete path taken by the current in a cathode-ray tube circuit.

Make a list of the important properties of the fluorescent material which is used to coat the screen of the tube.

Why, in practice, is a saw-tooth the most common of all time-base waveforms?

(C & G)

11.3 What do you understand by the striking and extinguishing voltages of a gas diode?

A battery is connected across a resistance of R ohms in series with a capacitance of C farads. A gas diode is connected across the capacitance. Draw a circuit diagram and a graph showing that the battery current has an alternating component. Derive an expression for the fundamental frequency of this alternating component.

(C & G)

11.4 Define the *grid control ratio* of a gas-filled triode valve (thyratron). State the factors on which the ratio depends and give a typical value for a small thyratron.

Show, with the aid of a basic circuit diagram, how a small thyratron may be used to generate an approximately linear sawtooth voltage waveform. Explain how the amplitude and frequency of the waveform may be controlled, and give a suitable method of injecting a synchronizing voltage. State one possible application of such a generator.

(C & G)

11.5 Make careful sketches to show the appearance of the following waveforms when displayed on a cathode-ray oscilloscope:

(a) a sinewave of frequency 1 kHz and 2 V peak, to which has been added 50 per cent second harmonic. Use scales of 1 inch = 1 V and 1 inch = 0·5 ms;

(b) a sinewave of frequency 1 kHz and 2 V peak to which has been added 50 per cent third harmonic. Use scales as in (a);

In both cases assume that the oscilloscope provides a linear timebase.

11.6 Draw diagrams to illustrate the effects on the trace of a sinewave signal of (a) a non-linear timebase, and (b) lack of flyback blanking.

11.7 Describe how a single-beam cathode-ray oscilloscope could be calibrated and used to measure:

(a) a d.c. voltage.

(b) the r.m.s. value of a sinusoidal a.c. voltage.

(c) the frequency of a sinusoidal a.c. voltage.

(d) the phase difference between two a.c. voltages.

(C & G)

11.8 If a supply frequency, $f = 1/2\pi\sqrt{(R_1 R_2 C_1 C_2)}$ is applied across XZ in Fig. 11.17 then the voltage across YZ will be in phase with that supply. Describe a test, using an oscilloscope, which would confirm this fact and suggest how the circuit could be adapted to measure an unknown frequency.

(C & G)

11.9 Describe in detail, with the aid of a circuit diagram, the operation of a linear timebase employing the Miller integrator principle.

Discuss the effect the choice of component values will have on:

(a) the linearity of the sweep,

(b) the sweep time,

(c) the flyback time.

Sketch on a common time scale the waveforms you would expect to observe at the input and output, and also at critical points in the circuit.

(C & G)

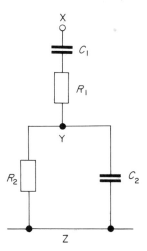

Fig. 11.17   Figure for question 11.8.

11.10 Draw a diagram to show how a crystal diode and a simple d.c. amplifier may be used as a valve voltmeter, and briefly describe its operation.

What limits the frequency range of the instrument you have described?

(C & G)

11.11 Why is the input impedance of an electronic voltmeter high?

How can an electronic voltmeter be used to measure r.f. current?

A series circuit ABC consists of a 100 Ω resistor between points A and B, and a coil of effective resistance r and inductance of 0·001 H between points B and C. An electronic voltmeter reads 5 V when connected across AB and 43·9 V when across BC. If the supply frequency is 100 kHz, determine the values of (a) the circuit current, and (b) the effective resistance r of the inductor.

(Based on a C & G question)

11.12 State Thevenin's Theorem.

A signal generator has its output terminals connected to a variable resistor in parallel with a valve voltmeter. The indication of the voltmeter as the resistance is varied is shown in the table below. Determine the e.m.f. of the generator and the internal resistance.

V (μV)	33	50	67
R (Ω)	25	50	100

What is the maximum power available from the generator?

(C & G)

11.13 Draw the circuit diagram of a valve voltmeter and briefly describe its operation.

What precautions must be taken when using such a voltmeter in conjunction with a signal generator to determine the resonant frequency of a tuned amplifier?

A peak reading voltmeter has been calibrated with a sine-wave signal and is scaled in r.m.s. volts. What will the meter indication be when a square-wave signal of +1 V peak amplitude is applied?

(C & G)

11.14 A 1 kHz square wave of 50 V peak-to-peak is applied to the circuit in Fig. 11.14(b). Sketch the output voltage waveform to scale, given that $C = 0.2\ \mu F$ and $R = 100\ k\Omega$; the forward conduction resistance of the diode is 150 Ω, and its reverse blocking resistance is infinite.

11.15 Sketch the output waveform if the diode in question 11.19 is replaced by a 5 V Zener diode.

11.16 Enumerate the general principles of *three* types of digital voltmeters, and describe *one* in detail.

11.17 Draw a circuit diagram and explain the operation of a thermionic valve or transistor multivibrator and show how it is synchronized to an external signal.

How can multivibrators be used to effect a frequency division of 100:1?

(C & G)

11.18 Explain how a cathode-ray oscilloscope may be used to measure the amplitude of an alternating voltage, and explain how it may be calibrated on direct current.

Discuss the advantages and disadvantages of this method of voltage measurement as compared with the use of a valve voltmeter.

(C & G)

11.19 (a) Draw the circuit and explain the operation of a diode voltmeter which indicates the peak value of an alternating signal.

(b) For a sinusoidal input signal, draw carefully typical waveforms of (i) the voltage across the reservoir capacitor, (ii) the voltage across the diode. Indicate on these sketches the effects of (iii) a high value of diode forward resistance, (iv) a relatively low value of reservoir capacitance.

(c) Draw a typical diode characteristic and explain how this affects the use of the instrument for measuring low voltages.

(C & G)

11.20 Describe the relative advantages of magnetic and electric deflection in cathode-ray tubes used (a) in television receivers, (b) in oscilloscopes.

Explain how a circular trace may be produced on a cathode-ray tube with electrostatic deflection. How may this be used to compare two frequencies, one many times the other?

(C & G)

11.21 A digital voltmeter (D.V.M.) often forms the basis of a data processing system. Draw a block diagram of a simple D.V.M. and explain briefly the circuit operation.

(C & G)

12. Switching circuits

12.1 The basis of engineering logic

Consider the statement 'X is either a cat or a horse, and X is not a cat. What is X?' This is a simple statement of logical fact, and one way of deducing the answer is to present the facts in the form of a symbolic equation. It is clear that the conventions of symbolic logic or *Boolean algebra*— named after Boole, its inventor—will differ in some respects from conventional algebra. In the following paragraphs, a general insight into the subject is given.

Generally speaking, we require a simple 'true' or 'untrue' answer to any engineering question that we may pose. Suppose that we assign the logical value of unity to an answer which is true, and the logical value zero to one which is untrue. By this simple means, we can readily detect the truth or otherwise of a logical proposition in, say, an electronic logic circuit simply by using a voltmeter which indicates the voltage level at a point in the circuit. If the two voltage levels that are available in the circuit are +10 V and zero, and if we let

$$\text{logic '1'} = +10 \text{ V}$$

$$\text{logic '0'} = 0 \text{ V}$$

then we say we are working in *positive logic,* since the higher of the two potentials represents logic '1'. Most circuits utilizing n-p-n transistors work in positive logic, since it is convenient to think of a signal existing when a positive potential can be detected.

Many early logic systems operated in what is known as *negative logic,* in which the lower of the two potentials is taken as logic '1'. With the above potentials, logic '1' in negative logic would be represented by zero voltage, while logic '0' would be represented by +10 V. Positive logic and negative logic are both employed in present-day engineering systems, and it is often convenient to mix both types of logic levels in order to simplify system design*.

* See *Logic Circuits* by N. M. Morris, published by McGraw-Hill (1969).

In order to reduce logical statements to meaningful symbolic equations we must be able to express such words as AND, NOT, and OR in symbolic form. In the following, circuits are developed which operate in positive logic and satisfy the logical requirements of these statements.

12.2   The AND gate

The most direct approach to the operation of logic circuits is by simple relay circuits, an example of which is shown in Fig. 12.1. Logic circuits are also described as *gates,* since they permit free flow of information when they are *open,* and stop the flow when they are *closed* or *inhibited.*

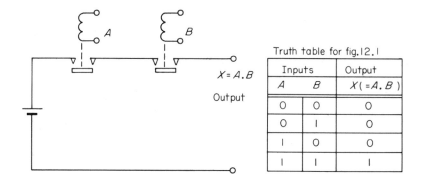

Truth table for fig. 12.1

Inputs		Output
A	B	$X(=A.B)$
O	O	O
O	I	O
I	O	O
I	I	I

Fig. 12.1   A relay AND gate with its truth table.

Under normal operating conditions, the coils of relays A and B are either energized at their full voltage, or they have no voltage applied to them. In the former case, we say that the signal has unity logical value (full voltage), and in the latter case it has zero logical value. Either coil may be energized at one of the two logic levels, so that there are four possible energization states for the two relays. All possible combinations of the states of inputs A and B are grouped together in the input columns of the *truth table* in Fig. 12.1; the name 'truth table' is derived from the fact that a logic '1' signal represents a state of 'existing' or one which is 'true', while logic '0' represents a state of 'not existing' or one which is 'untrue'.

When either or both of the relay coils are de-energized, one or more sets of contacts remain open and the output voltage is zero. This corresponds to the first three rows of the truth table. Only when both A AND B are energized simultaneously does an output exist, satisfying the fourth row of the truth table. Thus, the gate is closed when either A or B has zero value, and is open only when $A = B = 1$. An inspection of the rows of the truth table for the AND gate shows that the numerical value in the output column has the same value as the mathematical product of the two numbers in the input columns. Accordingly, the logical expression for output X is given by

$$X = A \cdot B$$

where the 'dot' (.) represents the logical AND function. Although the *logical product function* has the same value as the *arithmetic product function* of the variables, great care must be exercised in the use of this relationship, since the two functions do not mean the same thing. Also, the equality sign in the logical equation means logical equality, which is not necessarily the same thing as arithmetic equality.

If several inputs A, B, C, D, etc. are used, then the output of a series relay circuit is the product of the inputs, viz:

$$X = A \cdot B \cdot C \cdot D \ldots$$

A positive logic diode-resistor AND gate is shown in Fig. 12.2(a). The input voltage levels are set by the state of the input switches A and B. When both

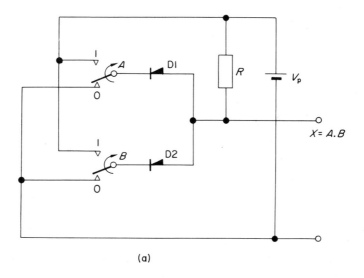

(a)

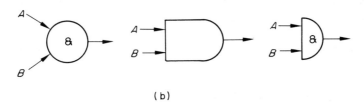

(b)

Fig. 12.2   (a) A semiconductor AND gate, and (b) some representative circuit symbols.

input lines are at the zero level, both diodes are forward biased via resistor R, and each carries a current of approximately $V_p/2R$. The output voltage in this condition is the voltage appearing across a forward biased diode, which is generally small. For most practical purposes, the output voltage may be taken to be zero, giving conditions which correspond to the first row of the truth table in Fig. 12.1.

By switching input B to the logic '1' level (with $A = 0$), diode D2 is reverse biased and the current flowing through it falls to zero; diode D1 remains forward biased and continues to pass current. Since the p.d. across D1 remains at a low value, the output signal is still at the logic '0' level. If the states of the input signal levels are reversed so that $A = 1$, $B = 0$, the roles of the two diodes are reversed, and the output remains at the zero level. This satisfies the second and third rows of the truth table of the AND gate.

When both input lines are connected to V_p (i.e., $A = B = 1$), the return path to the negative pole of the battery is broken, and the current flowing in resistor R falls to zero. As a result, the output potential rises to the logic '1' level, so satisfying the fourth row of the truth table. A number of circuit symbols are used to represent the AND gate, and three of the more popular symbols are shown in Fig. 12.2(b).

12.3   The OR gate

A relay version of the OR gate is shown in Fig. 12.3(a), which consists of two parallel-connected relays. When both relays are de-energized ($A = B = 0$), the output is zero; when either relay or both relays are energized the output voltage is + V_p, or logic '1'. A diode-resistor version of the circuit is shown in Fig. 12.3(b) and some popular circuit symbols in (c). The '1' inside the symbols signifies that an output exists when any one of the inputs is energized. It is left as an exercise for the reader to verify that the diode-resistor circuit satisfies the truth table given in the figure.

The first three rows of the OR gate truth table are satisfied by the arithmetic equation

$$X = A + B$$

and it is for this reason that the 'plus' sign is used to represent the logical OR function. It is only upon investigation of the fourth row of the truth table that any difference between the arithmetic sum function and the logical sum function can be seen. It is observed that, in this case, $1 + 1 = 1$! Again the reader is cautioned to be careful when dealing with the numerical value of a logical equation.

The function generated by the circuits described here is sometimes known as the INCLUSIVE-OR function, since the input combinations which result in an output from the circuit include the case where both inputs exist simul-

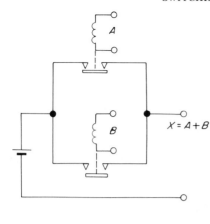

Inputs		Output
A	B	$X(A+B)$
0	0	0
0	1	1
1	0	1
1	1	1

$X = A + B$

(a)

$+V_p$

$X = A + B$

(b)

A
B

A
B

A
B

(c)

Fig. 12.3 (a) A relay OR gate with its truth table. A diode-resistor OR gate is shown in (b), and representative circuit symbols in (c).

taneously. If the gate has N input lines $A, B, C, \ldots, M, N$, the output is expressed in the logical form

$$X = A + B + C + \cdots + M + N$$

It can also be expressed in the form

$$X = A \vee B \vee C \vee \cdots \vee M \vee N$$

where the 'vee' (v) represents the INCLUSIVE-OR function. In some instances, this symbol is preferred to the 'plus' sign as it eliminates any difficulty which may arise when dealing with circuits which involve arithmetic addition.

The maximum number of input signals that can be accommodated by the gate is known as the *fan-in* of the gate, and can have a large value in some circuits.

12.4 The NOT gate

A NOT gate gives the *logical complement* of the input signal. It is said to *negate* or *invert* the input signal. Since there are only two possible signal levels, the following relationships hold good.

$$0 = \text{NOT } 1 = \bar{1}$$
$$1 = \text{NOT } 0 = \bar{0}$$

Variables which are complemented are symbolized by placing a bar over the variable.

A relay version of the NOT gate is shown in Fig. 12.4, and comprises a relay with normally-closed contacts. When the relay is de-energized ($A = 0$), the output signal is logic '1' ($= \bar{A}$); on energizing the relay ($A = 1$), the contacts open and the output falls to logic '0' ($= \bar{A}$).

A transistor version of the circuit is shown in Fig. 12.4(b). When the input line is taken to earth ($A = 0$), the resistor chain $R_1 R_2$ ensures that the emitter junction is reverse biased and the transistor is cut off. In this event, the

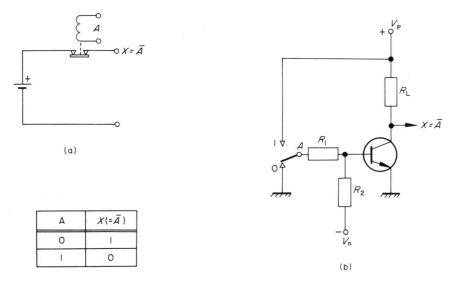

A	$X(=\bar{A})$
0	1
1	0

Fig. 12.4 (a) A relay NOT gate with its truth table. (b) The basic resistor-transistor NOT gate.

collector current is zero and the output voltage is at its highest level ($X = 1 = \bar{A}$). The output voltage is dependent upon the current drawn by the load which is driven by the NOT gate. The maximum number of similar gates that may be connected to the output terminal without causing faulty operation is known as the *fan-out* of the gate.

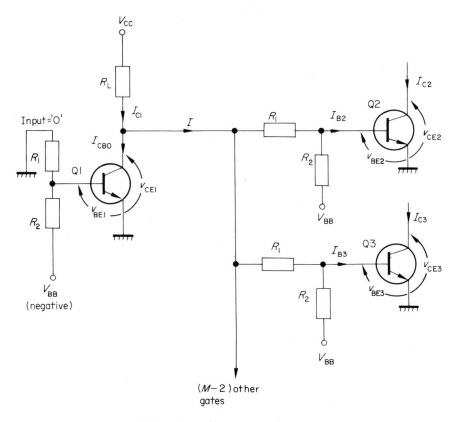

Fig. 12.5   Design procedure for a NOT gate.

The values of R_1 and R_2 are so chosen that when a logic '1' signal is applied to the input of the gate, the base potential becomes positive and the transistor is forced into its saturated state. The p.d. across the transistor under this condition is between about 0·15 V and 0·7 V; this voltage is taken to be the logic '0' level. A simplified design technique for a NOT gate with a fan-out of M is developed below.

The worst operating condition for this circuit occurs when the input signal is zero, and the transistor has to drive all the connected transistors into saturation. This condition is illustrated in Fig. 12.5, in which Q1 is cut off, and its collector current is equal to the leakage current I_{CBO} (since the emitter junction is reverse biased and $I_B = 0$); this results in a 'high' output potential. Current I_{C1} flowing in the load resistor of Q1 must then be adequate to drive all the connected transistors into the saturated state.

The relative values of R_1 and R_2 are calculated as follows. The reverse bias applied to the emitter junction of Q1 is

$$v_{BE1} = V_{BB} \frac{R_1}{R_1 + R_2}$$

or

$$\frac{V_{BB}}{v_{BE1}} = \frac{R_1 + R_2}{R_1} = 1 + \frac{R_2}{R_1}$$

hence,

$$R_2 = R_1 \left(\frac{V_{BB}}{v_{BE1}} - 1 \right) \tag{12.1}$$

Thus, if $V_{BB} = -10$ V, and v_{BE1} is to be -0.5 V when the input voltage is zero, then

$$R_2 = R_1 \left(\frac{-10}{-0.5} - 1 \right) = 19 R_1$$

In practice, a value of R_2 which is slightly greater than the calculated value would be chosen for reasons which are given later.

When Q1 is cut off, the current shown as I must be sufficient to hold Q2, Q3, $\ldots$, QM in the saturated state. Assuming that all the transistors are identical, and that:

(a) $v_{BE2} = v_{BE3} = \ldots = v_{BEM} = v_{BEsat} \simeq 0$
(b) $v_{CE1} = v_{CEsat} \simeq 0$
(c) the current flowing in the bias resistors R_2 in each of the driven stages is small compared with the base current
(d) $I_{CB0} \simeq 0$

then,

$$I_{C2} = I_{C3} = \ldots = I_{CM} \simeq \frac{V_{CC}}{R_L}$$

and

$$I_{B2} = I_{B3} = \ldots = I_{BM} \simeq V_{CC}/h_{FEsat} R_L$$

Since there are M similar stages

$$I = M I_{B2} = M I_{B3} = \ldots = M I_{BM}$$
$$= M V_{CC}/h_{FEsat} R_L \tag{12.2}$$

as the leakage current in Q1 is very small, then $I \simeq I_{C1}$ and

$$I = \frac{V_{CC}}{R_L + (R_1/M)} = \frac{MV_{CC}}{MR_L + R_1} \qquad (12.3)$$

Equating eqs. (12.2) and (12.3) yields

$$h_{FEsat} R_L = MR_L + R_1$$

Provided that three of the four quantities in this equation are known, then the fourth can be calculated. From the equation, we can deduce that

$$R_1 = R_L(h_{FEsat} - M) \qquad (12.4)$$

$$M = h_{FEsat} - \frac{R_1}{R_L} \qquad (12.5)$$

It is usual to select a value for R_1 which is less than the value given by eq. (12.4), for reasons which are given later.

The output voltage corresponding to logic '0' in this circuit is v_{CEsat}, which is 0·15 V to 0·7 V, typically. The 'high' output voltage corresponding to logic '1' is approximately

$$V_{CC}\left(\frac{R_1/M}{(R_1/M) + R_L}\right) = \frac{V_{CC}}{1 + (MR_L/R_1)} \qquad (12.6)$$

The fan-out of the gate is generally less than the theoretical value, due to a number of factors. Firstly, the value of h_{FEsat} of the 'worst' transistor in any production batch is much lower than the value which is 'typical' of that type of transistor. Secondly, the base-emitter saturation voltage v_{BEsat} has a finite value, which reduces the effective base current below the value calculated above. A third reason is that if one of the transistors has a value of v_{BEsat} which is much below that of the other transistors, that transistor draws a much greater proportion of the available current (I in Fig. 12.5) than do the other transistors, so reducing the current available to saturate other driven stages. This is known as *current hogging*. Fourthly, some of the current flowing in the input resistor (R_1) of each driven stage flows to the bias source via R_2, reducing the available base current further.

For these reasons, the value of h_{FEsat} used in calculations must be the lowest quoted for that type of transistor; this ensures satisfactory operation of any circuit using transistors of a given type. In addition, the value of R_1 should be below that calculated from eq. (12.4) to ensure that adequate base current is available, even with a transistor with the lowest possible value of h_{FEsat}. The value of R_1 should also account for its tolerance variations; if 10 per cent tolerance resistors are used, the value chosen should be at least 10 per cent lower than the theoretical figure. Similarly, the value of R_2 should be greater than that predicted by eq. (12.1) to ensure that the base current of the driven stages is not reduced significantly.

Example 12.1: Design a resistor-transistor NOT gate using an n-p-n transistor in which $h_{FEsat} = 10$, $V_{CC} = +10$ V, $V_{BB} = -10$ V, the maximum collector current is not to exceed 10 mA, the reverse voltage applied to the emitter junction to hold it in the cut-off condition is to be -0.5 V, and the design fan-out of the gate is 5. With the maximum number of gates connected to the output, estimate the minimum output voltage corresponding to the logic '1' condition.

Solution:

$$R_L = \frac{V_{CC}}{I_{C(max)}} = \frac{10 \text{ V}}{10 \text{ mA}} = 1 \text{ k}\Omega$$

From eq. (12.4),

$$R_1 = 1(10 - 5) = 5 \text{ k}\Omega$$

A resistor with a preferred value of 3.9 kΩ would be suitable. From eq. (12.1),

$$R_2 = 3.9((-10/-0.5) - 1) = 74 \text{ k}\Omega$$

A resistor with a preferred value of 82 kΩ would be used. The minimum output voltage corresponding to the logic '1' level is

$$\frac{10}{1 + (5 \times 1/3.9)} = 4.38 \text{ V}$$

12.5   The NOR gate

The logic NOR function is derived from the statement

$$\text{NOR} = \text{NOT OR} = \overline{\text{OR}}$$

that is, it is equivalent to an inverted OR function. A block diagram which represents the logic operations involved in a two-input NOR gate is shown in Fig. 12.6(a). The truth table of the function is given in Table 12.1, and it is observed that the output is logic '0' whenever either input signal is logic '1'. In the general case of a gate with N inputs, the output is zero when any of the input lines is energized, and is logic '1' only when all the input signals are zero. The output from the gate is described by either of the equations

$$X = \overline{A + B + \ldots + M + N}$$

$$X = \overline{A \vee B \vee \ldots \vee M \vee N}$$

A practical form of *diode-transistor logic* (DTL) NOR gate is illustrated in Fig. 12.6(b). When any input line is energized at the logic '1' level, the diode in that line becomes forward biased, causing the transistor to be switched into the saturated state, maintaining the output of the gate at logic '0'. This satisfies

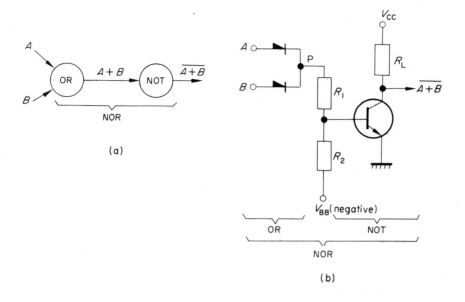

(a)

(b)

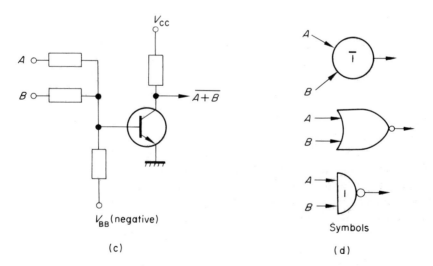

(c)

Symbols

(d)

Fig. 12.6   (a) Generation of the NOR function using an OR gate and a NOT gate. (b) A diode-transistor logic (DTL) NOR gate, and (c) a resistor-transistor logic (RTL) NOR gate. Some representative circuit symbols are shown in (c).

the last three rows of the truth table. When all the input lines are at the logic '0' level, the bias supply V_{BB} and the resistor chain R_1R_2 reverse bias the emitter junction of the transistor, causing the output voltage to be 'high', i.e., logic '1'.

Table 12.1

Truth table of the NOR gate

Inputs		Output
A	B	$\overline{A + B}$
0	0	1
0	1	0
1	0	0
1	1	0

The number of input lines converging at point P is known as the *fan-in* of the gate; the fan-in affects the fan-out in the following manner. A worst-case operating condition of a NOR element occurs when the output of the gate is logic '1', when it must provide sufficient current to saturate all the connected transistors. If, at the same time, all the other input lines of the driven gates are connected to transistors in the saturated state, then the driving stage must also provide the reverse leakage current flowing through the input diodes. Thus, if a DTL NOR gate has a fan-in of 10 and a fan-out of 8, then each of the driven gates has nine other diodes connected to it; the driver stage must be capable of providing sufficient current to saturate 8 transistors in addition to providing the reverse leakage current through $8 \times (10 - 1) = 72$ diodes.

It is possible in certain cases to increase the fan-in by connecting additional input diodes to point P in Fig. 12.6(b) (see also Fig. 12.7), subject to the limitation given in the paragraph above. The number of input and output lines that may be connected to the gate is dependent upon the type of *packaging* employed in its construction. For example, in the range of cylindrical metal canisters of the T0-5 variety, twelve leads is the maximum number that can normally be accommodated. A popular type of packaging for this type of work is the dual in-line pack, which is approximately 6 mm wide by 3 mm deep x 20 mm long (0·25 x 0·125 x 0·75 in), having fourteen connecting leads—seven on each long face with opposite leads being in line with one another for plugging into printed circuit boards. Owing to the limitation on the number of input lines, *fan-in expanders* containing only diodes are manufactured in the same package form as the gates as an economic method of increasing the fan-in of DTL gates.

The design procedure of the DTL NOR gate is generally similar to that of the NOT gate, with the exception that the leakage current of the input diodes must be accounted for. A simple way of assessing the effect of the fan-in of the driven gates on the fan-out of the driver gate is to estimate the total leakage current of the diodes, and to represent this in relationship to the current required to saturate a driven stage. Thus, if the driven stages have a total of

100 input diodes, each passing a reverse leakage current of 1 μA, the driver stage must provide 100 μA of output current in addition to that required to saturate the driven transistors. If the total leakage current approaches the current required to saturate a transistor, then the fan-out must be reduced to make allowance for this fact.

Example 12.2: Design a DTL NOR gate which uses the same transistor and power supplies as in example 12.1.

Solution: The calculation follows the same general lines as that of the NOT gate, giving R_L = 1 kΩ, R_1 = 3·9 kΩ, and R_2 = 82 kΩ. Now, the base current to cause saturation is 1 mA (I_C = 10 mA, h_{FEsat} = 10). This is very large when compared with the leakage current of modern semiconductor diodes, and a fan-in of at least 10 can be accommodated.

A *resistor-transistor logic* (RTL) NOR gate is shown in Fig. 12.6(c). This is generally similar to the DTL gate, other than that the diodes are replaced by resistors. The input resistive network must be designed so that a signal applied to any line holds all the driven transistors in the saturated condition, even if all the other input lines of the driven gates are at the logic '0' level. The design of RTL gates is more complicated than the design of DTL gates since the resistive input elements impose a much greater current drain on the gate than do the diodes in DTL gates. This factor imposes a severe limitation on the fan-in and fan-out capability of the gate; RTL gates do, however, have the merit of simplicity and only require three different resistor values. As a simple rule-of-thumb guide, the resistors used in the input circuits of RTL gates should have a value which is the next lower preferred value below that used in an equivalent NOT gate (see section 12.4).

The fan-in and fan-out capability of both types of NOR gate can be increased by various means; illustrative examples of some of the techniques used are shown in Fig. 12.7. The fan-in of the first stage is increased from 3 to 7 by the use of a fan-in expander unit comprising four diodes. The effective fan-in can be increased further by using a common collector resistor to two transistors (Q1 and Q2). In this connection, the output voltage is low whenever either transistor is saturated (since $v_{CEsat} \simeq 0$), so that a logic '1' input at any of the input lines $A, B, C, \ldots,$ $H, I, J,$ causes the output to be zero. It is not uncommon to use as many as ten to twenty transistors in this way.

The fan-out can be increased by the use of one or more emitter follower stages, providing an increase in the fan-out of the order of the h_{FE} of the transistor used in each emitter follower. However, this is not a very popular method.

So far we have discussed the effects of the fan-in of the driven stages on the fan-out of the driver stage, but we have not mentioned their effect on the switching speed of the gate. One reason for the bias supply V_{BB} is to reduce the switching time of the gate; when the transistor is being switched from the saturated state to the cut-off state, it is necessary to extract all the minority charge carriers from the base region in the shortest possible time. The base bias supply is one means of achieving this end. It has also been stated that the base bias supply effectively reduces the fan-out, since it reduces the current available to force the driven stage into saturation. Clearly, a reduced value of V_{BB} results in an increased fan-out at the expense of an increased turn-off time;

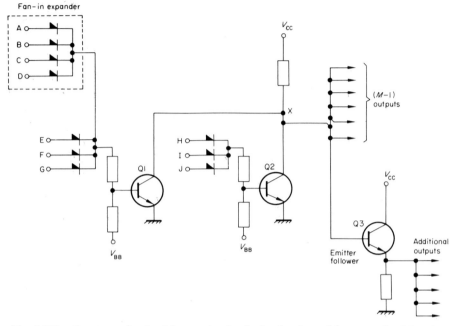

Fig. 12.7 Some methods of increasing both the fan-in and fan-out of NOR gates.

it is possible for gates to operate without a bias supply, the bottom end of R_2 being connected to earth, but these gates cannot operate at elevated temperatures and have a limited switching speed. Generally speaking, there is a trade-off between the fan-in, the fan-out, and the switching speed of the gate. At best, the performance is a compromise between these three factors, depending upon which are the most important.

The NOR gates in Figs. 12.6 and 12.7 are known as *current sourcing logic gates* since, when the transistor is cut off, the gate acts as a current source which drives all the connected gates into saturation.

12.6   The NAND gate

The logic NAND function is derived from the statement

$$\text{NAND} = \text{NOT AND} = \overline{\text{AND}}$$

A DTL NAND gate is shown in Fig. 12.8, and its operation is explained by considering each section separately. In the event that both input lines are at the logic '0' level, point P is at a small positive potential, and point Q is at a small negative potential due to the negative bias potential V_{BB} (in a p-n-p transistor circuit V_{BB} is a positive potential). This condition is maintained if either input line is at zero potential, since the diode in that line is forward biased. Only when both input lines are at the logic '1' level are the diodes reverse biased, and do not pass current. In the latter operating state, the potential of point Q is set by the values of R_1, R_2, R_3, V_{CC}, and V_{BB}; the circuit is so designed that point Q is at a positive potential when all the inputs are at the logic '1' level. From the fore-

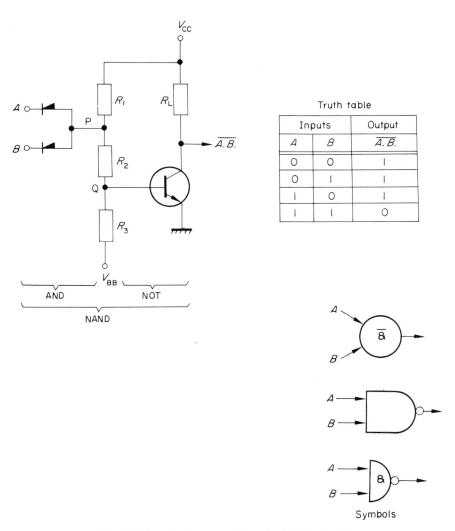

Truth table

Inputs		Output
A	B	$\overline{A.B}$
0	0	1
0	1	1
1	0	1
1	1	0

Fig. 12.8   A diode-transistor logic NAND gate.

going, it is seen that the potential at point Q is the positive logic AND function of the input signals. The signal is then complemented by the NOT section of the gate to give the NAND function overall.

A NAND gate with N inputs gives an output of logic '0' when all the input lines are at the logic '1' level, otherwise the output is logic '1'.

The circuit in Fig. 12.8 is described as a *current sinking logic gate,* since the driving gates must absorb or 'sink' the current flowing in R_1 in each of the driven gates when the driving transistor is saturated.

The circuit design of the DTL NAND gate is more complicated than its equivalent NOR gate, and a detailed design study will not be undertaken here, but approximate equations are given below. It should be understood that these equations are based on certain assumptions (not given here), which may not suit every application.

$$R_L \simeq 2V_{CC}/I_{CM}$$

$$R_1 = MR_L$$

$$R_2 = R_1/3$$

$$R_3 = 3R_1$$

A NAND gate designed using the above equations which uses a transistor with $h_{FEsat} = 25$, and $I_{CM} = 6$ mA with $V_{CC} = -V_{BB} = 6$ V gives the following results. $R_L = 2 \cdot 2$ kΩ (preferred value), $M = 5$, $R_1 = 10$ kΩ (preferred value), $R_2 = 3 \cdot 3$ kΩ (preferred value), $R_3 = 33$ kΩ (preferred value).

12.7   Function generation using NOR and NAND gates

The most important aspect of NOR and NAND gates to industrial users is that both types can generate any of the normal logic functions, e.g., AND, OR, and NOT functions. Circuits generating these functions are shown in Fig. 12.9.

Thus, a NOR gate or a NAND gate with a single input generates the NOT function of the input signal, shown in (a). Three NOR gates in the configuration in (b) generate the AND function, while the same configuration of NAND gates generates the OR function. The configuration in (c) allows the OR and AND functions to be generated by NOR and NAND gates, respectively.

12.7.1   The *S-R* flip-flop

A *set-reset* (*S-R*) *flip-flop* can be constructed from two NOR gates or four NAND gates, shown in Fig. 12.10(a) and (c), respectively. The flip-flop has two input lines shown as S and $R,$ and two output lines Q and $\bar{Q}$ which provide complementary output signals. When the S-line or *set line* is energized, the logic level of output Q becomes '1', and $\bar{Q}$ becomes '0'. When the R-line or *reset line* is energized, output Q becomes '0' and $\bar{Q}$ becomes '1'. Thus, a signal at the

	Circuit	Gates used	Function generated	Logical relationship
(a)	$P \longrightarrow \bigcirc \longrightarrow X$	NOR NAND	NOT	$X = \overline{P}$
(b)	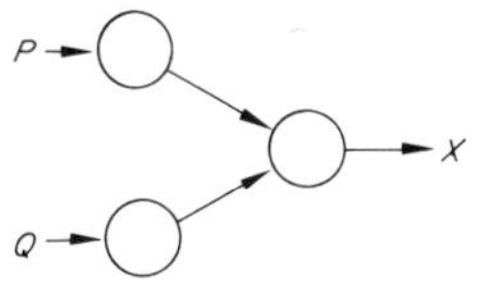 $P \rightarrow$, $Q \rightarrow$ gate $\rightarrow X$	NOR	AND	$X = P.Q$
		NAND	OR	$X = P + Q$
(c)	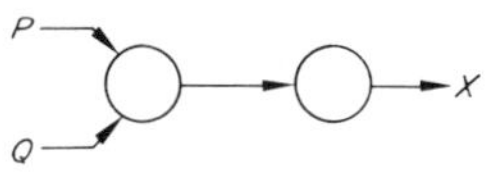 $P \rightarrow$, $Q \rightarrow$ gate $\rightarrow \bigcirc \rightarrow X$	NOR	OR	$X = P + Q$
		NAND	AND	$X = P.Q$

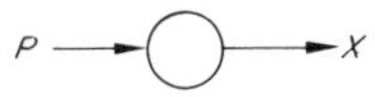

Fig. 12.9   Function generation using NOR and NAND gates.

S-line sets output Q to the '1' state, and a signal at the R-line resets output Q to zero.

The NOR circuit in Fig. 12.10(b) operates as follows. Assume that TR2 is saturated ($Q = 0$), and TR1 is cut off ($\overline{Q} = 1$). In the event that both input signals are zero, the logic '1' signal fed back to the base of TR2 from the collector of TR1 holds TR2 in the saturated state. This ensures that the voltage at the collector of TR2 is practically zero, since both input signals to TR1 (i.e., S and Q) are zero, TR1 remains cut off and its output is logic '1'. This is one of the two stable operating conditions with $Q = 0$, $\overline{Q} = 1$.

The application of a logic '1' signal to the S input line causes TR1 to saturate, and its output voltage falls to zero. The input conditions of TR2 have now changed, both inputs being zero. This causes TR2 to be cut off, forcing its collector potential to rise to the logic '1' level. This is the second stable operating state of the circuit, with $Q = 1$, $\overline{Q} = 0$. The signal applied to the S-line can now be reduced to zero, since the logic '1' signal fed back from output Q to the input of TR1 holds it in the saturated state. The flip-flop can be reset to its original state by applying a logic '1' signal to the reset input line of the circuit.

Both the R-line and the S-line may be energized simultaneously, resulting in both outputs being zero. This operating condition is generally not recommended,

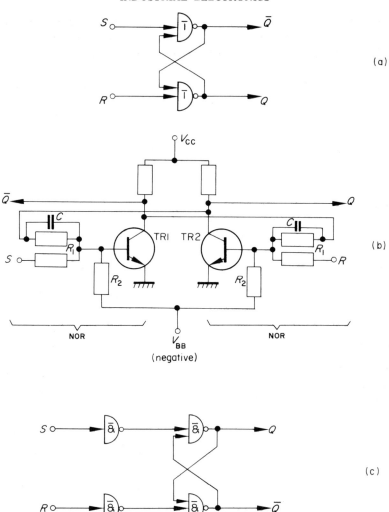

Fig. 12.10   Construction of an *S-R* flip-flop using (a) NOR gates, and (c) NAND gates. A practical *S-R* flip-flop using RTL NOR gates is shown in (b).

as the final output state is not only dependent on which input signal is the last to fall to zero but also upon the relative operating speed of each half of the circuit.

Capacitors *C* are used to improve the switching speed of the circuit; their function is to ensure that any change in voltage at the collector of one transistor is rapidly transmitted to the base of the other transistor. The value of capacitance used is of the order of a few hundred picofarad. Unfortunately, any electrical noise present at the collector is also transmitted in the same fashion,

and may cause spurious triggering of the flip-flop. This problem can be overcome by eliminating the capacitor and replacing the transistors with others of
a higher switching speed (i.e., transistors with a higher cut-off frequency).

The design of an S-R flip-flop similar to the one in Fig. 12.10(b) follows the
general procedure for the RTL NOR gate; if the fan-out of the flip-flop is to
be M, then it must be designed to drive (M + 1) similar gates since it has to
drive the opposite half of the flip-flop as well as the connected gates. When
using this type of circuit, care has to be taken not to apply an excessively un-
balanced load, otherwise the output voltage on the loaded side falls to a level
which prevents stable operation.

12.7.2   The 'trigger' flip-flop

Many forms of flip-flop have been developed by using the S-R flip-flop in conjunction with additional input circuitry*. The basis of the input circuitry
associated with the trigger flip-flop is the R-C circuit in Fig. 12.11(a). In this
circuit, the diode allows only negative-going signals to be passed to the output
terminals; positive-going signals cause the diode to be reverse biased. When the
voltage applied to input Y is zero (i.e., it is earthed), the leading edge of the
square wave applied to input X causes the capacitor to draw a charging current,
and a positive-going voltage spike is generated at point A, the duration of the
spike being dependent upon the time-constant of the network. Since a positive
voltage at A causes the diode to be reverse biased, the output from the network
is zero. When the signal at input X is reduced to zero the capacitor discharges,
taking the cathode of the diode to a negative potential, causing a negative-
going pulse to appear at the output terminal.

If input terminal Y is held at the logic '1' level, the cathode of the diode
never becomes negative whatever the change in logic level at input X, and the
output voltage remains zero. As a result, the output from the circuit in Fig.
12.11(a) is a negative-going pulse which is generated when the trailing edge
(a logic '1' to logic '0' change) of a square wave is applied to input X and input
Y is at zero potential, otherwise the output is zero.

The way in which the R-C pulse circuit is used with an S-R flip-flop is shown
in Fig. 12.12. Here two identical R-C networks are combined so that negative-
going pulses at A are transmitted to the base of TR2, and negative-going pulses
at A' are transmitted to the base of TR1. The combined R-C network is
described as a *pulse steering network.* Inputs Y and Z are energized at complementary logic levels, so that when one is at the logic '1' level the other is at
zero. The signal applied to input X is in the form of a series of rectangular
pulses in the manner of input X in Fig. 12.11.

Suppose that, initially, TR1 is cut off ($Q = 1$) and TR2 is saturated ($\overline{Q} = 0$).
If input Y is at the logic '0' level and input Z is at the logic '1' level, then a

---

* See *Logic Circuits* by N. M. Morris, published by McGraw-Hill, (1969).

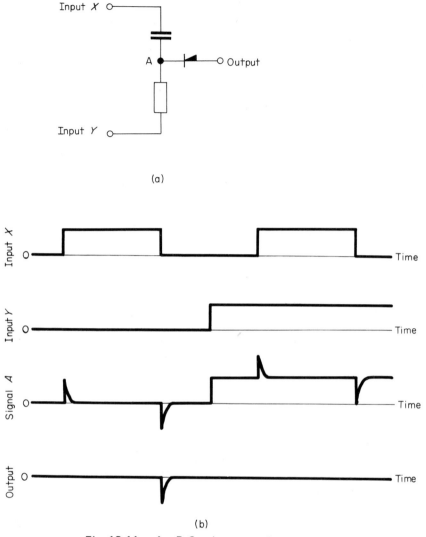

(a)

(b)

Fig. 12.11   An R-C pulse generating network.

negative-going pulse is transmitted to the base of TR2 whenever the trailing
edge ('1' → '0' edge) of the signal at input X occurs. No signal is applied to TR1.
The negative pulse has the effect of robbing TR2 of its base current, causing
it to be cut off; the output of TR2 then rises instantly to the logic '1' level.
Owing to the regenerative feedback action within the S-R flip-flop, TR1 is driven
into saturation, and its output voltage falls to zero.

If inputs Y and Z are maintained at the initial voltage levels, negative pulses
continue to be 'steered' to the base of TR2 at each trailing edge of the signal

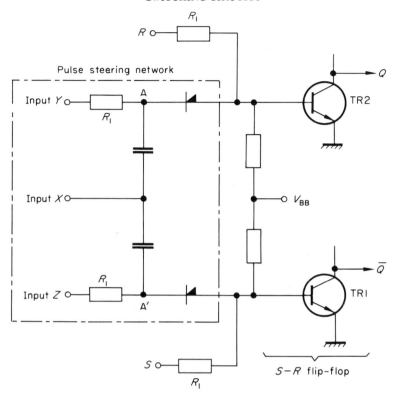

Fig. 12.12   A pulse steering network for a trigger flip-flop.

applied to input X. This has no further effect upon the state of the flip-flop, since the pulses continue to try to turn off a transistor which is already cut off.

If the voltage levels at inputs Y and Z are reversed so that Y = '1', Z = '0', then at the next '1' → '0' edge of the pulse applied to input X a negative pulse is applied to the base of TR1, turning it off. Regenerative action within the flip-flop causes TR2 to be saturated, and the outputs assume their original states (i.e., $Q = 0$, $\overline{Q} = 1$).

Alternative versions of the circuit are in use, but the general principle in all cases is the same insomuch that the conducting transistor is turned off upon the incidence of the trailing edge of the input square wave at X. The names of *toggle* and *Eccles-Jordan trigger circuit* are also given to this circuit.

Overriding 'set' and 'reset' to the trigger flip-flop can be exercised by applying signals to the S and R lines, respectively. This allows the flip-flop to be set to any desired state. In Fig. 12.12, all the resistors marked R_1 have equal values.

A *serial counter* or *ripple-through counter* operating in the pure binary code is shown in Fig. 12.13. It uses a number of the trigger flip-flops described above. Suppose that the initial states of the flop-flops are $A = B = C = D = 0$, where A

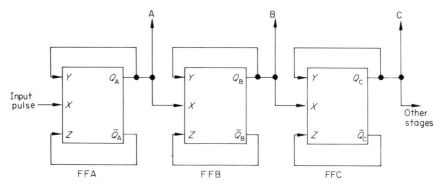

Fig. 12.13   A simple pure binary counter using trigger flip-flops.

is the output at the Q-line of FFA, B is the output of FFB, etc. Since $A = 0$, the Y input to FFA is zero, and the Z input of the same flip-flop is logic '1'. Upon the incidence of the trailing edge of the first input pulse, TR2 in FFA is turned off, causing output A to rise to the '1' level. This signal change is applied to input X of FFB but, as explained above, the leading edge ('0' → '1' edge) of the waveform applied to the trigger flip-flop does not alter the state of the flip-flop, and the output of FFB remains unchanged. When the trailing edge of the second input pulse arrives at X, a negative pulse is steered to the base of TR1 of FFA (since input Z is '0' after the first pulse), causing output A to fall to zero. This prepares the input circuit of FFA to steer the next negative pulse to the base of TR2.

The '1' → '0' change at output A also triggers FFB into operation, causing output B to rise to the logic '1' level. The counting process is seen to proceed in a serial manner in this flip-flop, and the input pulse is said to 'ripple-through'

Table 12.2

Counting sequence of a simple pure binary counter

Decimal value	Binary output		
	Flip-flop		
	C (4)	B (2)	A (1)
0	0	0	0
1	0	0	1
2	0	1	0
3	0	1	1
4	1	0	0
5	1	0	1
6	1	1	0
7	1	1	1

the counter. The complete sequence of events for the counter is given in Table 12.2, the decimal values or 'weights' of outputs A, B, and C are 1, 2, and 4, respectively.

12.8   The Schmitt trigger circuit

The Schmitt trigger circuit is a version of the long-tailed pair amplifier, and is used as a pulse height discriminator in which the output voltage from the discriminator abruptly increases when the input signal rises to a certain level, and abruptly falls when the input signal falls to a slightly lower level. The 'hysteresis' effect is necessary to ensure that the output voltage from the discriminator does not oscillate when a very slowly varying input signal is applied.

The basic circuit is shown in Fig. 12.14, and comprises an emitter coupled amplifier with a regenerative feedback network R_1R_2 between the transistors. When either transistor is conducting, the potential of the emitter connection is positive with respect to the common line. If the input signal potential is lower than that of the common-emitter point, the emitter junction of Q1 is reverse biased, and the transistor is cut off. The collector potential of Q1 is then at its highest value, and the ohmic values of R_L, R_1, and R_2 are such that Q2 is driven into saturation under this condition, and the collector potential of Q2 is at its lowest value.

When the input voltage exceeds the common-emitter potential, Q1 begins to turn on and its collector potential begins to fall. This has two immediate effects:

(a) The emitter potential begins to rise due to an emitter-follower action.
(b) The base potential of Q2 falls due to the feedback between the collector of Q1 and the base of Q2.

As a consequence of these actions, Q2 is very rapidly cut off and its collector potential rises to a high level.

The transition of the voltage level at the collector of Q2 is much faster than at the collector of Q1, and the output is normally taken from the collector of Q2 for this reason. The rise-time of the output waveform is further reduced by the use of the speed-up capacitor C which has a value of a few hundred pico-farad. A feature of the Schmitt trigger circuit is that both output signals are non-zero, and a simple way of reducing the lowest output potential to zero is by the use of the Zener diode circuit shown in broken lines. Alternatively, the output from Q2 can be used to drive a NOT gate to provide the dual features of increased fan-out and improved waveform.

The circuit can be further modified at the input to give adjustment of the turn-on voltage by adding the bias circuit shown in broken lines. The capacitor in the bias circuit is included for decoupling purposes.

The hysteresis effect can be modified to some extent by using resistors in the two collector circuits with different values. Normally, the resistance included in the collector circuit of Q2 is equal to or is less than that in the collector of Q1. To illustrate the operation of the circuit with widely differing resistor values, let us assume that the value of the resistance used in the collector circuit of Q2 is much lower than in the Q1 circuit. Initially, with a low input voltage, Q2 is saturated and the potential of the two emitters is high due to the low value of the collector circuit resistance of Q2. When the input voltage reaches a value which is equal to the potential of the emitters, Q1 turns on, and the

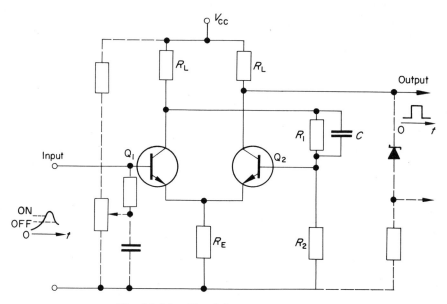

Fig. 12.14   The Schmitt trigger circuit.

current in R_E falls to a lower value due to the effect of the collector circuit resistance of Q1. Before the trigger circuit can be returned to its original state, the input signal level must fall below that of the emitter, which is a much lower value than that of the turn-on voltage due to the current difference. Generally speaking, the collector resistor of Q2 affects the turn-on voltage level and the collector resistor of Q1 affects the turn-off voltage level.

Example 12.3: Design a Schmitt trigger circuit which uses a transistor with the following parameters: $h_{FEsat} = 20$, $V_{BEsat} = 0.3$ V, $V_{CEsat} = 0.15$ V. The current drain is to be 5 mA, and the supply voltage is 10 V.

Solution:

Q1 OFF, Q2 ON
The general circuit conditions are illustrated in Fig. 12.15. It is first necessary to select a voltage at which the circuit should trigger. The voltage V_E should

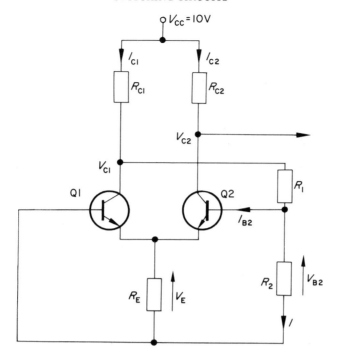

Fig. 12.15   Figure for example 12.3.

not be too large otherwise the output voltage change becomes small, and it must not be too small otherwise it may not be possible to provide an adequate reverse bias voltage at the base of Q2 when Q1 is saturated. Let us select $V_E = 2$ V.

Now,

$$V_{C2} = V_E + V_{CEsat} = 2 \cdot 15 \text{ V}$$

therefore,

$$I_{C2} R_{L2} = 10 - 2 \cdot 15 = 7 \cdot 85 \text{ V}$$

or

$$R_{L2} = 7 \cdot 85/5 = 1 \cdot 57 \text{ k}\Omega \quad \text{a value of } 1 \cdot 5 \text{ k}\Omega \text{ is suitable}$$

R_{L1} should have a value equal to or greater than $1 \cdot 5$ kΩ. For this example, let us assume that $R_{L1} = 1 \cdot 5$ kΩ.

To ensure that there is an adequate base drive to Q2 to maintain it in a saturated state, the current drawn by the resistor chain $R_1 R_2$ should be greater than I_{B2}.

Now,

$$I_{B2} = I_{C2}/h_{FEsat} = 5/20 = 0 \cdot 25 \text{ mA}$$

Let

$$I = 3I_{B2} = 0.75 \text{ mA}$$

hence,

$$I_{C1} = I + I_{B2} = 1 \text{ mA}$$

and

$$V_{CC} - V_{C1} = I_{C1} R_{L1} = 1 \text{ mA} \times 1.5 \text{ k}\Omega = 1.5 \text{ V}$$

therefore,

$$V_{C1} = 6 - 1.5 = 4.5 \text{ V}$$

Now, when Q2 is saturated,

$$V_{B2} = V_E + V_{BEsat} = 2 + 0.3 = 2.3 \text{ V}$$

and

$$V_{C1} - V_{B2} = (I + I_{B2}) R_1$$

therefore,

$$R_1 = (4.5 - 2.3)/1 = 2.2 \text{ k}\Omega$$

also

$$R_E = V_E/I_{C2} = 2/5 = 0.4 \text{ k}\Omega \quad \text{(say 390 } \Omega\text{)}$$

Q1 ON, Q2 OFF

The instant Q1 is triggered into saturation, a reverse bias voltage must be applied to the emitter junction of Q2. Let this have a value of 0·75 V.

$$V_{B2} = V_E + V_{BE2} = 2 - 0.75 = 1.25 \text{ V}$$

and

$$V_{C1} = V_E + V_{CEsat} \simeq 2 + 0.15 = 2.15 \text{ V}$$

Since $I_{B2} = 0$ in this operating state, the current through R_1 passes through R_2, and

$$\frac{R_1 + R_2}{R_2} = \frac{2.15}{1.25} = 1.72$$

or

$$R_2 = R_1/0.72 = 2.2/0.72 = 3.06 \text{ k}\Omega$$

(either 2·7 kΩ or 3·3 kΩ may be used)

12.9   Power switching circuits

So far, we have dealt only with low current switching circuits. In power electronics, currents of many thousands of amperes at high voltages have to be controlled. The most popular semiconductor device used to control this order of current is the thyristor, with the triac (bi-directional thyristor) increasing in popularity.

A simple thyristor circuit for half-wave control of a single-phase a.c. supply is shown in Fig. 12.16(a). The triac, inset (i), can replace the thyristor if a full-wave output is required. The control circuit provides either a series of pulses or a phase-shifted sinusoidal output; details of suitable circuits are given later.

The thyristor can be triggered into conduction whenever the anode is positive

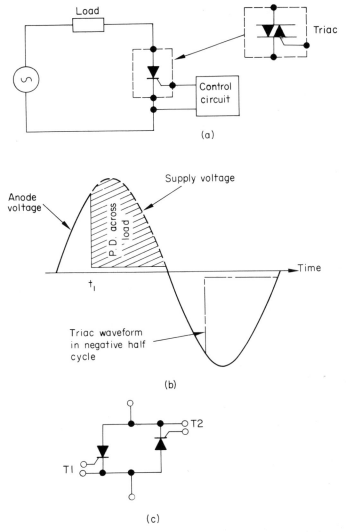

Fig. 12.16   A power switching circuit using a thyristor. The thyristor may be replaced by a triac (inset) if a full-wave output is required. Typical waveforms are shown in (b). An inverse-parallel connected pair of thyristors is shown in (c).

with respect to the cathode; when this happens the voltage across the thyristor falls to about 1 V, the remainder of the supply voltage being dropped across the load. In Fig. 12.16(b), the thyristor is triggered into conduction at time t_1; the waveform shown by the full line is the anode voltage of the thyristor, while the cross-hatched area is developed across the load. When using a triac as the controlling element, conduction can be forced in the negative half-cycle by applying either a positive or negative voltage to the gate of the triac. This is indicated by the chain-dotted waveform in Fig. 12.16(b).

A popular full-wave thyristor circuit is the *inverse-parallel connection* shown in Fig. 12.16(c). In this circuit, the two trigger signals T1 and T2 must be electrically isolated from one another, otherwise a short-circuit will develop across the supply. It is common to apply the gate pulse in this circuit through a double-wound transformer with two isolated secondary windings, one for each thyristor.

The applications of thyristors are many and varied, and range from logic circuits to the control of electrical machine systems with ratings of many megawatts. Details of these systems are available elsewhere*.

12.9.1   Phase-shift control

A simple form of R-C phase-shift control circuit is shown in Fig. 12.17(a), together with typical component values for use with a thyristor; the transformer output voltage in this case is 10-0-10 V. When resistor R is reduced to zero, point A is connected to point N and the voltage induced in the upper half of the transformer secondary winding is applied to the thyristor gate. Since the secondary voltage in this half of the winding is in phase with the primary voltage, the phase difference between the gate voltage and the anode voltage is zero, and the thyristor is triggered into conduction at the commencement of the positive half-cycle of the supply voltage. In this condition, the circuit between points N and L contains only capacitance C, and current I leads V_{NL} by 90 degrees.

As R is increased in value, the phase lead of current I on V_{NL} is reduced, as illustrated in the phasor diagram in Fig. 12.17(b). If the loading effect of the thyristor gate circuit on the transformer is small, then the locus of the voltage phasor of point A lies on a semicircle since the voltage V_R across resistor R is at right-angles to the voltage V_C across capacitor C. The phase-shift of the voltage V_{AM} is computed from triangle LAN as follows:

$$\tan \alpha/2 = \frac{V_R}{V_C} = \frac{IR}{I/\omega C} = \omega CR$$

or

$$\alpha = 2 \tan^{-1} \omega CR$$

* See *Control Engineering* by N. M. Morris, published by McGraw-Hill (1968).

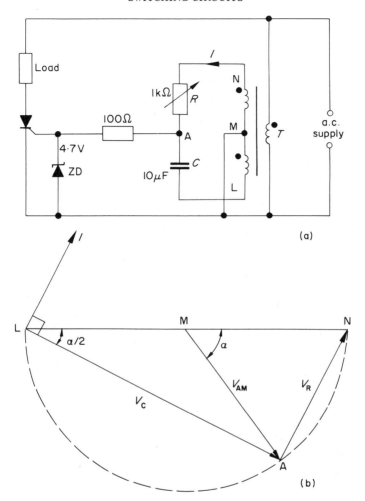

Fig. 12.17 (a) A thyristor circuit with a simple *R-C* phase-shift control circuit, and (b) shows the phasor diagram of the control circuit.

Thus, as R is increased in value the voltage V_{AM} is *phased-back* with respect to the supply voltage; as R is reduced in value, the phase lag of V_{AM} is reduced, and it is said to be *phased-forward*.

The 100 Ω resistor and the Zener diode shown in Fig. 12.17(a) act as a protective circuit for the thyristor gate; the diode limits the maximum forward voltage applied to the gate of the thyristor to the Zener breakdown voltage of the diode, the maximum reverse voltage applied to the gate being practically zero since the Zener diode then operates as a conventional forward biased diode. The 100 Ω resistor merely acts as a current limiting resistor.

12.9.2   Pulse generator circuits

Perhaps the simplest pulse generator is the unijunction transistor circuit in Fig. 12.18(a). This circuit operates on the same basic principle as the sawtooth generators described in section 11.7, as follows. If the capacitor is initially uncharged, the voltage at point X begins to rise in an exponential manner until it reaches the peak-point voltage of the transistor. At this instant of time, the

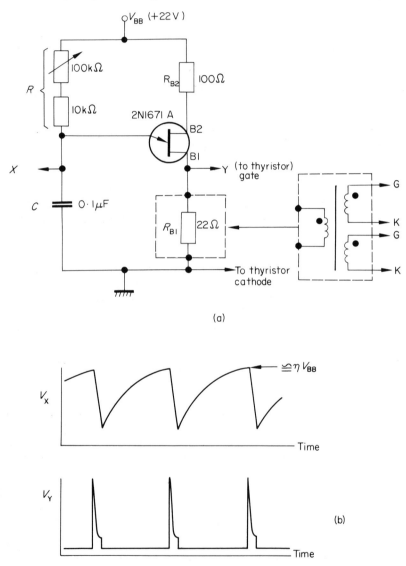

(a)

(b)

Fig. 12.18   (a) A relaxation oscillator circuit using a unijunction transistor, and (b) typical waveforms.

U.J.T. switches to its low resistance conducting mode and the capacitor is discharged through the 22 Ω resistor, causing a positive-going pulse to be generated at Y. The pulse repetition rate is controlled by the value of R, since this controls the time constant RC of the capacitor charging circuit, and the pulse width by R_{B1} since this affects the discharge time constant.

Assuming that the capacitor is initially uncharged, then the voltage at point X prior to breakdown is given by

$$V_X = V_{BB}(1 - e^{-t/RC})$$

where RC is the charging time constant of the resistor-capacitor circuit, and t is the time from the commencement of the waveform. Discharge occurs when V_X is equal to the peak-point voltage, which is taken to be equal to ηV_{BB} (where η is the intrinsic stand-off ratio of the U.J.T., and V_{BB} is the supply voltage). That is when

$$\eta V_{BB} = V_{BB}(1 - e^{-t/RC})$$

Hence, the periodic time is given approximately by

$$t = RC \ln 1/(1 - \eta) = 2\cdot 3 RC \lg 1/(1 - \eta)$$

A typical value for η is $0\cdot 55$, giving a periodic time of approximately $0\cdot 8RC$ s.

The function of resistor R_{B2} is to provide temperature stability, otherwise it has little effect upon the performance of the circuit. However, as a result of this resistor, a negative pulse is generated at base-two and can be used for other control purposes.

The pulse which is generated at point Y is of short duration, being typically 10 μs to 15 μs, and may be coupled directly to the gate terminal of the thyristor. As an alternative mode of connection, it may be coupled to the thyristor through an R-C network similar to that used between the stages of an R-C coupled amplifier (typical values being $C = 0\cdot 1$ μF, $R = 560$ Ω). Where complete isolation is required, a pulse transformer can be used, as shown in the inset in Fig. 12.18(a). The advantage of the pulse transformer is that a number of separate outputs can be used to trigger parallel (or inverse-parallel) connected thyristors.

A complete circuit showing the control and load sections is shown in Fig. 12.19. Resistor R_1 and the Zener diode form a voltage limiting circuit which performs two functions. Firstly, it limits the voltage applied to the control circuit in the positive half-cycle of the supply voltage to the breakdown voltage of the Zener diode. Secondly, during the negative half-cycle it restricts the voltage applied to the control circuit to about $- 0\cdot 8$ V, and provides a discharge path for the energy stored in capacitor C. As a result of the latter feature, the capacitor is fully discharged at the commencement of each positive half-cycle, ensuring that the first pulse generated by the U.J.T. occurs after a time interval of $(RC \ln 1/(1 - \eta))$ s after the commencement of the positive half-cycle. This also ensures that the thyristor is triggered into conduction at the same point in each positive half-cycle.

The circuit can be modified to be voltage controlled by replacing resistor R by a transistor (an ASY26 is suitable with the circuit shown). The pulse repetition rate is then controlled by the base-emitter voltage of the transistor. In this configuration, the capacitor charging current is a function of the transistor base current. If the base current is constant, then the waveform at the emitter of the U.J.T. is practically a linear ramp.

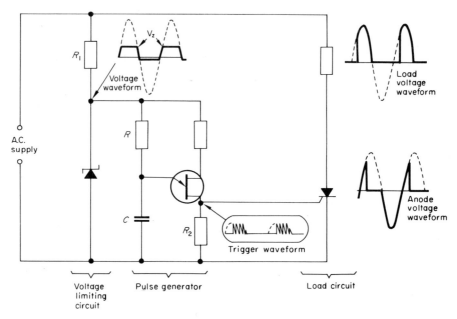

Fig. 12.19   A power switching circuit with details of the control circuit.

With a little modification, the circuit in Fig. 12.19 can be used to control a triac. All that is required is to replace the thyristor with a triac, and the supply to resistor R_1 and the Zener diode must be derived from a full-wave rectifier circuit. The latter feature allows pulses to be generated in both half-cycles.

In addition to the circuits described above, a multiplicity of other circuits including blocking oscillators, monostable multivibrators, and magnetic amplifiers has been used for thyristor and triac applications.

12.9.3   Mercury-arc convertor circuit

Mercury-arc convertors are ideally suited to certain types of drive, e.g., steel rolling mills, where large motors, often of 8000 hp or more have to be accelerated from standstill to full speed, and a short time later rapidly reversed. During the acceleration period, the convertor rectifies the alternating current and supplies

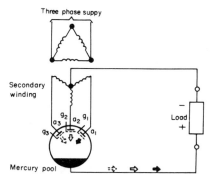

Fig. 12.20   Three-anode mercury-arc convertor.

energy to the mill motor; to produce the rapid deceleration required, the convertor inverts, allowing the kinetic energy of the motor and load to be converted into electrical energy, which is returned to the incoming supply lines. The latter process is known as *regenerative braking*. Other notable applications include traction, mine winders, and electrolytic processes.

As with thyratrons, the grids can only control the point of ignition of the arc to its own anode. The anode current falls to zero only if the anode potential is reduced to zero, or when the arc commutates to another anode at a higher potential. A three-anode convertor is shown in Fig. 12.20, the order of firing being a_1, a_2, a_3. The load voltage is dependent on the angle of conduction; using grid control, the point at which current flow commences in any anode is controlled by its grid circuit. Providing that the grid potentials are sufficiently negative, conduction can be completely prevented, resulting in zero load voltage.

As in the case of the thyratron, the grid voltage consists of a steady negative potential with either a sine wave or a series of pulses superimposed upon it, the

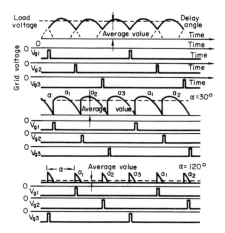

Fig. 12.21   Output waveforms for a three-anode convertor for various values of α.

maximum output occurring when natural commutation takes place between the anodes. Pulsed control is more precise than sinusoidal control since pulses can be generated at known times relative to the supply waveform; the capital cost of grid circuit equipment for pulsed control is greater than for a super-imposed sinusoidal signal. Output voltage waveforms for various firing delay angles with a resistive load are shown in Fig. 12.21.

Problems

12.1   What is meant by n-type and p-type impurity doping in a semiconductor device? Illustrate your answer with diagrams of the crystal lattice structure.

Explain the principle of operation of an n-p-n transistor and sketch its output characteristics. Show how it may be used as an electronic switch and indicate its limitations in this mode of operation.

(C & G)

12.2   Draw up truth tables for the NAND and NOR logic gates with two input lines.

Show that the logical NOT, AND, and OR functions can be generated using *either* a number of NAND *or* a number of NOR gates. Supplement your explanation with diagrams.

Explain, with the aid of diagrams, the operation of an *S-R* flip-flop constructed from NOR gates.

(C & G)

12.3   Describe the operation of a resistor-transistor NOR element.

12.4   Sketch the circuit diagram of an *S-R-T* flip-flop employing p-n-p transistors. Describe how the input 'pulse steering network' operates.

12.5   Explain, using suitable diagrams, the operation of diode logic circuits for performing the AND and OR functions, using (a) positive logic, and (b) negative logic.

Discuss briefly the relative merits and disadvantages of diode logic compared with other types of logic circuits.

(C & G)

12.6   Why are logic functions so named?

By means of (a) Boolean algebra and (b) truth tables, prove the following Boolean equations:

$$\overline{A \vee B \vee C} = \overline{A} . \overline{B} . \overline{C} \quad A \vee \overline{A} . \overline{B} = A \vee \overline{B}$$

Note: 'v' represents logic OR. '.' represents logic AND. '⁻' represents logic NOT.

(C & G)

12.7   Describe, with the aid of a suitable diagram, the operation of a bistable circuit using two p-n-p transistors. Discuss the minimum changes necessary to turn the circuit into

(a) A free-running multivibrator.
(b) A monostable circuit (one-shot multivibrator).

(C & G)

12.8   Sketch a circuit diagram and explain the action of a Schmitt trigger circuit using either valves or transistors.

Explain briefly with the aid of a block diagram the use of a Schmitt trigger circuit in industrial equipment.

(C & G)

12.9 Describe, with the aid of basic circuits and waveform diagrams, the two principal methods, viz. 'horizontal' (i.e. variable phase) and 'vertical' (i.e. variable bias), of controlling the firing of a gas-filled rectifier valve. State the advantages of applying a 'peaky' wave to the control grid of such a valve and give a method of producing such a waveform from one of sinusoidal shape.

(C & G)

12.10 A 6-anode mercury-arc rectifier is to be used to provide 6-phase rectification from a 3-phase supply. Sketch:

    (a) A circuit diagram showing the transformer, rectifier, and a resistive load;

    (b) The transformer phase voltage waveforms and the output voltage waveform for one complete cycle of supply voltage;

    (c) The output voltage waveform with grid control if the firing angle is retarded by 60°.

(C & G)

12.11 Explain the principle of operation of the thyristor. Sketch the static characteristic of a thyristor noting the salient features on the graph.

    Draw a circuit diagram to show how two thyristors can be connected in inverse parallel to supply controlled armature current to a d.c. motor, which has constant excitation.

(C & G)

12.12 (a) Sketch the circuit diagram for, and explain the operation of, a silicon controlled rectifier (thyristor) controlling the current in a load.

    (b) Draw a circuit diagram for an ignitron with external control and load circuits.

    State any protective devices that should be included in the circuits of both (a) and (b).

(C & G)

12.13 Explain the principle of the thyristor. Describe how thyristor convertors may be used to control the speed of an induction motor over a wide range.

    A thyristor convertor is to be used to control the speed of a d.c. motor and the maximum average current is to be 10 A. The supply is from a single-phase alternating supply of 220 V r.m.s. and the armature current is to be limited by connecting a resistor in series with the armature. Calculate the minimum value of this resistance if the armature resistance is 1 Ω.

12.14 Describe any one type of solid state controllable rectifier. Show how this could be embodied in a system for the control of the voltage or current delivered to a variable load, and briefly describe the system operation. State what safety features should be incorporated in the system.

(C & G)

12.15 Describe an ignitron rectifier and give a circuit diagram illustrating the use of such a device in any closed-loop application. Specify any protective devices which should be used.

(C & G)

12.16 Explain briefly why Boolean algebra is used in the design of modern digital systems.

    By means of (a) Boolean algebra and (b) truth tables, prove the following Boolean equations:

$$A \vee \overline{A} . B = A \vee B \qquad \overline{A} \vee \overline{B} \vee \overline{C} = \overline{A.B.C}$$

(C & G)

13. Regulated power supplies

13.1 Basic requirements

The regulated power supply is perhaps the most widely used piece of circuitry in electronics. Despite this fact, or perhaps because of it, it is probably the most misused equipment on the laboratory bench, having to sustain excessive overload conditions over very long periods of time. The advent of transistorized power supplies brought an awareness of these problems, and circuit designers have used great ingenuity to overcome them.

The basic block diagram of a regulated power supply is shown in Fig. 13.1. Broadly speaking regulated power supplies or *stabilized power supplies* can be reduced to four main areas:

(a) a.c. to d.c. supplies.

(b) d.c. to d.c. supplies.

(c) a.c. to a.c. supplies.

(d) d.c. to a.c. supplies.

Type (a) is used more than any other, and is the subject of this chapter. The remaining three groups are more specialized, and are dealt with in other texts.

In a.c. to d.c. supply systems, the unstabilized power supply is a simple rectifier and smoothing circuit. The output from the unstabilized power supply is applied to a controlling device which regulates the voltage to give a steady d.c. output. A proportion of the output signal is fed back, usually from a resistive type of network, and is compared with a reference signal. The difference between the two signals is then used to actuate the controlling element in such a way that it maintains the level of the output quantity at a constant value. The regulators described here are, clearly, negative feedback networks which contain an amplifier within the feedback loop.

338

These regulators can be further subdivided into two groups:

(a) Series regulators,
(b) Shunt regulators.

In a *series regulator,* the controlling device is connected in series with the load, and to regulate the output it must, at all times, absorb some of the supply voltage. Thus, if the series controlling device is a transistor and the unstabilized supply voltage is 20 V, then something like 5 V to 10 V would need to be dropped across the transistor. This allows the voltage across the transistor to be

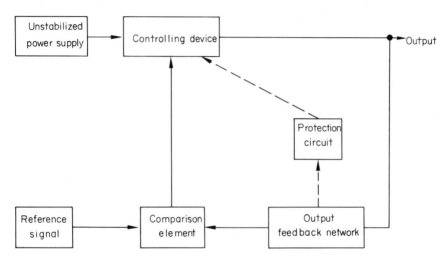

Fig. 13.1   A block diagram of a regulated power supply.

altered by the feedback signal to accommodate variations in the supply voltage and load resistance. In the case of series valve regulators, the voltage dropped across the valve is much greater, but they are capable of dealing with rather higher supply voltages than are transistors, a factor which is rapidly being overcome by improvements in semiconductor devices.

In a *shunt regulator,* the controlling device is in parallel with the load, and to effect output regulation it must pass current at all times. When the current through the controlling device falls to zero, regulating action ceases.

In more sophisticated power supplies, protection circuits are built into the network to limit one or more quantities to a safe level. For example, in the event of an excessive overload or a short-circuit being applied to a semiconductor voltage regulator, to protect the semiconductor regulating device the current must be switched off or reduced to a safe value within 10 μs to 100 μs. Fuses cannot provide this form of protection, and feedback techniques are used. The problem is not so serious in regulators using thermionic devices, and a thermal cut-out can be used. A feature of semiconductor devices is that when they fail catastrophically, they usually develop an internal short-circuit between the collector and emitter. In series regulators, this means that the unregulated power supply will be directly connected to the output terminals, possibly causing damage to the connected load. This can, in some instances, cause damage to large installations and it calls for additional means of protection. One method, known as *crowbar voltage protection*, employs a voltage-sensitive circuit which monitors the output voltage of the regulator; this circuit instantly applies a short-circuit to the output terminals whenever the voltage rises above a predetermined level. A thyristor can be used as a crowbar device. An inherent advantage of a shunt voltage regulator in the above case is that a short-circuit in the regulating device causes the output voltage to fall to zero instantly.

The reference voltage source in regulator circuits is usually either a Zener diode or a cold-cathode discharge tube. The design procedure for this type of circuit is considered in section 13.3.

13.2 Stabilization factor and output resistance

The degree of stabilization against output voltage variation offered by a constant-voltage regulated power supply is given by the *stabilization factor S,* where

$$S = \frac{\delta V_0}{\delta V_1} = \frac{\text{Change in output voltage}}{\text{Change in supply voltage}} \qquad (13.1)$$

the measurements being made with a constant value of load resistance. Strictly speaking, the stabilization factor should be computed on the basis that *only* V_0 *and* V_1 change. In practice, other quantities do change, e.g., the load resistance, the transistor or valve parameters, etc., but for our purposes they are considered to be constant.

In relatively simple regulators, S may have a value of 0·005 (a change of 5 mV/V), while sophisticated supplies provide values of S which can be smaller than 0·0002 (i.e., < 0·2 mV/V). Alternatively, the stabilization against voltage change may be defined as the inverse of S, when it is described as the *stability ratio*; in the above example, the stability ratios are 200:1 and 5000:1, respectively.

The *output resistance* of a power source is defined as

$$R_0 = -\frac{\delta V_0}{\delta I_0} = \frac{\text{Change in output voltage}}{\text{Change in output current}} \qquad (13.2)$$

the measurement being made with a constant value of input voltage. The negative sign in eq. (13.2) implies that an increase in load current results in a reduction in output voltage. The output resistance of a relatively uncomplicated voltage source for low voltage applications (e.g., up to 100 V) may be 20 mΩ, and in an equivalent sophisticated unit it may be 2 mΩ; in a 10 kV e.h.t. supply the output resistance could quite easily be 5 kΩ. The output resistance of a constant current source is, typically, many tens of thousands of ohms.

When a power supply is used to energize electronic amplifier circuits working in the audio-frequency and higher ranges, we may be more concerned with its output impedance than its output resistance. Typical values of output impedance are 0·1 Ω and 30 mΩ for a simple supply and a sophisticated supply, respectively.

Under resistive loading conditions, a small ripple normally appears at the output terminals of the regulator in addition to the steady output since its operation is never perfect. The ripple is expressed in millivolts or microvolts for constant voltage sources and milliamperes or microamperes for constant-current sources. Ripple and noise voltage are normally expressed in terms of their peak-to-peak values, and any potential user of the equipment should also be given information of the bandwidth of the instrument used to measure the ripple and noise. If the bandwidth of the measuring instrument is small, then high frequency spikes which may be detrimental to the load are concealed. Typical ripple figures to be expected from a simple supply and a relatively sophisticated low voltage supply are typically 10 mV peak-to-peak and 200 μV peak-to-peak, respectively; in a 10 kV unit the ripple voltage may be 3·5 V peak-to-peak.

13.3 Voltage reference sources

The simplest form of voltage reference source comprises a Zener diode and a current limiting resistor R_S, shown in Fig. 13.2(a). It is possible to use any Zener diode in this configuration, but if the output voltage is to be reasonably free from temperature drift the breakdown voltage should be in the region of 5 V; these diodes have the lowest voltage-temperature coefficient (see chapter 3) of the Zener diode range. Stabilized voltages at higher levels can be obtained by connecting 5 V diodes in series. One method of compensating for a positive voltage-temperature coefficient is to connect a silicon diode as shown in the inset in Fig. 13.2(a). As the temperature rises, the Zener breakdown voltage increases, and the forward breakdown drop across the forward biased diode reduces; with the correct combination of diodes, the net variation in output voltage with variation in temperature is very small. For output voltages in the region of 80 V to 150 V, it is often more convenient to use a cold-cathode discharge tube.

The circuit in Fig. 13.2(a) is, in fact, the simplest possible form of shunt voltage regulator. The basic design criterion of the shunt regulator circuit is that the Zener diode must always carry current. The moment that the diode stops carrying current, it assumes a reverse blocking state, and regulating action ceases. The current I drawn from the supply in Fig. 13.2(a) is

$$I = I_Z + I_L$$

and

$$I_Z = (V_0 - V_Z)/r_Z$$

where r_Z is the slope resistance of the Zener diode.

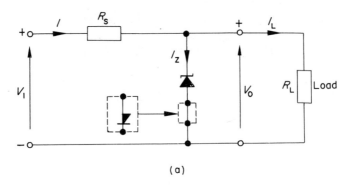

(a)

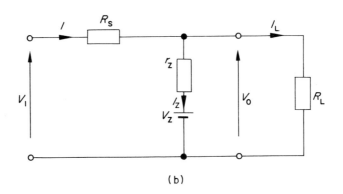

(b)

Fig. 13.2   A simple voltage reference source.

Hence,

$$I = \frac{V_0 - V_z}{r_z} + I_L \qquad (13.3)$$

The current I is also given by the equation

$$I = (V_1 - V_0)/R_S$$

Therefore

$$\frac{V_1 - V_0}{R_S} = \frac{V_0 - V_z}{r_z} + I_L$$

Grouping the terms yields

$$V_1 = \left(\frac{r_z + R_S}{r_z}\right)V_0 - \frac{R_S}{r_z}V_z + I_L R_S \qquad (13.4)$$

Since $I_L = V_0/R_L$, eq. (13.4) can be rewritten in the form

$$V_1 = \left(\frac{r_Z + R_S}{r_Z} + \frac{R_S}{R_L}\right)V_0 - \frac{R_S}{r_Z}V_Z \qquad (13.5)$$

If V_Z remains constant, the change δV_0 in output voltage due to a change δV_1 in input voltage is given by

$$\delta V_1 = \left(\frac{r_Z + R_S}{r_Z} + \frac{R_S}{R_L}\right)\delta V_0$$

therefore,

$$S = \frac{\delta V_0}{\delta V_1} = 1 \Big/ \left(\frac{r_Z + R_S}{r_Z} + \frac{R_S}{R_L}\right) \qquad (13.6)$$

The output resistance is deduced by rewriting eq. (13.4) in terms of V_0

$$\left(\frac{r_Z + R_S}{r_Z}\right)V_0 = V_1 + \frac{R_S}{r_Z}V_Z - I_L R_S$$

If, in the above equation, V_1 and V_Z are constant, then

$$\left(\frac{r_Z + R_S}{r_Z}\right)\delta V_0 = -R_S \delta I_L$$

where δI_L is the change of load current which causes the change δV_0 in output voltage hence,

$$\frac{\delta V_0}{\delta I_L} = -\frac{r_Z R_S}{r_Z + R_S} \qquad (13.7)$$

From eqs. (13.2) and (13.7),

$$R_0 = r_Z R_S/(r_Z + R_S) \qquad (13.8)$$

The output resistance can, alternatively, be deduced from the equivalent circuit of Fig. 13.2(b). If the internal impedance of V_1 is zero, then the effective impedance measured between the output terminals (with R_L disconnected) is r_Z in parallel with R_S, giving the result in eq. (13.8).

The design principles of gas discharge tube circuits are generally similar to that described above, but with the following exceptions:

(a) The supply voltage must be in excess of the *breakdown* voltage of the tube.

(b) The tube current must never be allowed to fall below the minimum current which is necessary to maintain the discharge.

In addition to the above, the maximum load current which may be drawn from a voltage reference source using either Zener diodes or gas discharge tubes should not exceed the rating of the device, otherwise the regulating device will be overloaded if the load becomes disconnected.

23*

Simplified design: Circuit design can be simplified by assuming that r_Z has such a small value that it can be neglected. In this event, the current in the current limiting resistor is

$$R_S \simeq (V_1 - V_Z)/I \qquad (13.9)$$

When R_L is disconnected ($R_L = \infty$) the Zener diode must be capable of absorbing the current given by eq. (13.9), hence

$$I \leqslant P_Z/V_Z \qquad (13.10)$$

where P_Z is the power rating of the Zener diode. The circuit ceases to act as a regulator when $I_Z = 0$, i.e., when $I_L = I_1$; this occurs when the minimum value of load resistance $R_{L(min)}$ is connected, giving

$$\frac{V_Z}{R_{L(min)}} = \frac{V_1 - V_Z}{R_S}$$

therefore,

$$R_{L(min)} = \left(\frac{V_Z}{V_1 - V_Z}\right) R_S \qquad (13.11)$$

As a matter of good practice, it is inadvisable to operate the Zener diode near to the zero current condition otherwise the output voltage will vary due to the curvature of the characteristic in the region of the 'knee'.

Example 13.1: Design a voltage reference circuit which uses a 5 V, 400 mW Zener diode with a slope resistance of 10 Ω. If the supply voltage is 11 V ± 2 V, calculate the maximum and minimum possible values of no-load output voltage if the resistor R_S is to be chosen from a range of preferred resistors with a 10 per cent tolerance; compute the stabilization factor of the circuit. What is the output resistance of the circuit? Also estimate the lowest value of load resistance that can be connected before the regulator fails to function correctly.

Solution: The maximum current that the diode can pass is

$$I_{Z(max)} = 400 \text{ mW}/5 \text{ V} = 80 \text{ mA}$$

The current in the circuit must be limited to this value under no-load conditions by r_Z and the minimum value of the current limiting resistor $R_{S(min)}$ when the maximum supply voltage is applied. That is when

$$r_Z + R_{S(min)} \geqslant (V_{S(max)} - V_Z)/I_{Z(max)} = (13 - 5)/80$$

$$= 0 \cdot 1 \text{ k}\Omega$$

giving

$$R_{S(min)} \geqslant 100 - 10 = 90 \ \Omega$$

A resistor of 100 Ω would be chosen, since the lowest value with a 10 per cent tolerance is 90 Ω. If R_S has its minimum value, then its rating is $0.08^2 \times 90 = 0.576$ W, and a rating of 1 W is appropriate.

With $V_{S(max)}$ applied and $R_{S(min)}$ in circuit, the maximum no-load output voltage is

$$V_{0(max)} = V_Z + I_{Z(max)} r_Z = 5 + (0.08 \times 10) = 5.8 \text{ V}$$

Under the same circuit conditions, with $V_{S(min)}$ applied

$$I_Z = (9 - 5)/100 \text{ A} = 40 \text{ mA}$$

and

$$V_0 = 5 + (0.04 \times 10) = 5.4 \text{ V}$$

hence,

$$S = \frac{\delta V_0}{\delta V_1} = \frac{5.8 - 5.4}{13 - 9} = 0.1$$

This value can also be obtained from eq. (13.6) by putting $R_L = \infty$.

Note: The minimum no-load output voltage occurs when R_S has its maximum value of 100 Ω + 10 per cent = 110 Ω, giving (with $V_{S(min)}$ applied) an output voltage of 5.36 V.

The maximum output resistance is evaluated by using $R_{S(max)}$ in eq. (13.8), giving

$$R_{0(max)} = r_Z R_{S(max)}/(r_Z + R_{S(max)}) = 10 \times 110/(10 + 110)$$
$$= 9.16 \ \Omega$$

The minimum value of R_0 occurs when $R_{S(min)}$ is in circuit, when $R_{0(min)} = 9 \ \Omega$.

The minimum value of load resistance that may be connected is calculated from eq. (13.11), as follows

$$R_{L(min)} = \left(\frac{V_Z}{V_{1(min)} - V_Z} \right) R_{S(max)} = \frac{5}{9-5} \times 110$$
$$= 137.5 \ \Omega$$

Approximate design procedure: Using eqs. (13.9) to (13.11), and assuming a nominal supply voltage of 11 V and that $r_Z = 0$

$$R_S = (11 - 5)/80 = 0.075 \text{ k}\Omega = 75 \ \Omega$$

Possible values for R_S are 82 Ω ± 10 per cent or 100 Ω ± 10 per cent. If the 82 Ω resistor is chosen, the lowest value of R_S is 73.8 Ω, which may result in the diode power rating being exceeded since it is less than the value calculated above. It is therefore prudent to choose a 100 Ω resistor.

The power consumption in R_S is $(11 - 5) \times 0{\cdot}08 = 0{\cdot}48$ W. A 1 W rated resistor would be selected.

The no-load output voltage on this basis is equal to the Zener breakdown voltage, i.e., 5 V, and the minimum load resistance that may be connected is

$$R_{L(min)} = \frac{5}{11 - 5} \times 100 = 83{\cdot}4 \ \Omega$$

The value of $R_{L(min)}$ calculated above is satisfactory for the nominal supply voltage, but if the supply voltage is known to have a minimum value, then this value should be used in the above calculation.

13.3.1 Two-stage voltage reference sources

A highly stabilized voltage reference source comprising two cascaded Zener diode circuits is shown in Fig. 13.3. The design principles of each stage are generally as outlined in section 13.3. The stability factor of the circuit is the product of the two stability factors of each half of the circuit, so that a stability factor of $0{\cdot}01$ can be achieved if the value for each half is $0{\cdot}1$. As a rule-of-thumb guide, the unstabilized input voltage should be of the order of three to four times greater than the output voltage; V_{Z1} should be about twice V_{Z2}. It is advisable in these circuits to use Zener diodes with breakdown voltages of the order of 5 V, for reasons already given, and for higher voltages than 5 V it is best to connect a number of 5 V diodes in series.

A variable output voltage between the values of V_{Z1} and V_{Z2} can be obtained if R_2 is replaced by a potentiometer, as shown in the inset in Fig. 13.3. The output is taken from the wiper of the potentiometer. This feature is obtained

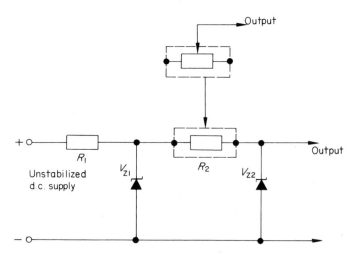

Fig. 13.3   A two-stage voltage reference source.

at a cost of a reduced stabilization factor and an increased output resistance, the
actual values depending upon the setting of the potentiometer wiper.

By using a larger number of cascaded stages, a voltage reference source with
the accuracy of a standard cell can be constructed.

In some applications, similar circuits using cold-cathode tubes are used.

13.4   Stabilizer with a shunt transistor

The current handling capacity of the stabilizer can be increased by using a transistor in conjunction with the Zener diode, as shown in Fig. 13.4(a). The emitter
current of the transistor is the sum of the base current (i.e., the Zener diode
current) and the collector current. Thus, the shunt circuit handles $(1 + h_{FE})$

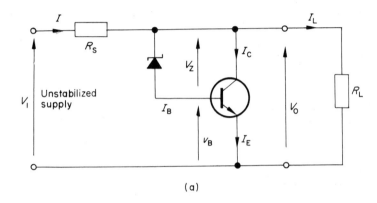

(a)

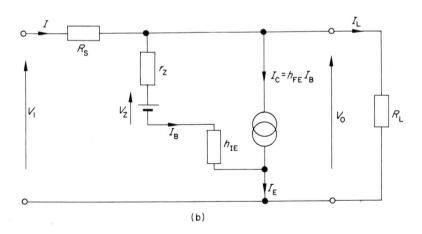

(b)

Fig. 13.4   A transistor shunt regulator.

times the current flowing in the Zener diode. The current handling capacity can be increased further by replacing the single transistor in Fig. 13.4(a) by a super-alpha pair of transistors.

Since the shunt transistor operates in the unsaturated region of its characteristics, the base-emitter voltage drop v_{BE} is of the order of 0·2 to 0·8 V. If the slope resistance of the Zener diode can be neglected, then

$$V_0 \simeq V_Z + v_{BE}$$

Since v_{BE} remains substantially constant over the working range of the transistor, the output voltage is slightly greater than V_Z. The base current is restricted by the current rating of the Zener diode, hence,

$$I_B \leqslant P_Z/V_Z$$

where P_Z is the power rating of the Zener diode.

The maximum load current should be limited to a value which is less than the emitter current of the transistor, otherwise both the diode and the transistor may be damaged if the load becomes disconnected. Hence,

$$I = I_B + I_C = (1 + h_{FE})I_B \leqslant (1 + h_{FE})P_Z/V_Z$$

and the value of R_S is calculated from the equation

$$R_S = (V_1 - V_0)/I$$

and it has a power dissipation of $I^2 R_S$.

The above analysis is adequate for basic design purposes, but that which follows allows the stabilization factor and output resistance to be predicted. The approximate equivalent circuit of the regulator is shown in Fig. 13.4(b), from which the base current is

$$I_B = (V_0 - V_Z)/(r_Z + h_{IE})$$

and

$$I_E = (1 + h_{FE})I_B = \frac{(1 + h_{FE})(V_0 - V_Z)}{r_Z + h_{IE}} \tag{13.12}$$

Now,

$$I = I_E + I_L$$

and

$$I = (V_1 - V_0)/R_S$$

therefore,

$$\frac{V_1 - V_0}{R_S} = \frac{(1 + h_{FE})(V_0 - V_Z)}{r_Z + h_{IE}} + I_L$$

Solving for V_1 gives

$$V_1 = V_0 \left\{ \frac{(1 + h_{FE})R_S}{r_Z + h_{IE}} + 1 \right\} - V_Z \frac{(1 + h_{FE})R_S}{r_Z + h_{IE}} + I_L R_S \qquad (13.13)$$

This equation is seen to have the same general form as eq. (13.4) for the Zener diode regulator.

Since $I_L = V_0/R_L$, eq. (13.13) may be rewritten as

$$V_1 = V_0 \left\{ \frac{(1 + h_{FE})R_S}{r_Z + h_{IE}} + 1 + \frac{R_S}{R_L} \right\} - V_Z \frac{(1 + h_{FE})R_S}{r_Z + h_{IE}} \qquad (13.14)$$

Accordingly, from eq. (13.14), the stabilization factor is

$$S = \frac{\delta V_0}{\delta V_1} = 1 \left/ \left\{ \frac{(1 + h_{FE})R_S}{r_Z + h_{IE}} + 1 + \frac{R_S}{R_L} \right\} \right. \qquad (13.15)$$

and from eq. (13.13) the output resistance is

$$R_0 = -\frac{\delta V_0}{\delta I_L} = R_S \left/ \left\{ \frac{(1 + h_{FE})R_S}{r_Z + h_{IE}} + 1 \right\} \right. \qquad (13.16)$$

Example 13.2: Estimate the maximum value of collector current and the power rating of a transistor suitable for use in a regulator of the type in Fig. 13.4(a), if the Zener diode has a rating of 5·6 V, 250 mW and has a slope resistance of 5 Ω. The value of h_{IE} of the power transistor is 15 Ω, and its h_{FE} is 20. What is the output voltage at the regulator terminals and the power rating of R_S if the maximum input voltage is 20 V? Estimate the maximum current which may be drawn from the circuit before the regulating action fails.

Calculate the output resistance of the regulator, and the stabilization factor of the circuit if $R_L = 100 \Omega$.

Solution: The maximum Zener diode current is

250 mW/5·6 V = 44·7 mA, hence I_{CM} = 44·7 x 20 = 894 mA,

and the maximum emitter current is 44·7 + 894 = 938·7 mA. The maximum current that may be drawn from the regulator is, therefore, 938·7 mA.

Now,

$$V_0 = V_Z + I_B(r_Z + h_{IE})$$
$$= 5 + 0·0447(5 + 15)$$
$$= 5·89 \text{ V}$$

The transistor rating must be at least 5·89 x 0·894 = 5·26 W. The value of R_S is calculated as follows

$$R_S = (20 - 5·89)/0·9387 = 15 \Omega$$

and it must have a rating in excess of $0.9387^2 \times 15 = 13.25$ W. From eq. (13.16),

$$R_0 = 15 \bigg/ \left\{ \frac{(1 + 20)\,15}{5 + 15} + 1 \right\} = 0.895 \ \Omega$$

and with $R_L = 100 \ \Omega$

$$S = 1 \bigg/ \left\{ \frac{(1 + 20)\,15}{5 + 15} + 1 + \frac{15}{100} \right\} = 0.0592$$

13.4.1 A high voltage shunt regulator

The output voltage of the regulator described in the previous section is limited by the collector-base breakdown voltage of the transistor. Higher output voltages are obtained by connecting the Zener diode in series with the emitter. A thermionic valve circuit using this principle is shown in Fig. 13.5, and its operation is described below.

The pentode, which has similar output characteristics to the transistor, may be used in any of the shunt regulator applications so far described. In Fig. 13.5, the cathode potential is held at $+V_T$ relative to the common connection, and the control grid potential is

$$V_G = R_2 V_S/(R_1 + R_2)$$

which is slightly less than V_T in order to maintain a negative grid-cathode potential. Any reduction in the supply voltage V_S causes V_G to fall, making the control grid of the pentode more negative with respect to the cathode than it was previously, causing the anode current to be reduced. In a correctly designed circuit, the reduction in p.d. across resistor R due to the reduction in valve

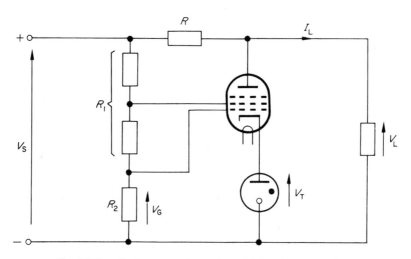

Fig. 13.5   A shunt regulator for a higher output voltage.

current just compensates for the reduction in the supply voltage, so maintaining a constant output voltage.

The circuit used here does not adequately stabilize the circuit against load resistance changes, since a reduction in load current due to a higher value of load resistance merely reduces the p.d. across R without altering the value of V_G. Some compensation in this direction can be obtained by connecting the resistor chain $R_1 R_2$ across the output terminals.

The value of the series resistor to give zero voltage regulation for small changes in input voltage can be predicted as follows. The change δi_A in anode current due to a change δv_G in grid voltage is

$$\delta i_A = g_m \delta v_G$$

Now, suppose that the supply voltage changes by δv_S. Since the cathode of the pentode remains at a substantially constant voltage, the change in grid-cathode voltage is equal to the proportion of δv_S that appears across R_2, hence

$$\delta v_G = R_2 \, \delta v_S / (R_1 + R_2)$$

The change in p.d. across R due to the change in current δi_A is $R \delta i_A$, and the change δv_L in the output voltage is

$$\delta v_L = \delta v_S - R \delta i_A = \delta v_S - \frac{R R_2 g_m \delta v_S}{R_1 + R_2}$$

$$= \delta v_S \left(1 - \frac{R R_2 g_m}{R_1 + R_2}\right)$$

The stability factor of the circuit is given by

$$S = \frac{\delta v_L}{\delta v_S} = 1 - \frac{R R_2 g_m}{R_1 + R_2}$$

Clearly, the voltage regulation is zero for small changes in V_S when

$$\frac{R R_2 g_m}{R_1 + R_2} = 1$$

Thus, if $R_1 = 90\,\text{k}\Omega$, $R_2 = 10\,\text{k}\Omega$, and $g_m = 5\,\text{mA/V}$, then R must have a value of $2\,\text{k}\Omega$ to give zero regulation with small changes of supply voltage.

13.5 Series voltage regulators

Series voltage regulators are simply emitter follower or cathode follower circuits with the control electrode (base or control grid) energized by a reference signal which is usually derived from a shunt regulator.

Figure 13.6(a) shows the simplest possible form of transistor series voltage regulator, in which the voltage reference source is a Zener diode. Two popular variations on the circuit are shown in the insets. Inset (i) shows the addition of a

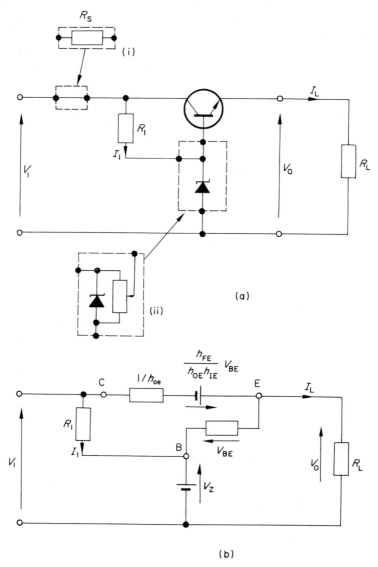

Fig. 13.6   The basic series regulator.

series resistor R_S, which acts as a current limiting resistor in the event of a short-circuit at the output. Inset (ii) shows one common method of controlling the output voltage; unfortunately, this modification tends to increase the voltage regulation. The output voltage of the basic circuit is $V_Z - V_{BE}$, and the modification in inset (ii) allows values below this to be obtained.

In simple circuits of this kind, a proportion of the ripple voltage at the input appears at the output terminals since a small ripple voltage is developed across the Zener diode. This can be minimized by shunting the diode with a capacitance of the order of $100 \, \mu F$. To reduce the output impedance (as distinct from the output resistance) of the circuit to high frequency fluctuations in load current, the output terminals can be shunted by a capacitance of the order of $0.1 \, \mu F$ to $1 \, \mu F$.

The design of the voltage reference source follows the lines already outlined. In order to maintain a substantially constant voltage across the Zener diode, the current through the diode should not change significantly over the range of base current required by the transistor. Thus, the maximum collector current I_{CM} to be handled by the transistor requires a base current of I_{CM}/h_{FE}. In a practical circuit, the current through the Zener diode should be about five times this value, and the maximum current through R_1 should be of the order of $6I_{CM}/h_{FE}$; whence,

$$\frac{V_1 - V_Z}{R_1} \geqslant \frac{6I_{CM}}{h_{FE}}$$

or

$$R_1 \leqslant \frac{(V_1 - V_Z)h_{FE}}{6I_{CM}}$$

If the terms in the above expression are subject to changes in tolerance, then $V_{1(min)}$, $V_{Z(max)}$, and the minimum value of h_{FE} should be used in the calculation.

In the following analysis, the slope resistance of the Zener diode is neglected, and the constant voltage equivalent circuit of the transistor is used. The transistor input parameter h_{RE} is neglected since it is usually very small. The results of this analysis can then be replaced by the valve equivalent equations, since $r_a = 1/h_{OE}$, and $\mu = h_{FE}/h_{OE}h_{IE}$ (see Table 9.2, page 236); the input resistance ($\equiv h_{IE}$) of the valve is infinitely large. Equating the supply voltage to the p.d.s and e.m.f.s within the main loop gives

$$V_1 = V_0 - \frac{h_{FE}}{h_{OE}h_{IE}}V_{BE} + \frac{I_L}{h_{OE}}$$

$$= V_0 - \frac{h_{FE}}{h_{OE}h_{IE}}(V_Z - V_0) + \frac{I_L}{h_{OE}}$$

$$= V_0\left(1 + \frac{h_{FE}}{h_{OE}h_{IE}}\right) - V_Z\frac{h_{FE}}{h_{OE}h_{IE}} + \frac{I_L}{h_{OE}} \tag{13.17}$$

The form of eq. (13.17) is seen to be generally similar to that of eqs. (13.4) and (13.13) for shunt voltage regulators. The output resistance of the regulator is, therefore,

$$R_0 = -\frac{\delta V_0}{\delta I_L} = \frac{1/h_{OE}}{1 + (h_{FE}/h_{OE}h_{IE})} = \frac{1}{h_{OE} + (h_{FE}/h_{IE})} \tag{13.18}$$

In many cases $h_{OE} \ll h_{FE}/h_{IE}$, and $R_0 = h_{IE}/h_{FE}$.

Since $I_L = V_0/R_L$, eq. (13.17) may be rewritten in the form

$$V_1 = V_0\left(1 + \frac{h_{FE}}{h_{OE}h_{IE}} + \frac{1}{R_L h_{OE}}\right) - V_Z\frac{h_{FE}}{h_{OE}h_{IE}} \qquad (13.19)$$

giving a stabilization factor of

$$S = \frac{\delta V_0}{\delta V_1} = 1\bigg/\left\{1 + \frac{h_{FE}}{h_{OE}h_{IE}} + \frac{1}{R_L h_{OE}}\right\}$$

$$= 1\bigg/\left\{1 + \frac{1}{h_{OE}}\left(\frac{h_{FE}}{h_{IE}} + \frac{1}{R_L}\right)\right\} \qquad (13.20)$$

When a series resistance R_S is included in the circuit, the relevant equations are

$$R_0 = \frac{R_S + 1/h_{OE}}{1 + (h_{FE}/h_{OE}h_{IE})}$$

$$S = 1\bigg/\left\{1 + \frac{h_{FE}}{h_{OE}h_{IE}} + \frac{1}{R_L}\left(\frac{1}{h_{OE}} + R_S\right)\right\}$$

Equations for equivalent valve circuits are

$$R_0 = (R_S + r_a)/(1 + \mu)$$

$$S = 1\bigg/\left\{1 + \mu + \frac{1}{R_L}(R_S + r_a)\right\}$$

The current that may be handled in the case of the transistor circuit can be increased by replacing the single transistor with a super-alpha pair. Alternatively, in the case of both valves and transistors, the current can be increased by connecting a number of devices in parallel; in the event of matched devices not being available, some method of forcing correct current sharing must be adopted.

Example 13.3: Design a series regulator similar to that in Fig. 13.6(a) where the maximum load current is to be 0·5 A and the supply voltage is 22 V ± 2 V. The transistor parameters are $h_{FE} = 50$, $h_{IE} = 40\ \Omega$, $h_{OE} = 0·02$ S; a 12 V Zener diode is used as the reference source. Estimate the output resistance of the regulator and its stabilization factor.

Solution:

$$I_{BM} = 0·5/(1 + 50) = 0·0098 \text{ A or } 9·8 \text{ mA}$$

The current through R_1 should, therefore, be approximately 6 x 9·8 = 58·8 mA (the rating of the Zener diode should be greater than 12 x (5 x 9·8) = 588 mW).
Hence,

$$R_1 \leqslant ((22 - 2) - 12)/0·0588 = 136\ \Omega$$

and a 120 Ω resistor with a tolerance of $\pm$ 10 per cent would be suitable. The rating of R_1 should be greater than

$$((22 + 2) - 12)^2/(120\,\Omega - (10\% \text{ of } 120\,\Omega)) = 12^2/108 = 1\cdot333\text{W}$$

From eq. (13.18)

$$R_0 = 1 \left/ \left(0\cdot02 + \frac{50}{40} \right) \right. = 0\cdot787\ \Omega$$

Since the output voltage from the regulator is approximately 12 V, then the minimum load resistance that can be connected is 12 V/0·5 A = 24 Ω. Under this condition,

$$S = 1 \left/ \left\{ 1 + \frac{1}{0\cdot02} \left(\frac{50}{40} + \frac{1}{24} \right) \right\} \right. = 0\cdot0155$$

13.6   Series regulators using amplifiers

In the simple series regulator of Fig. 13.6(a), the voltage $(V_Z - V_0)$ is used to directly control the series transistor. By amplifying this voltage difference before applying it to the series transistor, the output impedance and the stabilization factor are improved. One such circuit is shown in Fig. 13.7. The error voltage amplifier comprises transistor Q1 and resistor R_1, the input voltage to the amplifier being $(\beta V_0 - V_Z)$, where β is the proportion of the output voltage which is fed back.

The output voltage is set by adjusting the position of the wiper of RV1, thereby altering βV_0. Since the base-emitter voltage of Q1 is small, $V_Z \simeq \beta V_0$, or $V_0 \simeq V_Z/\beta$.

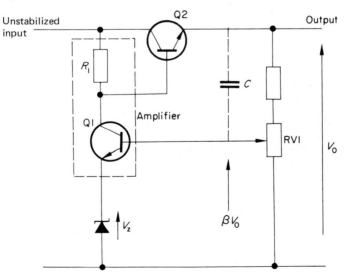

Fig. 13.7   An improved series regulator.

Since the emitter of Q1 is maintained at a constant voltage by the Zener diode, any change in output voltage is applied in attenuated form (depending on the setting of RV1) between the base and emitter of Q1. If the gain of the error amplifier is m (which has a negative value since the amplifier is phase-inverting), then the change in the collector voltage of Q1 is $m\beta\delta V_0$, where δV_0 is the change in output voltage. Therefore, a transient increase in output voltage causes the base voltage of Q1 to increase, and its collector potential to fall; this reduces the base voltage of Q2, so reducing the output current and voltage to its original level. The numerical value of the gain of the amplifier in Fig. 13.7 is approximately $R_1 h_{FE1}/h_{IE1}$, where h_{FE1} and h_{IE1} are parameters of Q1.

The equivalent circuit of the regulator in the figure generally resembles that shown in Fig. 13.6(b), the principal difference being that the 'battery' V_Z in Fig. 13.6(b) is replaced by one with a value $m(V_Z - \beta V_0)h_{FE2}/h_{OE2}h_{IE2}$, where the parameters given are those of Q2. The loop equation then becomes

$$V_1 = V_0 + \frac{I_L}{h_{OE2}} - m(V_2 - \beta V_0)\frac{h_{FE2}}{h_{OE2}h_{IE2}}$$

from which equation the output resistance is

$$R_0 = \frac{1/h_{OE2}}{1 + m\beta\dfrac{h_{FE2}}{h_{OE2}h_{IE2}}}$$

and

$$S = 1 \left/ \left\{ 1 + m\beta\frac{h_{FE2}}{h_{OE2}h_{IE2}} + \frac{1}{R_L h_{OE2}} \right\} \right.$$

To improve the high frequency performance of the circuit, a capacitor of small value may be coupled between the positive line and the base of Q1, shown by the dotted connection in the figure. This allows high frequency variations in output voltage to be transmitted directly to the base of Q2.

Valve versions of the circuit in Fig. 13.7 are in common use, the reference source in these circuits being either a Zener diode or a gas discharge tube, and Q1 is replaced by a pentode. Q2 is usually replaced by a triode with a low value of slope resistance (or a pentode connected as a triode). The equivalent equations for the valve circuits are

$$R_0 = r_a/(1 + m\beta\mu_2)$$

$$S = 1 \left/ \left\{ 1 + m\beta\mu_2 + \frac{r_{a2}}{R_L} \right\} \right.$$

where μ_2 and r_{a2} are parameters of the series valve, and m is the small-signal gain of the shunt amplifier circuit.

An improved version of the circuit is shown in Fig. 13.8, in which R_2 and Z_2 form a *pre-regulator circuit* which provides a relatively ripple-free supply to the shunt amplifier comprising Q1 and R_1. The value of resistor R_2 is calculated by the equation

$$R_2 \geqslant \frac{V_{1(\text{max})} - (V_{0(\text{max})} + V_{Z2})}{I_1 + I_2}$$

where I_1 is dictated by the maximum base current of Q3, and I_2 by the current rating of Z_2. An advantage accruing from the use of a pre-regulator is that supply voltage variations are not passed on to the collector of Q1, and thereby to the output circuit. A feature of the circuit in Fig. 13.8 is that the supply voltage must be higher than in the equivalent rated circuit in Fig. 13.7, due to the additional voltage required to operate the pre-regulator. Thus, if a 6 V or 7 V Zener diode is used in the pre-regulator, then V_1 should be at least 10 V to 15 V greater than V_0; in the series regulator in Fig. 13.7, V_1 is about 6 V to 8 V greater than V_0. The use of the Darlington-connected transistors in Fig. 13.8 allows a much greater output current to be handled than in the single-transistor circuit.

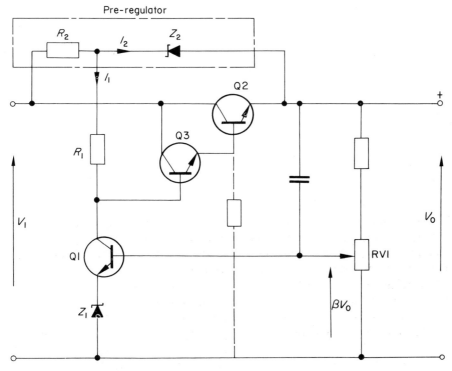

Fig. 13.8   A series regulator incorporating a pre-regulator for improved performance.

In Fig. 13.8, it is sometimes advisable to connect the base of Q2 to the negative line via a resistor (shown by the dotted connection), otherwise control of the output voltage may be lost at very low values of load current. This is due to the fact that the leakage current flowing out of the base of Q2 at very low values of load current may be greater than the emitter current of Q3, resulting in the base current of Q2 falling to zero. This problem may be overcome in one

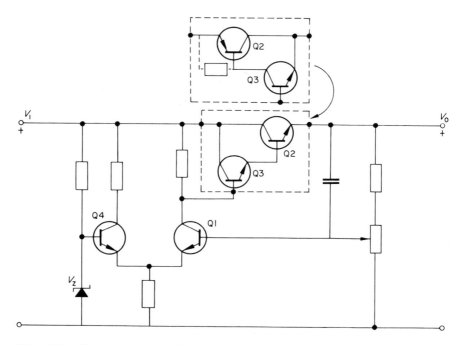

Fig. 13.9　The use of a long-tailed pair improves the temperature stability of the regulator.

of two ways. Firstly, a resistor may be connected between the base of Q2 and the negative line (mentioned above) to provide a path for the collector leakage current; secondly, the bleed current flowing through RV1 can be made sufficiently large to ensure that Q2 is always conducting.

In the circuits in Figs. 13.7 and 13.8, both V_Z and the base-emitter voltage of Q1 are temperature-sensitive, so that the output voltage of the regulator is also temperature-sensitive. The effects of a temperature change on Q1 can be minimized by replacing the simple amplifier by a long-tailed pair, as shown in Fig. 13.9. A change in temperature also affects the value of v_{BE} at any given current of both Q2 and Q3, but since these are both within the feedback loop any variations will be largely corrected for. As the temperature rises, the base-emitter voltage of Q4 in Fig. 13.9 falls and its collector current rises. Owing to the mode of operation of the long-tailed pair circuit, the collector current of Q1 should also be reduced. However, the effects of temperature change on Q1 are generally similar to those on Q4, since identical transistors are used, with the result that the two effects cancel each other out. This means that the collector voltage of Q1 remains substantially constant when the ambient temperature changes. Further improvement is effected by the use of two cascaded long-tailed pair stages, but this brings in its wake the possibility of instability and oscillation.

It is difficult to compensate for changes of V_Z with temperature, and its effects can be minimized by maintaining the Zener diode at a constant temperature in an 'oven'.

As an alternative arrangement, the super-alpha pair in the figure can be replaced by a complementary pair of hook-connected transistors shown in the inset. This latter circuit is less temperature-sensitive than the super-alpha pair, since there is only one base-emitter voltage drop between the collector of Q1 and the output. In the super-alpha pair, there are two base-emitter voltage drops, both of which are temperature-sensitive. The leakage current of Q2 in the inset is allowed for by connecting a resistor with a value of about 1 kΩ between the base and emitter of Q2 (shown dotted). In both versions of the circuit, ripple may be reduced further by connecting the base of Q3 to the negative line via a capacitor with a value of about 0·001 μF.

13.7 Series current regulators

In some applications, it is desirable to be able to maintain a constant current in a circuit, despite supply voltage variations and load resistance changes. A simple experimental circuit is shown in Fig. 13.10, which is designed to deliver a constant current of 10 mA into a load which has a resistance of any value between zero and about 2 kΩ. In the circuit shown, assuming that v_{BE} is small compared with V_Z, the voltage drop across the two resistors comprising R is equal to V_Z

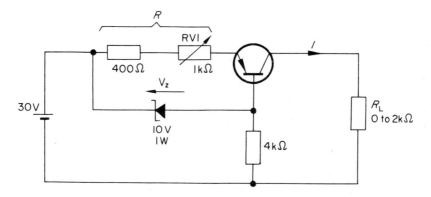

Fig. 13.10   A simple series current regulator.

that is,

$$V_Z \simeq IR$$

or

$$I \simeq V_Z/R$$

The load current is set up initially by means of RV1, so that with a 10 V Zener diode and a load current of 10 mA, R has a value of 10 V/10 mA = 1 kΩ. The power rating of resistor R must be greater than 10 V x 10 mA = 100 mW. The lower end of the 4 kΩ resistor may, alternatively, be connected to the collector of the transistor instead of the negative supply rail.

Under no-load conditions (R_L = 0), the transistor must absorb a voltage $(V_S - IR)$ = 30 − 10 = 20 V. The limiting condition for the circuit to maintain the load current constant occurs when the p.d. across the transistor falls to zero. This happens when the p.d. across the transistor is zero, and $IR_{L(max)}$ = 20 V, or when

$$R_{L(max)} = 20 \text{ V}/10 \text{ mA} = 2 \text{ k}\Omega$$

In order that the transistor can deal with all possible load conditions, it must have a maximum current rating greater than 10 mA, and a voltage rating greater than 20 V.

Since the voltage across a Zener diode is absolutely constant only when it carries a constant current, an accurately stabilized voltage source can be constructed by driving a multi-stage Zener diode circuit of the type in Fig. 13.3 by a constant current source of the type described here.

13.8 Protection of transistor power supplies

The problem of *overvoltage protection* has already been raised in section 13.1, and one solution to this is some form of 'crowbar' protection in which a voltage-sensitive circuit applies a short-circuit to the output whenever an overvoltage occurs. The overcurrent resulting from this form of protection is cleared by back-up protection such as a fuse or a current limiting circuit.

Overcurrent protection, to be effective, must limit the current to a safe value within about 10 μs to 100 μs of the application of the fault. Fuses do not provide this speed of response, and the first line of defence is some form of current-limiting circuit. Fuses or thermal cut-outs are provided in regulators to give back-up protection in the event of a serious failure in the regulator.

Characteristics relating to various forms of overcurrent protection circuits for constant voltage regulators are shown in Fig. 13.11. Curve A is for a *constant-current limiting circuit* which limits the load current to a constant value whatever value of load resistor is connected. Curve B is representative of *cut-out circuits*, in which the output voltage is suddenly reduced to zero upon the incidence of a fault. It is then necessary to locate and remove the fault before the supply can be reset. Between curves A and B lie a great variety of possible characteristics known as *re-entrant characteristics,* curve C being an example.

With cut-out protection, the regulator may suddenly be tripped out by the application of a transient load, e.g., a lamp load; this may cause inconvenience. This may be overcome by the introduction of a time delay in the tripping circuit, but this defeats the very essence of an electronic circuit.

Constant current protection overcomes this objection, and it maintains the supply until the overcurrent is removed. However, this form of protection results in operating conditions in the series regulator which are more severe than those encountered under normal operation. Consider a regulator which employs a 20 V unregulated supply, and provides an output of 12 V and a current of 1 A.

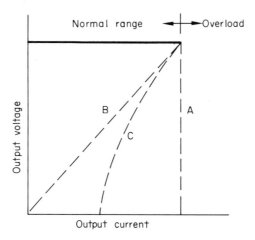

Fig. 13.11 Overcurrent protection characteristics.

Under full-load conditions the series transistor dissipates $(20 - 12) \times 1 = 8$ W. If the output current is limited to 1 A when a short-circuit is applied, the series transistor must support the whole of the unregulated supply voltage under short-circuit conditions and still carry the full-load current, giving a dissipation of 20 W.

A simple constant current protection circuit is shown in Fig. 13.12. In this circuit, a resistor R_1 and a *silicon* diode D are connected across the unregulated supply. The anode voltage of the diode is applied to the base of a *germanium* transistor; the reason for this combination of materials is that the forward voltage drop across the silicon diode is greater than the drop across the emitter junction of the transistor. At low values of load current, the p.d. across R_2 is small, and the transistor operates in its saturated region as a result of the larger forward p.d. across the diode. As the load current builds up, the p.d. across R_2 increases until the transistor operates in the unsaturated region, when it begins to support some of the supply voltage, so reducing the output voltage. The circuit is seen to be similar to the basic constant-current regulator in Fig. 13.10. The value at which the current is limited can be adjusted by shunting R_2 (which has a value of a few ohms) with a potentiometer with a much higher value of resis-

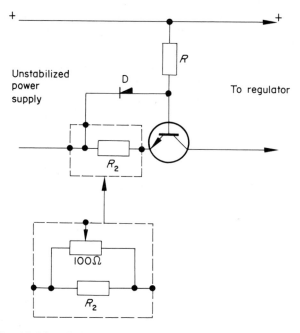

Fig. 13.12   A simple means of providing current limiting.

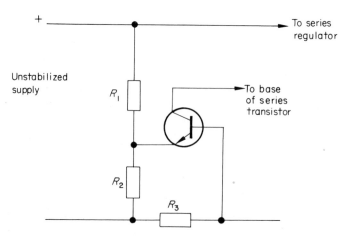

Fig. 13.13   One method of providing re-entrant protection.

tance than R_2 (say 100 Ω), the diode cathode being connected to the wiper of the potentiometer (see inset). The setting of the position of the wiper provides control of the point at which the transistor moves into the unsaturated region of its characteristics.

In re-entrant protective circuits, the output current and voltage are reduced simultaneously, so that the problem of excessive power dissipation that occurs in the series regulator is eliminated. A simple re-entrant protection circuit is shown in Fig. 13.13. The potential of the emitter of the transistor in the figure is fixed by the potential divider R_1R_2 which is connected across the unstabilized supply. When the load current is small, the p.d. across R_3 is less than that across R_2 and the emitter junction of the transistor is reverse biased. In this case, the collector current of the transistor is zero. When the load current becomes excessive, the p.d. across R_3 exceeds that across R_2, and collector current begins to flow. Since the collector of the transistor is connected to the base of the power transistor in the series regulator (Q3 in Figs. 13.8 and 13.9), it reduces the base drive to that transistor, and in so doing it reduces the output voltage and current.

While re-entrant protection overcomes one difficulty associated with constant current protection, it has its own problems with reactive loads and with loads with non-linear characteristics, e.g., lamp loads. In these applications, the initial rush of current may cause the output to be reduced to a level from which it is not possible to achieve normal operating conditions. In some power supplies, a hybrid arrangement of part constant-current and part re-entrant protection is used.

13.9   High voltage and high current supplies

In the voltage range 150 V to 450 V, regulators using valves can be used directly because of their superior high voltage capability. Alternatively, using semiconductors, a common technique is to use two voltage sources in series. One of the two supplies is a smoothed but unregulated voltage which provides the bulk of the output, and the other voltage source is a stabilized supply providing about 30 V to 50 V. A proportion of the total output voltage is then fed back to control the stabilized source. This is sometimes known as *piggy-back control*.

Where an e.h.t. supply is required it can be obtained from an audio-frequency oscillator, the output being raised to the desired level by a transformer and rectifier system. A proportion of the output voltage is fed back and is used to control the voltage regulator which supplies power to the oscillator. As a result, the amplitude of the oscillations is maintained at the desired level.

Thyristor and triac circuits are now commonly used to provide current at medium and high voltages. These devices can be used in one of several different ways. Firstly, they can be used to control the alternating voltage fed to the primary of the transformer which feeds a conventional rectifier and smoothing system. Secondly, they can be connected in the rectifier circuit on the secondary side of the transformer which supplies the load. In a third mode, they can be used as 'choppers' on the d.c. side of the circuit. In all these cases, a signal proportional to the output quantity is fed back to a pulse generator which triggers the thyristor or triac into conduction. Unfortunately, it is not possible to deal with these circuits in detail here, but information is available elsewhere.*

Problems

13.1 Draw the static characteristic of a Zener diode, and explain its principle of operation. Show how the diode can be used in a simple shunt regulator.

13.2 Derive an expression for the stabilization factor and output resistance of the shunt regulator in question 13.1 in terms of the component parameters; assume that the source resistance is zero.

13.3 In a stabilizer of the type in question 13.1, the diode has a breakdown voltage of 10 V and a slope resistance of 10 Ω, and the series resistance is 100 Ω. Calculate the output resistance of the regulator, and hence compute the change in output voltage when the load current is changed by 50 mA.

13.4 Calculate the stabilization factor of the circuit in question 13.3 with a load resistance of 500 Ω. If the supply voltage changes by 5 V, what is the change in output voltage?

13.5 If, in the shunt stabilizer in the above questions, the Zener diode draws a current of 20 mA under no-load conditions, what is the value of the load resistance which must be connected to reduce the diode current to 5 mA? State any assumptions made in the calculations.

13.6 Design a Zener diode shunt regulator to supply 0·2 A at 15 V from a 50 V unstabilized supply. Assume that the slope resistance of the diode is zero. State the ratings of the Zener diode and the series resistance.

13.7 A cold-cathode gas-filled stabilizing tube is to provide a supply of 80 V across a 3·2 kΩ resistor from a nominal 220 V d.c. supply. The tube has a striking voltage of 105 V and over the current range of 5–45 mA the tube voltage can be considered constant at 80 V.

Calculate the series resistance necessary to limit the tube current to 45 mA with the load disconnected and the supply voltage at its nominal value.

Draw a circuit to show the load connected and determine:
(a) The maximum and minimum values of supply voltage between which stabilization will be satisfactory.
(b) The minimum supply voltage to ensure that the tube will strike.

(C & G)

13.8 Explain the operation of a neon-tube voltage stabilizer and draw a typical voltage/current characteristic for the tube.

A neon tube and a resistor are connected in series to provide a 150 V stabilized output from a 200 V d.c. supply. If the resistance of the load is 15 kΩ and the current through the neon tube is 10 mA, determine the value of the series resistor and the power dissipated in it.

If the input voltage rises to 220 V, determine:
(a) The voltage drop across the series resistor.
(b) The power dissipated in the series resistor.
(c) The current flowing in the neon tube.
Assume ideal operation.

(C & G)

---

* See *Control Engineering* by N. M. Morris (McGraw-Hill, 1968).

13.9 A shunt regulator uses the transistor circuit in Fig. 13.4(a). If R_S = 50 Ω, the Zener diode slope resistance is 10 Ω, and the relevant transistor parameters are h_{FE} = 49, h_{IE} = 10 Ω, calculate (a) the output resistance of the regulator, (b) the change in output voltage when the load current changes by 1·5 A, the supply voltage remaining constant, and (c) the change in output voltage when the supply voltage changes by 5 V, the load resistance being 10 Ω.

13.10 The circuit in Fig. 13.14 is that of a series valve voltage regulator; the characteristic equations of the valves used are

$$\text{for V1} \quad i_A = (1\cdot25v_A + 5\cdot25v_G - 67) \times 10^{-3}$$
$$\text{for V2} \quad i_A = (0\cdot02v_A + 2v_G) \times 10^{-3}$$
$$\text{for V3} \quad v_A = 500\, i_A + 85$$

where v_A and v_G are in volts and i_A is in amperes.

The output from the regulator is to be 50 mA at 350 V. Calculate a suitable value for R, and determine the ratio R_1/R_2 given that $(R_1 + R_2) \gg R_L$ and that the current through V3 is 2 mA. The supply voltage is 500 V.

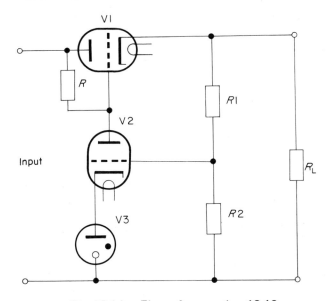

Fig. 13.14   Figure for question 13.10.

13.11 Draw the complete circuit diagram of a series regulator valve unit, suitable for interposing between an unregulated supply and its load for the purpose of controlling and regulating the output voltage. Include a control of output voltage and insert suitable component values, assuming a supply of 500 V and that the required output is 50 milliamperes at 300 volts. Explain the action of the circuit, stating the purpose of each component. State the advantages of such a unit over one using a series resistor in conjunction with a shunt connected gas-filled (e.g., neon) regulator tube.

(C & G)

Solutions to numerical problems

Chapter 1

1.6 0·00854 mA/V; 60 V; 5·55 mA.
1.7 (i) (a) 444 Ω, (b) 333 Ω; (ii) (a) 266 Ω, (b) 202 Ω.
1.9 27·5 V.
1.11 $E_S = 10$ V, $r = 100$ Ω; $I_S = 0·1$ A, $R_S = 100$ Ω.
1.13 $g_m = 4$ mA/V; $r_a = 5$ kΩ.
1.14 3·33 kΩ; 10; 3 mA/V.
1.15 $\mu = 19·5$; $g_m = 2·75$ mA/V
1.18 333 kΩ; 333 kΩ; 666 kΩ; 666 kΩ; 2930.
1.19 $\mu = 1200$; $r_a = 200$ kΩ.
1.24 $C_{gk} = 1·5$ pF; $C_{ga} = 1·5$ pF.
1.25 $g_m = b + 2\,cv_G$; 9.

Chapter 2

2.5 100 MΩ; 100 Ω.
2.6 -15 V.
2.9 (i) 20; (ii) 260 V; (iii) $-$ 3 V.
2.12 9·9 Ω; 31·42 A.

Chapter 3

3.9 4·7 V.
3.12 0·68 V, 0·52 V; 5 mA.

Chapter 4

4.6 99; $-$ 0·988.
4.7 $h_{ie} = 1800$ Ω; $h_{re} = 8 \times 10^{-4}$; $h_{fe} = 82$; $h_{oe} = 128$ μS.
4.8 $r_e = 1$ Ω; $r_b = 1626$ Ω; $r_d = 8·7$ kΩ; k = 77.

Chapter 5

5.4 45 V.
5.5 22·5 V.

Chapter 6

6.2 (i) 31·8 mA; (ii) 34·6 per cent; (iii) 1·21.

6.3 100 Hz.

6.4 57·3 V; 63·7 mA; 3·64 W; 200 V; 200 V.

6.9 (c).

6.10 5 kΩ; (a) 221 V, (b) 196 V.

Chapter 7

7.3 (a) 6·54 mA; (b) 6·54 V; (c) 42·8 mW; (d) $h_{fe} = 101, h_{oe} = 87 \ \mu S$.

7.4 (a) 250; (b) 300; (c) 400; 270.

7.5 $R_L = 3 \ k\Omega; R_B = 300 \ k\Omega$.

7.6 − 135; 6075.

7.7 − 22·5.

7.9 $S = k(1 + h_{FE})$.

7.10 $R_L = 6 \cdot 25 \ k\Omega; R_3 = 80 \cdot 6 \ k\Omega^*; R_4 = 11 \cdot 4 \ k\Omega^*; C_2 > 50 \ \mu F$.

7.13 (a) 985 Ω; (b) 39·5 Ω.

7.14 4·34 V; 57 mW.

7.15 (a) 280 V; − 18·5; (b) 120 V, − 22·7.

7.16 (i) 6·05 mA, 147 V; (ii) 42·5 V; (iii) − 15; (iv) $P_{ac} = 72 \cdot 25$ mW.

7.17 12·8 V; 12·3 V.

7.20 80 mV.

7.21 $(1 + \mu)R_L/(R_L + r_a + (1 + \mu)R_S)$, where R_S is the source resistance.

7.22 0·75 V.

7.23 (a) 1 kΩ, 50; (b) − 1665, 41 600.

7.24 (a) 6 mA; (b) 5 mA; (c) 1 mA; (d) 100 V.

7.25 − 21·8.

7.26 8·92; 178·5.

Chapter 8

8.1 (a) 30 dB; (b) 20 dB; (c) 10 dB; (d) 0 dB; (e) − 10 dB; (f) − 40 dB.

8.2 10·2 dB.

8.3 32·36 mV.

8.4 − 9·8 dB; 16·2 mV.

8.5 (a) 40 dB; (b) 20 dB; (c) 10 dB; (d) − 10 dB.

8.6 270 Hz, 3·2 kHz; 370 Hz, 3 kHz; 3·34 V.

8.7 11·8 V.

8.11 0·25 A, 1 A; 50 W, 50 W; 4:1.

8.13 2250 Ω.

8.17 44·1, 0·978; (a) − 17·1 V; (b) 39·3; (c) − 47·1.

8.18 $V_{CC} = 14 \ V; I_C = 1 \cdot 8 \ A; R_B \simeq 490 \ \Omega$; 23 V, 4 V (for a ± 20 mA base current swing).

8.20 63·6 mA; 40·5 mW.

Chapter 9

9.1 1/100; 300.

9.3 1/50; (i) 20 per cent; (ii) 11·2 per cent.

* These values are based on the assumption that the current drawn by R_3 and R_4 is ten times greater than I_B.

9.4   A reduction of 8·34 per cent or 0·76 dB.
9.5   100; (a) 20 per cent, (b) 2·5 per cent.
9.6   2 kΩ; 0·05 Ω; 0·5 per cent.
9.7   1/60; (i) 33·3 per cent; (ii) 25 per cent.
9.8   (a) 1510 Ω; (b) 510 Ω; $\beta = 0.184$.
9.10 29·8 dB or a numerical gain of 30·9.
9.11 40 dB (100); 40·8 dB (109·6).

Chapter 10

10.3   33·3.
10.7   22·5 kHz.
10.20 (i) − 8·34; (ii) 0·9 kΩ; (iii) 60 kΩ; (iv) − 7·15.

Chapter 11

11.11 (a) 0·05 A; (b) 614 Ω.
11.12 102 µV, 52·5 Ω; 49·5 pW.
11.13 0·707 V.
11.14 The output voltage rises to 50 V, falling to 47·5 V after 0·5 ms. The output voltage then instantaneously changes to a reverse voltage of 2·5 V, after which it falls to zero with a time constant of 30 µs.
11.15 Output rises to 5 V, falling to 4·75 V after 0·5 ms. The output instantaneously reverses to a voltage of 35·25 V, falling to zero with a time constant of 30 µs.

Chapter 12

12.13 8·9 Ω.

Chapter 13

13.3   9·1 Ω; 455 mV.
13.4   0·0892; 0·446 V.
13.5   670 Ω.
13.6 R_S = 175 Ω, 7 W; 15 V, 3 W diode.
13.7   3.15 kΩ; (a) 298 V, 173 V; (b) 207 V.
13.8   2·5 kΩ, 1 W; (a) 70 V; (b) 1·96 W; (c) 18 mA.
13.9   (a) 0·397 Ω; (b) 0·595 V; (c) 0·038 V.
13.10 81·7 kΩ; 3·14.

Index

Printed by William Clowes & Sons Ltd., London, Beccles and Colchester.